Fruit Breeding

Fruit Breeding

Edited by

M.R. Dinesh
Principal Scientist
Indian Institute of Horticultural Research
Hessaraghatta Lake Post, Bengaluru
Karnataka - 500 089

NEW INDIA PUBLISHING AGENCY
New Delhi – 110 034

NEW INDIA PUBLISHING AGENCY
101, Vikas Surya Plaza, CU Block, LSC Market
Pitam Pura, New Delhi 110 034, India
Phone: + 91 (11)27 34 17 17 Fax: + 91(11) 27 34 16 16
Email: info@nipabooks.com
Web: www.nipabooks.com

Feedback at feedbacks@nipabooks.com

General Editor (Series) Dr. K.V. Peter

ISBN: 978-93-83305-51-3

Composed, Designed and Printed in India

DEDICATION

Dr. M.R. Dinesh dedicates this book FRUIT BREEDING to his sister Mrs. Y. Jayalakshmi N Sharma and brother-in-law Mr. Y.K.N. Sharma for their encouragement and support throughout.

Dr. K.V. Peter dedicates this book FRUIT BREEDING to Prof. P.I. Peter Chairman NoniBiotech, Chennai who introduced the term "Fruit is the healthy food for wellness and welfare". He brought to light Noni (*Morinda citrifolia* L) (Family-Rubiaceae) from the status of an uncommon and underexploited fruit to the status of a fruit for health care and wellness. Let his mission to validate therapeutic actions of Noni and create wellness awareness among people to provide health security by 2020 be materialized.

डा. एस. अय्यप्पन
सचिव एवं महानिदेशक
Dr. S. AYYAPPAN
SECRETARY & DIRECTOR GENERAL

भारत सरकार
कृषि अनुसंधान और शिक्षा विभाग एवं
भारतीय कृषि अनुसंधान परिषद
कृषि मंत्रालय, कृषि भवन, नई दिल्ली 110 001

GOVERNMENT OF INDIA
DEPARTMENT OF AGRICULTURAL RESEARCH & EDUCATION
AND
INDIAN COUNCIL OF AGRICULTURAL RESEARCH
MINISTRY OF AGRICULTURE, KRISHI BHAVAN, NEW DELHI 110 001
Tel.: 23382629; 23386711 Fax: 91-11-23384773
E-mail: dg.icar@nic.in

Foreword

India is the second largest grower of fruits after China, with a production of 74.8 million tons from a area of 6.38 million ha. Mango, citrus, guava, banana, papaya, pomegranate, jackfruit, sapota, custard apple, ber and jamun are the common fruits grown in addition to a number of under-utilized and under-exploited fruit crops. Fruits yield larger quantities of food per unit area as compared to cereals. For instance, paddy yields 3 tons/ha, while it is 22 tons for banana; 45 tones for pineapple and 40 tones for grapes. Less area is required to obtain energy/adult/year (11,00,000 kcal) from banana (0.03 ha.) or mango (0.16 ha) than from growing wheat (0.44 ha).

Given the present context, food security is complete only with nutritional security through consumption of fruits and vegetables. Micronutrients malnutrition, a threat to vulnerable India, can be minimized through consumption of fruits affordable to common man. Fruit consumption is as low as 10g/day/person in Nagaland and Manipur to 70g/day/person in Chandigarh. Per capita consumption of fruits in India is only 65g against a minimum of 120g. With the present level of consumption, the annual requirement of horticulture process will be 90 million tons in 2016-17 and 98 million tons 2020-21. Globally, fruit production in India is 11.7/ha compared to 16.1 tons in Brazil, 22.2 tons in USA, 22.4 tons in Indonesia and 12.8 tons in Turkey. Efforts are made to improve productivity of fruit crops through breeding. In spite of the perennial nature and high heterozygosity, work carried out at different Institute and Universities have resulted in several new varieties in different fruit crops viz., *Amrapali* and *Mallika* in mango *Lalith* and *Arka Kiran* in guva, *PKM* varieties in sapota, *Arka Prabhath*, Pusa varieties and CO series in papaya and *Arka Sahan* in custard apple. Several Institute and Agricultural Universities *viz.*, Island Central Agricultural Research Institute, Port Blair; Central Institute for Arid Horticulture, Bikaner; Central Institute of Sub-Tropical Horticulture, Lucknow; Central Institute of Temperate Horticulture, Srinagar; Indian Institute of Horticultural Research, Hessarghatta; National Research Centre for Banana, Tiruchirapalli; National Research Centre for Citrus, Nagpur; National Research Centre for Grape, Pune; National Research Center for Litchi, Muzzafarpur; National Research Centre for Pomegranate, Solapur; All India Co-ordinated Research Projects on Fruits, Arid Zone Fruits and Department of Fruit Science attached to the Faculty of Agriculture/Horticulture and Fruit Research Farms of State Governments are doing commendable work on improvement of fruit crops.

Feeling the needs at national and international perspectives, the Education Division of ICAR has revised and updated the UG and PG syllabi in Horticulture and fruit breeding is offered as a compulsory/elective course. The present edited book on "Fruit Breeding" has 20 chapters, 9 chapters dealing the fundamentals and 11 chapters on individual fruit crops, as edited by Dr. M.R. Dinesh, Principal Scientist from IIHR, Bangalore and Dr KV Peter, Former Vice-chancellor, Kerala Agricultural University.

I congratulate the editor for this useful compilation

Dated the 6th December, 2013, New Delhi

S. Ayyappan

Acknowledgement

The editor express profound gratitude to Dr S Ayyappan Director General and Dr N Krishna Kumar Deputy Director General(Horticulture), ICAR New Delhi for the academic guidance and FOREWORD to the book. Dr M R Dinesh acknowledges Dr. C.P.A. Iyer, Ex Director, CISH, Lucknow, his guide and teacher and Late Dr. M.D. Subramanyam, Sr. Scientist (Retd), Division of Fruit Crops, IIHR under whom he learnt the science and art of fruit breeding. He also acknowledges Dr. S.D. Shikhamany, Ex Director, IIHR and Dr. G.S. Prakash, Ex-Head, Division of Fruit Crops who encouraged him in undertaking the arduous work of fruit breeding by providing the right atmosphere for working and the facilities. He acknowledges the Director, Indian Institute of Horticultural Research, Bangalore and the eminent scientists who contributed valuable chapters to the edited book. He Acknowledges Miss Bhuvaneshwari for typing entire the manuscript.

Dr. K.V. Peter places on record his gratitude to Prof. P.I. Peter Chairman Noni Biotech, Chennai for the facilities extended and permission to engage in this academic pursuit. He also acknowledges Dr. Kirti Singh, Chairperson, World Noni Research Foundation for the guidance and blessings.

Dr. K.V. Peter acknowledges the smiles and handwaves of his grandson Master Antony Ajay Peter which enthuse him to be in this area of academic journalism.

The editors express gratitude to the staff of New India Publishing Agency, New Delhi for the excellent printing of the book.

Editor
M.R. Dinesh

General Editor (Series)
K.V. Peter

Contents

Contributors

A. Rekha
Principal Scientist
Division of Fruit Crops
Indian Institute of Horticultural Research
Hessaraghatta Lake P.O., Bengaluru – 560 089, Karnataka, India
arekha@iihr.ernet.in

Awtar Singh
Principal Scientist
National Research Centre Litchi
Muzaffarpur – 842 002, Bihar, India
awtar_saini@yahoo.co.in

Bharathi Nirujogi
Asst.Professor
Department of Horticulture
Agricultural college,Naira
Srikakulam – 532 185, Andhra Pradesh, India

C. Vasugi
Sr. Scientist
Division of Fruit Crops
Indian Institute of Horticultural Research
Bengaluru – 560 089, Karnataka, India
vasuc@iihr.ernet.in

G.L. Veena
Ph.D Scholar, School of PG Studies
Bengaluru
University of Horticultural Sciences
Bagalkot – 587 103, Karnataka, India
veena.lgowda@gmail.com

K.V. Ravishankar
Principal Scientist
Division of Biotechnology
Indian Institute of Horticultural Research
Hessaraghatta Lake P.O., Bengaluru – 560089
Karnataka, India
kvravi@iihr.ernet.in

K. Sreedevi
Sr. Scientist
Division of Entomology
Indian Agricultural Research Institute
New Delhi – 110012, India
kolla.sreedevi@gmail.com

Kanupriya
Scientist, Division of Biotechnology
Indian Institute of Horticultural Research
Hessaraghatta Lake P. O.
Bengaluru – 560 089, Karnataka, India
kanu@iihr.ernet.in

M.R. Dinesh
Principal Scientist
Division of Fruit Crops
Indian Institute of Horticultural Research
Bengaluru – 560089, Karnataka, India
drmrdinesh@gmail.com

Prakash Patil
Principal Scientist
Project Coordinator's (Fruits) Unit
Indian Institute of Horticultural Research
Hessaraghatta Lake P.O., Bengaluru – 560 089
Karnataka, India
pnp@iihr.ernet.in

P. V. Rami Reddy
Principal Scientist
Division of Entomology and Nematology
Indian Institute of Horticultural Research
Hessaraghatta Lake P. O.
Bengaluru – 560 089
Karnataka, India
pvreddy2011@gmail.com

R. Venugopalan
Sr. Scientist
Section of Economics & Statistics
Indian Institute of Horticultural Research
Hessaraghatta Lake P.O.
Bengaluru – 560 089, Karnataka, India
venur@icar.org.in

S.S. Hiwale
Principal Scientist
Central Horticultural Experiment Station
(CIAH-ICAR)
Godhra Baroda Highway
Vejalpur – 389 340, Gujarat, India
sshiwale@yahoo.com

S. Simi
Assistant Professor
AICFTP
Department of Pomology and Floriculture
College of Horticulture
Kerala Agricultural University
PO KAU Thrisur – 680 656, Kerala, India
simishibu@gmail.com

T. Saktivel
Sr. Scientist
Division of Fruit Crops
Indian Institute of Horticultural Research
Hessaraghatta Lake P.O.
Bengaluru – 560089
Karnataka, India
sakthivel@iihr.ernet.in

T. Radha
Professor
Department of Pomology and Floriculture
College of Horticulture
Kerala Agricultural University
PO KAU Thrisur – 680 656
Kerala, India
radhathekoot@gmail.com

1

History of Fruit Breeding

T. Radha and S. Simi

Origin of fruit culture can be traced back to 8000-5000 years ago and the horticultural practices like irrigation, vegetative propagation, selection etc. developed about 5000-3000 years back. The sophisticated fruit culture might have originated before 3000-2000 years. Basically all the cultivated fruits are the gift of diversity and not results of systematic breeding. Most of them have spread to different parts of the world through various media from the primary centres of origin.

Centres of origin of major fruit crops are as follows:

Mediterranean basin : datepalm, fig, grape, olive, pomegranate
East Asia : banana, citrus, mango, peach, kiwi fruit, persimmon
Central Asia: pome fruits, apricot
Europe : plum, raspberry
North and South America: avocado, black berry, papaya, pineapple, strawberry

Domestication: The process of domestication involved the identification of certain useful wild species combined with a process of selection that brought about changes in their appearance, quality, and productivity. Spontaneous hybridization between wild races and cultivated clones was critical to the early domestication of fruits. Selection from sexual recombinants still remains as the dominant force in the domestication process as well as in modern fruit breeding. The isolation of elite selections, combined with mass multiplication and plantings created a situation where mass selection and recurrent selection could operate naturally. In the late Neolithic and Bronze Ages (6000 and 3000 BCE), the ancient Mediterranean fruits (date, olive, grape, fig, and pomegranate) were domesticated. Citrus, banana, various pome fruits (apple, pear, quince) and stone fruits (almond, apricot, cherry, peach, and plum) were domesticated in Central and East Asia and reached the West in antiquity. A number of fruits and nuts were domesticated only in the 19th and 20th centuries (blueberry, blackberry, pecan, and kiwifruit). Though not purposefully done, some genetic changes as given below were associated in a number of fruit crops during the process of domestication:

*Breakdown of dioecy in fig, grapes, papaya and strawberry

*Loss of self incompatibility in cherry

*Parthenocarpy and seedlessness in apple, pear, banana and plantains, citrus, fig, grapes, loquat, pineapple, persimmon

*Allopolyploidy in banana and plantain, blackberry, raspberry, strawberry, European plum

*Triploidy in banana and plantain, apple, pear

*Tetraploidy in raspberry, kiwifruit, blackberry, blueberry

*Hexaploidy in kiwifruit, European plum

*Octaploidy in strawberry

*Ease of vegetative propagation in apple rootstocks, datepalm

* Loss of spines, thorns and pubescence in pineapple, citrus, apple, peach, pear

Fruit breeding

Conventional plant breeding can be defined as systems for selection of superior genotypes from genetically variable populations derived from sexual recombination and in the modern sense can be defined as purposeful genetic improvement of plants. Fruit breeding was elevated as an organized activity during 19th century with reference to mass selection efforts in strawberry and pear. Thomas Andrew Knight (1759–1838) was the first to improve fruits by selection from genetic recombination derived from inter-pollinations of clones and he literally initiated the fruit breeding works. He was responsible for the release of a number of improved fruit cultivars (apple and pear, red_currant, strawberry, cherry, nectarine and plum). In the U.S. fruit breeding became a part of research at the state and federal experiment stations and a number of important breeding programs were initiated throughout the country. Fruit breeding also became an activity of the private sector. Luther Burbank (1849–1926) was the first to consider fruit breeding as a commercial endeavour.

Although fruit breeding has been a major activity since the early 20th century, the results were not even and varied from negligible to extraordinarily successful. Progress from breeding a number of fruit crops, however, has shown significant advances in the second half of the 20th century and selections from controlled crosses became significant in many crops. Many of the fruit cultivars are ancient and based on grower-selected seedlings and somatic mutations. Advances in molecular genetics may be the solution for the limitations to conventional fruit breeding based on sexual recombination by increasing selection efficiency using molecular markers and by transgenic technology. Progress has already been achieved in papaya (virus resistance) and apple (fire blight resistance), fears of consumer resistance remain a problem. Introduction has played a major role in fruit improvement.

Levels of fruit breeding can be summarised as the following techniques:

Primitive: vegetative propagation of unique clones, mass culture of superior clones, selection of chance seedlings and propagation of clonal variants.

Conventional breeding: recombination of elite clones, inter-specific recombination, back cross breeding and mutation breeding.

Biotechnological breeding: embryo rescue, protoplast fusion, marker assisted selection and transgenic breeding.

GM Crops: Genetically modified crops (GMCs, GM crops, or biotech crops) are plants used in agriculture, the DNA of which has been modified using genetic engineering techniques. In most cases the aim is to introduce a new trait to the plant which does not occur naturally in the species. Farmers of various countries have widely adopted GM technology during the period between 1996 and 2011. As of 2011, 11 different transgenic crops were grown commercially on 395 million acres (160 million hectares) in 29 countries. In papaya, ring spot virus resistance was the major issue addressed through genetic engineering.' Sun Up' and 'Rainbow' varieties were the results of this technique. As of 2013, an apple that has been genetically modified to resist browning, known as the Nonbrowning Arctic Apple produced by Okanagan Speciality Fruits, is awaiting regulatory approval in the US and Canada.

India

India is an important centre of origin and diversity of many horticultural crops including fruit crops like mango, citrus and banana. In 1890 introduction of grape variety Anab-e-Shahi from Middle East by Abdul Baquer Khan which revolutionized grape cultivation in the country can be considered as the earliest work of fruit crop improvement. Fruit research in the country was started in the departments of Botany of the six Agricultural Colleges established in 1905 at Pune, Coimbatore, Lyallpur, Nagpur, Sabour and Kanpur. Establishment of a Division of Horticulture in IARI in 1956-57 and creation of a separate division of Fruits and Horticultural Technology during 1982 boosted the research on different aspects of fruit crops. In 1968, Indian Institute of Horticultural Research(IIHR) at Bangalore and the Central Mango Research Station at Lucknow in 1972 were established and systematic research on fruit crops such as mango, grapes, banana, papaya, guava, sapota, citrus, pineapple etc. were initiated. Plant introduction facilitates further breeding programmes in crops. In India, National Bureau of Plant Genetic Resources (NBPGR) plays the role of the nodal Centre for the introduction and maintenance of germplasm of agricultural and horticultural crops including fruit species. The 'State Agricultural Universities', the 'All India Coordinated Fruit Improvement Project', different 'National Research Centres' under ICAR and other national institutes and sub centres were responsible for breeding work on all the important fruit crops and have come out with considerable achievements in this field as evident by the number of specific selections and hybrids evolved for commercial cultivation.

Milestones in fruit breeding

The significant inventions, research results, variety releases and other achievements in the field of fruit breeding is presented in the following table on a chronological order for ready reference .

Year	Major findings/results	Scientists/Authors
1905	First breeding work on pine apple was started in Florida	Webber
1905	Fruit research in India was started in the Departments of Botany at six Agricultural Colleges at Pune, Coimbatore, Lyallpur, Nagpur, Sabour and Kanpur	
1905	The varietal improvement programme in pomegranate was started at Ganeshkhind Botanical garden, Pune	
1907	Guava improvement work was initiated at Ganeshkhind Fruit Experiment Station, Pune	
1908	Earliest report on fig breeding	Swingle
1910	High yielding mono embryonic seedling of the variety Mulgoba was selected in Florida and named as Haden	
1911	The first to initiate planned hybridization work on mango at Pune	Burns and Prayag
1912	The earliest attempt made in West Indies to develop new cultivars by combining the good qualities of Indian mango with indigenous types by artificial crossing	Brooks
1912	Systematic evaluation work on citrus started in India at College of Agriculture, Pune	

(Contd.)

Year	Major findings/results	Scientists/Authors
1915	Kinnow mandarin, the first generation hybrid between King and Willow leaf mandarin developed at California Citrus Experiment Station	Frost
1917	Breeding programme for sweet cherry improvement was started at the Fruit Growing Institute in Plovdiv, Bulgaria	
1920	Possibility of radiation breeding in crops was discovered initially using X rays	
1922	Banana breeding programmes through hybridization started in Trinidad, West Indies	
1927	Developed Sardar guava through a Selection from Allahabad Safeda	Cheema and Deshmukh
1928	Large scale introduction of grapes in a planned manner was initiated at Lyallpur	Lal Singh
1933	Commercial cultivar of pineapple Cabezona was identified as a natural triploid with 75 chromosomes	Collins
1938	Described about 30 heritable mutant forms in Smooth Cayenne pineapple fields	Collins and Kerns
1945	Report of polyploidy in papaya	Hofmeyer
1946	Systematic introduction and exchange of plant genetic resources of horticultural plants started by Division of Botany, IARI, New Delhi	
1946	Scientists first discovered that DNA naturally transfers between organisms	
1946	Two promising mango hybrids – Mahmud Bahar and Prabha Shankar were evolved at Sabour	Sen *et al.*
1947	Adventive embryony in mango reflects the effect of one or more recessive genes	Leroy
1948	Significant variation among trees of the same clone in a mango orchard with respect to fruit shape, size, colour and quality was ascribed to bud mutations	Naik
1949	Breeding work on banana started in Tamil Nadu	
1949	Apricot genebank was established in Romania	
1950	Induced polyploidy in Musa using colchicine treatment	Vakili
1950	Stated that mango is an allopolyploid	Mukherjee
1952	Reported for the first time that the seedless guava are autotetraploids based on somatic chromosome number	Kumar and Ranade
1952	The Fruit Experimental Station at Kallar began breeding work on jack with a view to developing short, compact, many-branched trees, precocious and productive	
1953	Induction of tetraploidy in *Annona squamosa*	Islam
1954-58	First hybridization work on citrus in India was carried out at Fruit Research Station, Kodur	
1954	Preliminary work on hybridization in sapota was attempted at Agricultural College and Research Institute, Coimbatore	Sambamurthi and Ramalingam

(*Contd.*)

Year	Major findings/results	Scientists/Authors
1954	No inbreeding depression occurs in papaya during selfing of gynodioecious population	Hamilton
1955	Reported scoring technique in banana to indicate the relative contribution of two wild species for constitution of a given cultivar	Simmonds and Shepherd
1956	Published a monograph on Citrus classification	Bhattacharya and Datta
1958	FHIA was established by the United Fruit Company in La Lima, Honduras, which initially focused on dessert banana works.	
1958	All the present commercial mango varieties were developed from open pollinated seedlings	Bhujanga Rao and Rangacharlu
1958	Cultivar Suvarnarekha was identified as an isolated chance seedling	Naik *et al.*
1958	Mango hybrids Swarna Jehangir, Neeluddin, Neeleshan and Neelgoa were released from the Fruit Research Station, Kodour	Bhujanga Rao and Rangacharlu
1958	Hybridisation work in grapes was started at IARI, New Delhi to develop seedless varieties with early maturity, high yield and better quality seedless varieties with resistance to biotic stress.	
1958	Classified papaya flowers into eight categories	Storey
1959	First gynodioecious variety of papaya Coorg Honey Dew was selected from the local type Honey Dew	Aiyappa and Nanjappa
1960	Reported heterosis in the cross between Phillipines x Solo varieties of papaya	Dai
1960	Important mango varieties such as Dashehari, Langra, Samar Bahisht, Chausa and Ratul were the result of chance selection	Singh
1960	Developed a botanical key to the genera and species and the major characteristics of five cayenne clones for selection of appropriate parents for pineapple breeding	Collins
1961	Reported dual factor hypothesis for the flowering of banana	Chailakhyan
1961	Classified the forms of Citrus in detail	Tanaka
1962	Reported the presence of self incompatibility in Dashehari mango	Singh, Majumdar and Sharma
1963	Reported successful interspecific hybridization in the Genus *Mangifera*	Mukherjee
1965	Pollen sterility can be induced by applying growth regulators in grapes which can replace emasculation during production of male sterile lines.	Iyer and Randhawa
1966	First tried mutation induction using ionizing radiation in mango	Siddiqui , Mujeeb and Vasti
1966	Systematic work for breeding of uniform papaya varieties was started at IARI, Pusa, Bihar	
1968	IIHR was established and started grapes breeding programme to develop superior varieties for table, raisins, wine and juice.	

(*Contd.*)

Year	Major findings/results	Scientists/Authors
1968	Numerical taxonomy for Citrus classification was attempted	Singh and Singh
1969	Found pre- selection criteria for dwarfness in mango	Majumdar, Mukherjee and Chakladar
1969	In papaya only bonafide varieties in existence were Solo and Bush of Hawaii and Hortus Gold of South Africa	Storey
1970	Pomegranate type GBG-I released for cultivation in 1936 was renamed as Ganesh	
1970	Pusa Seedless grapes variety was released from IARI	
1970	Described Muttam Varikka, a selection in jackfruit varikka type from Kerala	Srinivasan
1970	Improvement in mango hybridization technique – caging technique was discovered	Sharma and Singh
1971	All India Coordinated Fruit Improvement Project (AICFIP) was launched by ICAR to conduct research on banana, citrus, pineapple and papaya	
1971	Mallika, world's first mango regular bearing hybrid released for commercial cultivation, from IARI	
1972	TNAU released Papaya variety CO-1 evolved through sibmating Ranchi and sapota hybrid CO-1	
1972	The citrus rootstock hybridization was started at Central Horticultural Experiment Station, Chethalli	
1972	Systematic breeding programme for developing an ideal rootstock variety for citrus was started at CHES, Chethalli	
1972	The Central Mango Research Station was established to conduct research on different aspects of mango	
1974	Sapota variety CO-2, selection from Baramasi was released from TNAU	
1975	Spontaneous hybridization between wild races and cultivated clones was critical to the early domestication of fruit crops	Zohary and Spiegel Roy
1976	Breeding of acidlime for resistance to canker was started at Tirupathi and Parbhani	
1978	All India Coordinated Research project on arid zone fruits was started	
1979	TNAU released papaya variety CO2 as pureline selection from a local type	
1979	Reported that when Totapuri was crossed with Alphonso, the resultant F_1 hybrids had beaked fruits of inferior quality	Iyer and Subramaniam
1979	Central Horticulture Experiment Stations were established at Gujarat and Ranchi to work on the fruit crops of western India and eastern India respectively	
1979	Semi-vigorous mango hybrid Amrapali was released from IARI	Negi and Randhawa
1980	Four grape hybrids Arkavati, Arka Kanchan, Arka Shyam and Arka Hans were released from IIHR	
1980	Released Oroblanco an early maturing seedless grape fruit type citrus	Soost and Cameron

(Contd.)

Year	Major findings/results	Scientists/Authors
1980	Sapota variety PKM-1, a selection from Guthi released from Periyakulam, TNAU	
1981-82	IBPGR mounted several Commissions to collect ten fruit tree species in several countries to build up the genetic resources	
1981	Four superior clones of Dashehari mango variety were located	Singh and Chadha
1981	First report of organogenesis- induced roots from callus that was initiated from mango cotyledons on MS medium supplemented with Kinetin and NAA	Rao, Sin, Kathagoda and Hutchinson
1981	Reported for the first time the use of isozymes to note the genetic variations in mango cultivars	Gan, Zaini and Idris
1981	Developed a dwarf mutant line by treating papaya seed with 15K gamma rays	Ram and Majumdar
1981	Attempted inter-specific hybridization in genus *Carica*	Subramanyam and Iyer
1981	Mango hybrid PKM-1 and sapota selection PKM -1 were released from TNAU	
1982	A separate division of Fruits and Horticultural Technology was established at IARI, new Delhi	
1982	First to report induction of embryogenic mango cultures from nucellar tissues of some poly embryonic cultivars after 1-2 months from ovule cultures excised from 40-60 days old fruitlets	Litz , Knight and Gazit
1983	Regular bearing mango hybrid variety Ratna was released from Dapoli	
1983	Pusa Nanha, variety of papaya was developed by mutation breeding	Ram
1983	Hybrid papaya varieties CO-3 and CO-4 were released from TNAU	
1983	Pomegranate variety CO-1developed through clonal selection was released from TNAU	
1983	Large number of crosses were effected by synchronizing the flowering, storage of pollen and identification of hybrids using marker gene as a part of rootstock breeding in Citrus at CHES, Chethalli	Iyer *et al.*
1983	Floral biology of most of the citrus types were well studied and reviewed for successful hybridization	Iyer *et al.*
1984	Banana hybrid CO-1 developed through multiple crossing was released from TNAU	
1984	Somatic embryogenesis was found to be successful for multiplication of jamun plants	Litz
1985	In mango red skin colour was reported to be dominant and influenced by duplicate genes	Sharma and Majumdar
1985	Biennial bearing character of a mango variety is dominant over regular bearing	Sharma and Majumdar
1985	In mango, purple colour of new leaves and panicles and beak characters of fruits will be of help in identifying the hybrid seedlings in the nursery	Sharma and Majumdar

(*Contd.*)

Year	Major findings/results	Scientists/Authors
1985	Isolated high yielding clones of Langra mango variety	Singh Gorakh, Rao and Mishra
1985	Melogold, an early maturing seedless hybrid grapefruit type was developed	Soost and Cameron
1985	Papaya CO-5 was released as a selection of Washington variety from TNAU	
1985	NRC on citrus was established at Nagpur	
1985	International network for improvement of banana and plantains (INIBAP) was established in France	
1985	Analyzed the GE interaction and stability of several papaya varieties developed by MARDID's breeding programme using structural relationship analysis proposed by Tai, 1971	Chan
1986	Indian mango cultivars, Kerala Dwarf, Janardhan Pasand, Manjeera, Amrapali and Creeping are useful as source for dwarfness	Iyer and Subramanyam
1986	Mango hybrids Neelphonso, Neeleshwari and Neeleshan were released	
1986	Papaya CO-6, selection from a giant variety was released from TNAU	
1986	Papaya varieties Pusa Dwarf, Pusa Delicious and Pusa Majesty were released from IARI	
1986	The hybrid embryo resulting from inter-specific cross of *Carica papaya* and *C. cauliflora* was successfully rescued by embryo culture	Yung
1986	Cariflora a dwarf papaya variety tolerant to PRSV became popular in Florida	Conover *et al.*
1987	The Hongkong wild Kumquat, *Fortunella hindsii* was first reported as tetraploid	Soost
1987	Described two breeding systems for papaya – the Hawaiian system and the Queensland system	Aquilizan
1987	MARDI Backcross Solo, later named as Eksotika was released through backcross breeding in papaya	
1987	Reverted 'sexually ambivalent male' (SAM) has been used for development of inbred lines in breeding of dioecious papaya varieties	Aquilizan
1987	Coorg Honey Dew developed as a popular selection from Honey Dew at IIHR	Morton
1987	Apple variety KKL-1, selection from Parlin's Beauty was released from TNAU	
1988	Mutation breeding was initiated in mango through physical and chemical mutagens	Sharma and Majumdar
1988	Identified two promising clones of acid lime, Vikram and Pramalini in Maharashtra	Chakrawar *et al.*
1988	Stable transformation of papaya has been achieved through the use of various DNA transfer technologies	Pang and Sanford

(Contd.)

Year	Major findings/results	Scientists/Authors
1989	Establishment of the International Musa Testing Programme (IMTP) Phase I was started to evaluate germplasm from the FHIA breeding programmes in Honduras	
1990	Agrobacterium mediated transformation of mango embryonic cultures	Litz, Mathews, Bharathan and Narayan
1990	Selection of somaclonal variants in banana was started at TBRI, Taiwan	Hwang and KO
1990	Kagzi Kalan, lemon variety released from IARI	
1990	PAU Selection- Selection from two strains of baramasi lemon released from Ludhiana	
1990	Mango variety PKM-2 and acid lime variety PKM-1 were released from TNAU through selection procedures	
1990	Nileswar Dwarf variety is useful as source for dwarfness in mango	Singh
1990	Jai Devi, an acid lime variety released from Priyakulam, Tamil Nadu	
1990	Seedless Santra was selected with commercial potentiality from Nagpur	Chadha and Singh
1991	Used flow cytometry to determine the nuclear content of mango	Arumuganathan and Earle
1991	Higher phenolics in the apical bud is associated with a reduction in vigour and dwarfing in mango	Iyer
1991	Higher diversity is found among cross pollinating species than in self pollinating species	Hamrick *et al*.
1992	Regular bearing 'seedless' mango variety Sindhu with papery stone was released from Dapoli	
1992	Banana improvement was started formally in response to the severity of damage by panama wilt	
1992	First androgenic banana plants were obtained from the wild clone Long Tavoy of *Musa acuminata*	Bakry and Horry
1992	Mango selection Paiyur-1, jackfruit selection Palur-1 (PLR-1) and sapota hybrid PKM-2 were released from TNAU	
1992	Three mango hybrids Arka Aruna, Arka Puneet, Arka Anmol were released from IIHR	Iyer and Subramanyam
1992	One guava selection Arka Mridula and one guava hybrid Arka Amulya were released from IIHR	Subramanyam, Iyer and Dinesh
1992	Arka Neelamani, a black seedless grape variety released from IIHR	
1992	Pollen of mango cultivars were successfully cryopreserved for two years	Rajasekharan and Ganeshan
1993	*Mangifera* genus has been divided into two subgenera – Mangifera and Limus	Kostermans and Bompard
1993	IPGRI initiated the project on 'Promoting conservation and use of tropical fruit species in Asia	

(Contd.)

Year	Major findings/results	Scientists/Authors
1993	A number of mango varieties were introduced from other countries for use as donor parents for good colour	Singh and Rana
1993	Use of RAPD genetic markers to identify mango cultivars and to determine genetic relationships among cultivars	Schnell and Knight
1993	First transformation of papaya using *Agrobacterium* was reported	Fitch *et al.*
1993	In papaya hybridization using female flowers will obviate emasculation and the seed production is also four times higher compared to that with hermaphrodites	Chan and Mak
1993	FHIA-01 and FHIA -02 were released as dessert bananas and FHIA -03 as cooking variety	
1993	Citrus breeding started- Trifoliate orange was used for imparting resistance	
1994	Sapota hybrid PKM-3 was released from TNAU	
1994	Resistance of transgenic papayas like Rainbow seemed to indicate strain specificity	Tenant *et al.*
1994	Cryopreservation can be applied for long term conservation of banana seeds	Bhat *et al.*
1994	Grape hybrid, Arka Magestic developed by crossing Angur Kalan and Black Champa was released from IIHR	
1994	Grape hybrid, Arka Soma developed by crossing Anab-e-Shahi and Queen of Vineyards was released from IIHR	
1994	Grape hybrid, Arka Trishna developed by crossing Bangalore Blue and Convent Large Black was released from IIHR	
1994	Grape hybrid, Shweta Seedless developed by crossing Anab-e-Shahi and Thompson Seedless was released from IIHR	
1994	Kagzi Lime selection, Sai-Sharbati released from Rahuri	
1994	Pollen genepools can provide sources of resistance for exchange with other breeders	Poulos
1994	The first reliable protocol reported for transformation of citrus relative *Poncirus trifoliata* by co cultivation of etiolated epicotyl segments with *Agrobacterium tumefaciens*	Kaneyoshi *et al.*
1995	Mango hybrid Arka Neelkiran was released from IIHR	Dinesh, Iyer, Subramanyam and Yadav
1995	ACC synthase and alternative oxidase are the first mango genes that have been cloned and have potential horticultural interest to the control of mango ripening	Cruz-Hernandez and Gomez-Lim
1995	BSR-1 selection of amla was released from TNAU	
1995	The first transgenic banana plants were produced through particle bombardment of embryogenic suspension cultures	Sagi *et al.*
1995	PCR-RFLP analysis of cpDNA can be successfully used for phylogenetic analysis of fruit crops like *Prunus* sp	Badenes and Parfitt
1995	Described elaborately the methodology involved in pollen cryopreservation	Grout and Roberts

(Contd.)

Year	Major findings/results	Scientists/Authors
1996	'Golden Delicious' apple was proved to be a pre potent parent for producing many important new cultivars from breeding works	Janick *et al.*
1996	Characterization of citrus genome through analysis of restriction fragment length polymorphisms (RFLPs) has been carried out	Liou *et al.*
1996	IMTP – International Musa Testing Programme- Phase II was started	
1996	A banana mutant Novaria was derived from gamma radiation of the clone Grand Naine	Mak, Ho and Tan
1996	PPI-1 jackfruit selection was released from TNAU	
1996	Grape varieties Pusa Urvashi and Pusa Navrang were released from IARI	Villegas *et al.*
1996	New hybrid papaya Sinta with moderate tolerance to PRSV was released	
1997	Direct plantlet regeneration was induced from higher embryos and somatic embryogenesis was induced in immature zygotic embryos of some popular mango varieties	Ghosh
1997	Avocado variety TKD-1 and papaya hybrid CO-7 were released from TNAU	
1997	A model for how to plan and establish a genetic reserve has been developed	Maxted *et al.*
1998	Mutagenic variety Osa Gold Pear was popularized in Japan	
1998	A gynodioecious advanced generation papaya hybrid Arka Surya was released from IIHR	Dinesh, Iyer, Subramanyam and Yadav
1998	BRS-1 Banana hybrid developed by crossing Agniswar and Pisang Lilin was released from BRS, Kannara, KAU	
1998	BRS-2 Banana hybrid developed by crossing Vannan and Pisang Lilin was released from BRS, Kannara, KAU	
1998	Lalit, a selection from local guava type was released from CISH	
1998	Pineapple hybrid, H7 developed by crossing Kew with Ripley Queen released from BRS, Kannara, KAU	
1998	Pineapple hybrid, H8 developed by crossing Kew with Ripley Queen released from BRS, Kannara, KAU	
1998	Sun Up-first genetically engineered commercial variety of papaya developed in 1991 and Rainbow- a hybrid between SunUp and Kapoho Solo have been successfully commercialized in Hawai, representing the first practical application of transgenic fruit crop.	Gonsalves
1998	First attempt for the identification of molecular markers that coded for sex in papaya	Somsri *et al.*
1999	Mango variety Konkan Ruchi was released from Dapoli for pickling purposes	
2000	Mango hybrid Ambika developed from a cross between Amrapali and Janardhan Pasand was released from CISH	Rajan *et al.*

(Contd.)

Year	Major findings/results	Scientists/Authors
2000	Mango hybrid Sonpari with Alphonso as female parent was released from AES, Paria, Gujarat	
2000	Hybrid sapota CO-3 was released from TNAU	
2000	Grapefruit mutant variety, Rio Star was popularized in USA	
2000	Pummelo is the female parent of sour orange, sweet orange and grapefruit while sour orange is the maternal parent of *Citrus limon*. Citron is the male parent in the origin of many hybrid species of Citrus including limes and lemons	Nicolosi *et al.*
2000	FAO/IAEA database includes 48 mutant cultivars of fruit crops including 10 in apple and 8 in sweet cherry	
2001	Fruit set in mango panicle is less than 1 % in general	Usman *et al.*
2001	Identified Kesar, Ratna and Goamankur as pollinizer varieties for Alphonso mango to increase pollination and fruit set in Alphonso.	
2001	Mango variety Pusa Peetambar was released from IARI	
2001	Sekaki, the foot long papaya was introduced to Malayasia	Chan
2001	Report of the first successful PRSV resistant Tainung-2 papaya through replicase mediated resistance	Chan *et al.*
2002	A GM hybrid papaya was developed from cross between Kamiya and Rainbow F2	Fitch *et al.*
2002	Poamoho Gold - a back cross 1 hybrid of Kapoho and Rainbow F2 was developed	Fitch *et al.*
2002	Mango varieties Pusa Arunima and Pusa Surya from IARI were released	
2002	Cryopreservation in banana plants has been shown as a powerful tool for the sanitation of varieties infected by viruses	Helliot *et al.*
2002	Transgenic plants of table grape cultivars 'Silcora' and 'Thompson Seedless' have been produced via organogenesis	Mezzetti *et al.*
2003	Custard apple variety APK(Ca)-1 and sapota selection PKM (Sa)-4 were released from TNAU	
2003	Exceptionally excellent selections of pineapple, CO-2 and MD-2 (Gold pineapple) from Hawaii were developed and characterized	Janick
2004	Jackfruit variety Konkan Prolific, Karonda variety Konkan Bold and Jamun variety Konkan Bahadoli were released from Dapoli	
2004	Guava variety Pusa Srijan, useful as an extremely dwarf rootstock was released from IARI	
2004	A high density genetic map of papaya using 54 F2 plants derived from Kapoho and Sun Up cultivars was constructed with large number of markers	Ma *et al.*
2004	Polyembryony is genetically inherited and thought to be influenced by a single gene in mango	Aron *et al.*
2005	Udhayam- banana variety evolved by selection- ABB Pisang Awak sub group from NRC for Banana, Trichy	
2005	Off season guava variety TRY(G)-1 was released from TNAU	

(*Contd.*)

Year	Major findings/results	Scientists/Authors
2005	Successfully produced antibody reactive antigen in banana leaves and fruits	Sunil Kumar *et al.*
2005	Possibility to increase the proportion of the more bioavailable glucoside form using interspecific hybridization in cranberry	Vorsa and Palashock
2006	Most developed and sustained pomegranate breeding programmes in India, China and Israel. In China 50 improved cultivars have been selected and are being grown	Feng *et al.*
2006	Balaji, acid lime variety developed by pure line selection released from Periyakulam	
2007	Intergeneric hybridization between Carica papaya var. Arka Surya & *V. Cauliflora* was successfully carried out	Dinesh *et al.*
2007	Jackfruit selection PLR(J) 2 and sapota selection PKM(Sa) 5 were released from TNAU	
2007	Lemon variety Konkan Lemon was released from Dapoli	
2007	First genome sequences produced for a fruit crop-two high quality draft genome sequences reported for grapevine	Jaillon *et al.* and Velaco *et al.*
2008	Arunika, mango variety developed from a cross between Amrapali and Vanraj was released from CISH	
2008	Manilla tamarind variety PKM (MT)-1 was released from TNAU	
2008	Citrus somatic hybrid with resistance to Citrus Tristeza Virus infection was successfully evolved	Mourao Filho *et al*
2009	Suvarna, the first mango hybrid having Alphonso as female parent was released from Dapoli	
2009	One advanced generation papaya hybrid with pink pulp colour high TSS and medium sized weighing on an average about 1200g was released	Dinesh
2009	Banana variety Konkan Safed Velchi was released from Dapoli	
2009	One pink pulp F_1 guava hybrid Arka Kiran was released from IIHR	
2010	In peach and apple there appear to be one or two major genes which allow the anthocyanins development in the flesh	Sekido *et al.*
2010	Mango variety Konkan Raja was released from Dapoli	
2010	Swarna, a jack selection was developed at UAS, Bangalore	
2011	Mango hybrids Pusa Lalima and Pusa Shreshth were released from IARI	
2012	Three new somatic hybrids, including an allotetraploid hybrid and two cybrids (2n and 4n), were obtained by protoplast fusion of 'Valencia' sweet orange (*Citrus sinensis* L. Osbeck) + 'Femminello' lemon (*C. limon* L. Burm.).	Abbate, Tusa, Fatta Del Bosco, Tonia, Renda, Ruberto
2013	Genetically modified apple to resist browning known as Non browning Arctic apple produced by Okanagan Speciality Fruits (awaiting approval in the US and Canada)	
2013	Single plant selection as a methodology for selecting integeneric *hermaphrodite* plants was suggested	Dinesh *et al.*

Further Reading

Beiquan, M. 2011. Transgenic crops Challenges and Oppurtunities. Scorza, R. (Ed.) CRC press, London

Chadha, K.L. and Pareek O.P. 1993. Advances in Horticulure Vol 1. Fruit crops. Malhotra Publishing House, New Delhi. p. 1661-2127

Chadha, K.L. and Pareek O.P. 1993. Advances in Horticulure Vol 4 Fruit crops. Malhotra Publishing House, New Delhi 514 p.

Janick, J. and Moore, J.N. 2011. Fruit Breeding Vol. 1 Tree and tropical fruits. Wiley India Pvt. Ltd., New Delhi 616 p.

Janick, J. and Moore, J.N. 2011. Fruit Breeding Vol. 2 Vine and small fruits. Wiley India Pvt. Ltd., New Delhi 477 p.

Kumar, N. 2006. Breeding of Horticultural Crops – principles and practices. New India Publishing Agency, New Delhi 202 p.

Shukla, A.K. and Vashishta, B.B. 2004. Fruit Breeding Approaches and Achievements. International Book Distributing Co., Lucknow, UP 342 p.

2

Biodiversity of Fruit Crops

M.R. Dinesh and G.L. Veena

The concept of origin of cultivated plants was first developed by A de Candolle and the geographic centres of variability were developed by Vavilov. He identified Asia as a major centre with "Indian centre" of North East region as primary or secondary centre of origin for many crop plants. Plant genetic resources are of great importance as they form the basic raw materials to meet the current and future needs of crop improvement programmes. A wider genetic base, thus, assumes priority in plant breeding research aimed at developing new varieties for increased crop production (Paroda, 1991). This diversity comprises of native landraces, local selections, elite cultivars and wild relatives of crop plants. The collection and conservation of this diversity in a systematic manner is the primary responsibility of all plant genetic resources institutes/centres. The mention of use and cultivation of fruits can be seen in epics like 'Ramayana'. Plant genetic resources are thus our heritage, which need conservation for posterity. During the long period of domestication, utilization and cultivation, a wide array of fruit crop variability got generated by natural means and through both conscious and unconscious selection. Huge wealth of variability also got generated/adapted and diversified by crop introductions in the exotic environment or through migration of human population.

Mankind has used only about 5,000 plant species worldwide to meet food and other needs, this number is just a fraction of the total world flora. With population growth, we are increasingly dependent on most productive plants. Today, only about 150 plant species are important in meeting the food (calories) needs of humans worldwide. Hence, there is greater dependence on a fewer plant species; 20 to 30 in global context (Harlan, 1975). This gradually, has resulted in the loss of native genetic resources which are otherwise essential as building blocks of genetic diversity. It is estimated that there are about 500 species of tropical fruit trees in Asia Pacific Oceania region, which include 30 families and 59 genera (Arora, 1998). In Southeast Asia alone, there are 120 major fruit species and 275 minor fruit species (Verheij and Coronel 1992). In Asia, 50-60 species belong to the most important indigenous fruits (Arora and Rao 1996). Citrus, mango, banana, rambutan, jackfruit, litchi and durian occupy 80% of total fruit production in the region.

Wild species and diversity

The role of wild species in the fruit improvement programme is increasingly becoming important as the donor source for many of the disease and pest resistance. However, it is to be mentioned here that in most of the perennial trees, evaluation of wild species or the indigenous germplasm has not been done extensively either by morphological or by molecular means. Some areas may be richer in biodiversity than other areas, and some species may also have more variation than others in a particular area.

Conservation is very important, because many species are becoming extinct and many others are threatened and endangered. The diversity of some fruits is well collected, while for other fruits relatively little has been done (Arora, 1994). Gaps in collections are found both between species and between regions. This is especially true for both underutilized species and wild crop relatives, where big gaps are noted. Kostermans and Bompard (1993) indicated that *Mangifera blommesteinii,M. leschenaultii*, *M. superba* and *M. paludosa* are in real danger of extinction. High genetic erosion has been noticed for jackfruit, *Citrus* spp. and *Litchi chinensis* in a survey carried out by the International Centre for Underutilized Crops (ICUC) and IPGRI (Haq, 1994). It is to be mentioned here that collection and utilization of wild species are not an easy task, as they require specific climate and do not so easily get acclimatized to the *ex situ* conditions on introduction.

Biodiversity of fruits

The North East region is centre of diversity for several important native fruits including mango, jackfruit and citrus. The plant genetic resources represent a sum total of the diversity which come from wild species and primitive forms, accumulated through evolution and natural selection, plant introduction, migration and domestication and also materials developed by artificial selection and breeding. The region had remained isolated for a long time and even today the accessibility is poor to many parts of this region. This undisturbed condition with suitable environmental conditions and variable altitudes is one of the major reasons for genetic diversity of the region through natural selection and evolution although the primary causes being the agro-climate, which are wet tropics with rain forests.

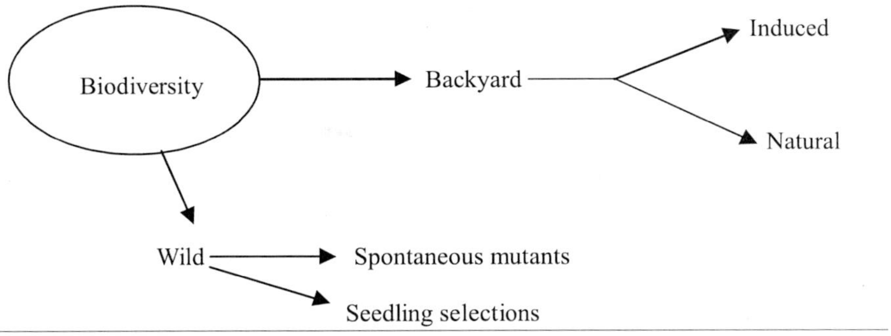

Fig 1: Biodiversity of Tropical Fruit Species

Biodiversity can be located either in the wild or in the backyard. With regard to many of the tropical fruit species, the variability can be traced in wild, wherein many species grow naturally even today viz., the occurrence of *Mangifera sylvatica* in the North-Eastern parts of India or *M. andamanica* in Andaman group of islands. In the wild, this diversity has come about mainly because of spontaneous mutants over a period and due to the dispersal of seeds and seedling population. In the backyard mainly the seedling populations have been the cause of diversity as noticed in the case of fruits like mango and jackfruit. Diversity due to natural means has come about because of seed dispersal as in pickling types of mango viz., Appemidi types in Uttara Kannada district of Karnataka or be it the varietal wealth found in the Western Ghat regions.

Characteristic features of tropical fruit tree diversity

The main causes for the tropical fruit diversity in India be it mango or an underutilized fruit like jamun, whether in the wild or in the cultivated types has been;

1. the presence of high heterozygosity
2. cross pollination
3. seed propagation
4. absence of vegetative propagation in the earlier days
5. indiscriminate multiplication.

Unlike other crops, wherein there is a need to create variability, in the case of tropical fruit species, it is the management of diversity, which is the more challenging task. In fact in crops like mango, the varietal diversity itself is considered as a hindrance to the improvement (Naik, *et. al.,* 1958).

The main centres of diversity for fruits in India are as follows;

Region	Species
Western Himalayas	*Elaeagnus hortensis, Ficus palmata, Fragaria indica, Moms spp., Prunus acuminata, P. cerasiodes, P. cornuta, P. napaulensis, P. prostrata, P. tomentosa, Pyrus baccata, P. communis, P. kumaoni, P. pashia, Ribes graciale, R. nigrum, Rubus ellipticus, R. moluccanus, R. fruticosus, R. lasiocarpus, R. lanatus, R. niveus, R. reticulatus, Zizyphus vulgaris.*
Eastern Himalayas	*Fragaria indica, Morus* spp., *Myrica esculenta, Prunus acuminata, P. cerasiodes, P. cornuta, P. jenkinsii, P. napaulensis, Pyrus pashia, Ribes graciale, Rubus lineatus, R. ellipticus, R. lasiocarpus, R. moluccanus, R. reticulatus.*
North-Eastern region	*Citrus assamensis, C. ichangensis, C. indica, C. jambiri, C. latipes, C. macroptera, C. media, C. aurantium, Docynia indica, D. hookeriana, Eriobotrya angustifolia, Mangifera sylvatica, Musa accuminata/M. balbisiana* complex, *M. manii, M. nagensium, M. sikkimensis, M. superba, M. velutina, Pyrus pyrifolia, P. pashia, Prunus cerasiodes, P. cornuta, P. jenkinsii, Ribes graciale, Rubus ellipticus, R. moluccanus, R. reticulatus, R. lasiocarpus, Myrica esculenta.*
Gangetic plains	*Aegle marmelos, Cordia myxa, C. rothii, Emblica officinalis, Grewia asiatica, Morus* spp.; *Phoenix* spp.; *Syzygium* spp.; *Zizyphus nummularia and other* spp.; and *Manilkara hexandra* (more in north-western plains).
Indus plains	Meagre occurrence of *Syzygium*,rich variaton in *Carissa congesta.*
Western peninsular tract	*Artocarpus heterophyllus, A. lakoocha, Garcinia indica, Diospyros* spp., *Ensete superba, Mangifera indica, Mimosops elengii, Spondias pinnata, Vitis* spp., *Zizyphus oenoplia, Z. rugosa, Rubus ellipticus, R. lasiocarpus, R. moluccanus.*

Arora and Nayar, 1984

Mango

Mango is native to India. Mukherjee (1949, 1985) opined that this genus might have originated in the region covering Burma, Siam, Indo-China and Malayan peninsula. The genetic diversity of mango available in India is very rich and at present more than one thousand vegetatively propagated varieties exists in the country. Clonal selection, selections from chance seedlings and breeding efforts have resulted in identification of many elite improved varieties of mango for commercial cultivation in the country.

All these varieties belong to one species *Mangifera indica*. Apart from *M. indica*, India is also reported to be the home of four other species viz., *M. andamanica*, *M. khasiana, M. sylvatica* and *M. camptosperma* (Mukherjee *et al.,* 1985). The species of *Mangifera* occur mainly as complex biotic community in tropical humid forests, sub-tropical rain forests and tropical dry forests/woodlands of Indo-Malayan biogeographic realm (Mukherjee, 1985).

The *Mangifera* germplasm can be classified into two categories

1. ***Seedling races***: This group includes both wild and cultivated types. Under this category the cultivated ones come under the monoembryonic types. The polyembryonic types are seen generally in the Western Ghats of Peninsular India.
2. ***Horticultural races***: They include varieties, which when grown under different agro-climatic conditions and propagated vegetatively from the parent material have shown clonal differences. Varieties like Alphonso, Dashehari and Langra are noticed to have clones resembling them in a few of the morphological characters. Yadav and Singh (1985) opined that mango varieties of Northern and Southern regions belong to two different eco-geographic regions.

Centres of mango diversity and distribution in India

In India, seven centres of mango diversity have been recognized, one being a sub-centre (Yadav and Rajan, 1993). These are the places where maximum diversity has been noticed for species as well as varietal diversity. They are:

1. Humid Tropical region-Manipur, Tripura, Mizoram and S. Assam
2. Chota Nagpur Plateau-Trijunction of Orissa, Bihar and Madhya Pradesh
3. Santal Paraganas in Bihar
4. South Madhya Pradesh adjoining Orissa and Andhra Pradesh
5. Dhar Plateau of Madhya Pradesh adjoining Gujarat and Maharashtra
6. Humid Tropical South Peninsular India
7. Andaman and Nicobar Islands.

Varietal diversity

In India about thousand varieties of mango are grown. Most of these varieties have arisen as chance seedlings. Each mango-growing region in India grows a different variety. Although thousand varieties are grown, one characteristic feature has been the growing of hardly twenty-five on a commercial scale in different states. Most of the commercial varieties have arisen as a result of selection of seedling types for different fruit characteristics like colour, taste, flavour, size and bearing habit. In India, there have been three main centres of development of mango i.e. Lucknow - Saharanpur belt of Uttar Pradesh, Murshidabad area of West Bengal and Hyderabad area of Andhra Pradesh. Most of these varieties have specific characteristics, require specific climate for optimum performance and have strong regional consumer preference. Although, growth in mango is genetically controlled, the environmental interaction has brought about the change in growth pattern under different agro climatic conditions, which also has contributed for its biodiversity.

Citrus

The North East hilly region is rich in fruits, vegetables and flowers, especially orchids. It is considered as a centre of origin of mandarins and a few other citrus fruits. Sixteen species of Citrus, 52 varieties, and seven natural hybrids of Assam were described by Bhattacharya and Dutta as early as in 1956. They also reported two species of sub genus *Eu citrus* viz., *C. indica* and *C. assamensis* and three species of sub genus of Papeda viz., *C. ichangensis, C. latipes* and *C. microptera* which grow at high altitudes. *C. indica* is considered to be the most primitive species of citrus and probable progenitor of cultivated species. It is found in various fruit forms in NEH region. Diverse forms of Pummelo, Sour Orange, Rough Lemon, Sour Pummelo, Adajamir Sweet Lime etc. are found in this region.

Varietal diversity

Mandarin is concentrated in Maharashtra (Nagpur, Amaravathi, Wardha and Yavatmal) in North East region of India (Assam, Arunachal Pradesh and Meghalaya), limited areas of Karnataka (Kodagu), Tamil Nadu (Nilgiris, Palani and Shevroy hills) and Kerala (Wynad). Sathpura hills of Madhya Pradesh adjoining Vidharbha region of Maharashtra also grow good quality mandarins. Kinnow Orange, a hybrid of King × Willow Leaf Mandarin has recently spread in North West India, especially in Punjab, parts of Himachal Pradesh, Uttar Pradesh and Rajasthan. The main advantage of cultivation of introduced varieties / hybrids is that they also add to the varietal diversity by throwing spontaneous mutants over a period of time.

Sweet oranges are well adapted to arid tropics and sub tropics. They are commercially grown in Andhra Pradesh, Maharashtra, Punjab and parts of Tamil Nadu, Rajasthan and Uttar Pradesh. Sweet orange cultivar in Andhra Pradesh is Sathgudi, whereas in Western and Central India, sweet orange cultivar

Mosambi is popular. In North Western India, the cultivars Malta, Jaffa and Valencia are popular.

Acid lime is grown on a commercial scale in Tamil Nadu, Andhra Pradesh, Maharashtra and Karnataka states. Lemons are grown commercially only on a limited scale. Eureka lemon in some regions and Assam lemon in North Eastern India are popular.

Banana

Bananas are one of the ancient fruits cultivated by man. It could be assumed that the fruit has evolved with the civilization (Krishnamurthi and Seshadri 1958) and found in Indus valley as early as in 327 B.C. Apart from its mention in Valmiki's Ramayana, it also finds a mention in Kautilya's Arthshastra and ancient Tamil classic Silappadikaram. These evidences suggested the early existence of banana in India. The wild *Musa acuminata* occurs in Assam, Myanmar, Siam, Indo-China, the Malayan peninsula and Archipelago and the Philippines.

The north eastern region of India - including the states of Assam, Arunchal Pradesh, Tripura and Mizoram, lie at a point where *Musa balbisiana* from the Indian subcontinent meets *Musa acuminata* from Southeast Asia. These two species, as well as other wild relatives have mingled to form a distinctive concentration of genetic diversity, occurring in the semi-evergreen, sub-tropical forests of the hill slopes. Further sources of diversity occur in the valleys and plains where bananas are common as a backyard crop. The habitat of wild bananas is shared with tribal groups who practice a form of shifting cultivation. Unfortunately, a growing number of sites where wild bananas once grew are now denuded. Wild species found in this part of India include *Musa acuminata, Musa balbisiana,* several species from the section Rhodochlamys as well as *Ensete glaucum*.

The diploid and triploid *acuminata* cultivars were taken by man to the native areas of *balbisiana*, which resulted in natural hybridization and formation of hybrid progeny with the genomes: AA, AAB, and ABB. It is thought that subsequent dispersal of edible bananas out of Asia was brought about by man. Secondary diversification with-in the groups of cultivated bananas are the result of somatic mutations.

Eumusa and Rhodochlamys are found in Assam region of India and Thailand whereas Callimusa and Rhodochlamys in Borneo, its surrounding islands, and Indonesia. Australimusa is largely found in Malayan islands, and Indonesia. It is also found in Assam, Indo-China, Malayan and Papua New Guinea, which is a primary centre of cultivated AA types. *M. balbisiana* occurred in Sri Lanka, India, Myanmar, Siam and Malaya where the A × B hybrids have evolved.

Region/ State	Cultivars
Andhra Pradesh	Amrit Pani (Rasthali,AAB),Thella Chekkarakeli (AAA), Karpura Chekkarakeli (Poovan, ABB), Monthan (ABB), Robusta (AAA)
Assam	Bhimkol, Manohar (ABB), Chini Champa (AB), Malbhog (AAB), Jahaji (AAA),BorJahaji (AAA), Kanch Kol (ABB),
Bihar	Alpan (AAB), Chini Champa (AB),Basrai (AAA),Kothia/ Muthia (ABB),Bateesa (ABB),Malbhog (Rasthali,ABB).
Gujarat	Dwarf Cavendish (AAA), Lacatan (AAA), Harichal (AAA)
Karnataka	Dwarf Cavendish (AAA), Robusta (AAA), Poovan (AAB), Rasbale(AAB, Rasthali), Marabale (Pome,AAB),Monthan (ABB), Elekki Bale (AB, Ney Poovan)
Kerala	Nendran (ABB, plantain), Palayankondan (AAB, Poovan), Kunnan (AB), Rasthali (AAB), Monthan (ABB) and Red Banana (AAA)
Maharashtra	Basrai (Dwarf Cavendish, AAA), Robusta (AAA), Safed Velchi (AB), Monthan (ABB), Rajeli (Plantain, AAB)
Tamil Nadu	Virupakshi (Pome, AAB), Poovan (AAB), Rasthali (AAB), Nendran (AAB), Monthan (ABB), Dwarf Cavendish (AAA), Robusta (AAA), Peyan (ABB)
Bengal and Orissa	Champa (AAB), Morthaban (AAB, Rasthali), Amrit Sagar (AAA), Giant Grover (AAA), Lacatan (AAA), Monthan (ABB)

Singh, 1996

Grape (*Vitis* spp)

European grape *V. vinifera* is considered to have originated primarily between Caspian and Black sea, and also considered a hybrid between two American spp *V. vulpine* and *V. labrusca*, which resembles *V. parviflora* and *V. lanata* found in Himalayan region. This region may be considered as a secondary centre of origin. Famous Indian medicine scholars, Sushruta and Charaka in their medical treatises entitled 'Sushruta Samhita' and 'Charaka Samhita', respectively, written during 1356-1220 BC, mentioned the medicinal properties of grapes. Kautilya in his 'Arthashastra' written in the fourth century BC mentioned the type of land suitable for grape cultivation. Native spp. resembling *Vitis lanata* and *Vitis palmata* grow wild in the northwestern Himalayan foothills. Indigenous varieties known as 'Rangspay', 'Shonltu White' and 'Shonltu Red' are grown in Himachal Pradesh even today.

Cultivated grapes are believed to have been introduced into the north of India by the Persian invaders in 1300 AD, from where they were introduced into the south (Daulatabad in Aurangabad district of Maharashtra) during the historic event of changing the capital from Delhi to Daulatabad by King Mohammed-bin-Tughlak. Ibn Batuta, a Moorish traveller who visited Daulatabad in 1430 AD, reported to have seen flourishing vineyards in South India.

Species	Region	Salient Characters
V.riparia	North Westeran Himalayan region	Small berries, purple black in colour, cold hardy, early flowering
V.lanata	Himalayan region	Purple black berries known for crack resistance, and plant resistant to diseases
V.barbata	Parts of Assam, Khasi hills and Bengal	
V. parviflora	North West Himalayas from Kashmir to Nepal	Small berries, delicately flavoured
V.tomentosa	Greater part of Deccan peninsula	

The spp *V. vinifera*, has not originated in India. However, it is very interesting to note that Thompson Seedless, which belongs to the spp *V. vinifera,* and which is grown in several parts of the world as well as in India has contributed for the diversity in varieties; several mutants have come about in the grapes due to its extensive cultivation over the years.

Guava

Guava is an important fruit crop of India. It is said to have originated from tropical America and seems to have been growing from Mexico to Peru. It is widely distributed all over the equatorial regions of tropical and sub-tropical climate. Guava is reported to have been introduced during the seventeenth century into India.

It has gained considerable prominence on account of its high nutritive value, availability at moderate prices, pleasant aroma and good flavour. It is one of the commonest fruit liked by the rich and the poor alike and is popularly known as the 'apple of tropics'. At present, it is grown throughout the length and breadth of the country from sea level to 1300 m altitude and is so much acclimatized that it appears to be native to India. The most important guava growing states are Uttar Pradesh, Bihar, Madhya Pradesh and Maharashtra.

The genus *Psidium* of Myrtacease family comprises of about 150 species of small trees and shrubs. About 20 species have edible fruits, of which the most commonly cultivated is the common guava i.e., *Psidium guajava* L. It has been reported that the value of the wild *Psidium* species mainly lies in their utility as rootstocks for regulation of vigour, fruit quality and resistance to pests and diseases.

Varietal diversity

Guava is mainly a self-pollinated crop but cross-pollination is also common. This has resulted in large variability in the seedling population from which promising genotypes have been selected in different agro-climatic regions of the country. In India different workers in different regions have described guava varieties. The main centre of variability in guava has been the Allahabad area in Uttar Pradesh. The promising cultivars of different states are as follows:

States	Cultivars
Andhra Pradesh	Allahabad Safeda, Anakapalli, Banarasi, Chittidar, Hafsi (Red Fleshed), Lucknow-46, Sardar, Seedless, Smooth Green, Smooth White.
Assam	Amsophri, Madhuriam, Safrior Payele.
Bihar	Allahabad Safeda, Chittidar, Hafsi (Red Fleshed), Harijha, Seedless.
Maharashtra	Dharwar, Dholka, Kothrud, Lucknow-24, Sardar,
Gujarat	Nasik, Seedless, Sindh.
Tamil Nadu	Anakapalli, Banarasi, Bangalore, Chittidar, Hafsi, Nagpur Seedless, Smooth Green.
Uttar Pradesh	Allahabad Safeda, Apple Colour, Chittidar, Red Fleshed, Banarasi Surkha, Sardar, Mirzapuri Seedless.
West Bengal	Behrampur and cvs. of Uttar Pradesh.

In India, due to seed propagation, varietal diversity is seen for guava, but species diversity is not seen.

Papaya

The papaya (*Carica papaya* L.) is one of the most important fruit crops valued for its rich nutrient content. It is a rich source of Vitamin A (2020 I.U), Vitamin B1 (40 mg), Vitamin C (46mg), protein (0.5%) and mineral matters (0.4%). Papaya is native to tropical America, its place of origin is said to be in Southern Mexico and Costa Rica. It was taken to Manila by Spanish in the mid 16th century, reached Malacca shortly afterwards. It was introduced into India during 16th century. It is grown both in tropical and sub-tropical parts of the world. Wild diversity is not reported for papayas in India.

Varietal diversity

In India, varietal diversity is seen for papaya. The variability seen is more because of the open pollination and indiscriminate multiplication using these seeds. In papaya there are two basic types of varieties. Those varieties, which are dioecious, produce only female and male plants, and 'gynodioecious' that produce both female and hermaphrodite plants.

Some of the varieties that are grown in different states are hereunder

State	Cultivars
Andhra Pradesh	CO2, CO5, Sunrise Solo, Taiwanese lines
Bihar	Pusa Dwarf, Pusa Majesty, Pusa Nanha, Pusa Giant, Pusa Delicious and Ranchi.
Karnataka	Coorg Honey Dew, Washington, Sunrise Solo, CO2, Surya and Taiwanese lines.
Maharashtra	Washington, CO2, Pusa Delicious, Pusa Majesty, Ranchi and Taiwanese lines
Orissa	CO2, Coorg Honey Dew, Washington, Ranchi, Pusa Dwarf and Pusa Delicious.
Tamil Nadu	CO2, CO3, CO4, CO5, CO6, CO7 and Coorg Honey Dew.
Uttar Pradesh	Coorg Honey Dew, Pusa Dwarf, Pusa Delicious, CO3 and Barwani Red

Sapota

Sapota (*Achras zapota* L.) is a popular dessert fruit belonging to the family Sapotaceae. It is believed to have originated in tropical America, taken to Philippines by the Spanish and from there has spread to other countries (Purseglove, 1968).

In India it is grown in the states of Andhra Pradesh, Gujarat, Tamil Nadu, Karnataka and Orissa. About 30 varieties are reported in India at various places. A number of locally grown genotypes identified include Bhuripatti, Morabba, Kalipatti, Turipatti, Golepatti, Singapuri, Khabari and Chhumukia type. Among these, a genotype identified in Navsari, in Gujarat locally known as 'Morabba' bears fruits of bigger size and superior quality in comparison to Kalipatti, a local genotype grown in about 80% area of Gujarat. It is a selection from grafted plants collected from nursery located in Golwal. It may have originated as bud mutant and is now being propagated vegetatively. Another somatic mutation having desirable characteristics of the fruit was identified in Paria (Rai, 1995). Wild diversity is not observed for sapota, as it has been grown over the years by using grafts.

Biodiversity of underutilized fruits

In India various native fruits, such as aonla (*Emblica officinalis)*, bael fruit (*Aegle marmelos*), jackfruit (*Artocarpus heterophyllus*), jamun (*Syzygium cuminii*), karonda (*Carissa congesta*), Kokum (*Garcinia indica*) and phalsa (*Grewia subinaequalis*) with lot of diversity in a wide range of agro-ecological situations throughout the tropics, subtropics and temperate regions, which could be grouped as underutilized. Some of these fruits yield juice with excellent flavour, which can be converted into blended beverages and these could play an important role in meeting the demand for nutritious, pleasantly flavoured and attractive natural food of high therapeutic value. Encouraging local people to produce these fruits can help in significantly checking uncontrolled harvesting from the wild and assisting in the retention of the various species in their native habitats where they perform the best.

Jackfruit

Artocarpus is a genus of small to large evergreen trees, distributed from Sri Lanka and India to South China and through Malaysia to Solomon Islands. Nine species are recorded in India. The spp, *A. heterophyllus* Lam. is grown for their edible fruits, and *A. chaplasha* Roxb.,*A. hirsutus* Lam. and *A. lakoocha* Roxb., are important timber trees.

A. chaplasha Roxb, is found distributed in the moist deciduous and evergreen forests of the sub-Himalayan tracts from Nepal eastwards to West Bengal, Assam and Tripura. In West Bengal and Assam, it occurs in moist types of mixed deciduous and evergreen forests. In Andaman and Nicobar islands it is an important constituent of evergreen and deciduous forests.

A. cummunis J.R. & G.Frost, commonly known as bread fruit is found mainly in West coast and Western Ghats, Wynad, in the Nilgris, Lower Plains, the Courtallam hills and the Annaimalais. There are two distinct varieties in this species. One is a seeded type and the other entirely seedless. The seeded types are found in a wild state in its native and are of little economic value. It is not useful in culinary preparations but the seeds, which resemble chestnut, are relished when roasted or boiled.

A.heterophyllus Lam commonly called as jackfruit is one of the most popular fruits of South India. The tree is indigenous to the evergreen forests of the Western Ghats at altitudes of 450-1200 m, but seen growing throughout other hotter parts of India too. Because of seed propagation, the existing population of jack comprises innumerable trees differing from each other in fruit characters of shape, size and quality.

A. hirsutum Lam is commonly found in the evergreen forests of Western Ghats from Konkan southwards and is fairly common in North Kanara and Kodagu in Karnataka to Kerala where it is an important tree. It requires heavy rain fall, not less than 174 cm annually and thrives well on lateritic soils at the foot of the Ghats. The tree can stand shade, but thrives the best with a fair amount of light. It does equally well in the open and withstands exposure to sun after the first a few years.

A. lakoocha Roxb, is commonly known as monkey jack. In its wild state, it is chiefly found in the moist or deciduous forests along the banks of streams and along the site of moist ravines. It thrives the best in deep laterite soils and generally comes to bear after about eight years. It is commonly cultivated throughout the greater part of India as a shade or ornamental tree. It is perhaps one of the foremost among neglected but useful trees. It is found distributed in evergreen, semi-evergreen and moist deciduous forests upto an altitude of 1800 m in eastern and northern India. On the west coast it is found from Konkan southwards to Kerala, and in Tamil Nadu. It is also found growing in many localities in Andaman Islands.

Aonla

Aonla or Indian gooseberry (*Emblica officinalis* Gaertn.) is considered as a wonder fruit for health conscious population. It is being grown in India for more than 3500 years. Sushruta, the father of ancient medicine (during 1500 BC-1300 BC), has mentioned about its usefulness in 'Ayurveda' in great detail. It belongs to family Euphorbiaceae and is one of the important indigenous fruits of Indian subcontinent. In different parts of India, it is known by different vernacular names such as Amla or Aonla in Hindi, Dhatri, Dhatriphala, or Amalaki in Sanskrit, Amla or Amalaki in Bengali and Oriya, Nelli in Malayalam and Tamil, Amlakamu in Telugu, Amolphal in Punjabi (Gurmukhi) and Aonla, Myrobalan or Indian gooseberry in English (Pathak, 2003). The plant and fruit of aonla are regarded sacred by 'Hindus' and have great mythological significance.

The aonla tree is native to tropical Southeast Asia, particularly central or southern India, Pakistan, Bangladesh, Sri Lanka, Malaya, Southern China and to Mascarene Islands. Seedling trees are of common occurrence in the mixed deciduous dry forests of India, ascending from sea level (western and Eastern Ghats, Aravali and Vindhya hills) to 1300 m above mean sea level, from northwest Himalayas (Jammu & Kashmir, Himachal Pradesh, Uttranchal) to eastern Himalayas in Assam, Meghalaya, Mizoram, Manipur and Tripura (Pathak, 2003). The natural distribution of wild aonla is found on the Himalayas, Chota Nagpur, Bihar, Orissa, West Bengal, North Circars, Deccan, Karnataka and in Western Ghats (Rawat and Uniyal 2003).

In India, the homeland of aonla, domestication first began in Varanasi (earlier known as Benaras) district of Uttar Pradesh with the initiative of Maharaja of Kashi. Banarasi, a superior genotype was selected from the wild aonla trees available in large numbers in the nearby Vindhyan hills. Authentic information regarding its cultivation dates back to 1881-82 in the Pratapgarh district of Uttar Pradesh.

The wild aonla germplasm is mostly confined in the mixed forests with sloppy topography and sometimes even difficult to approach. A rich genetic diversity of aonla exists in northeastern region of India, particularly in lower Assam, Maghalaya, Mizoram and Tripura (Yadav *et al.,* 2001). Aonla grows abundantly in the forest of Khasi and Garo hills of Meghalaya and locally known as "Sohmylleng" (Pandey *et al.,* 1993). The natural population of aonla in west Khasi hills (Nongkhyllum, Rajaju, Khonjoy area) of Meghalaya warrants *in situ* conservation, which may even be declared as gene sanctuary for this species (Hore 1998). Mizoram is homeland of wild aonla and star gooseberry (*Phyllanthus acidus*), which has potential as dwarfing rootstock for aonla. Wild Star gooseberry trees are found in forests of Kolasib, Thingdawl and Champhai in Mizoram. Madhya Pradesh forests have rich diversity of aonla. Jharkhand and adjoining areas of Chhatisgarh have rich diversity of aonla in the native forest. The important sites in Jharkhand are Lali forest near Ranchi, Dalma range of Jamshedpur, Theo Ghat forest of West Singhbhoom, Tiamara valley area between Ranchi and Jamshedpur, Ramgarh area of Hazaribagh, Parasnath hills of Girideeh, Kodemera and Jaomi areas of Bihar border, Simdega and Netarhat forest areas of Gumla, Belta forest of Daltonganj, Palamu and Garhwa of Jharkhand and adjoining areas of Sarguja and Ambikapur districts of Chha tisgarh and Sahdol district of Madhya Pradesh. The Belta forest (Daltonganj), Netarhat range in Gumla and Dalma range of Jamshedpur have comparatively high plant population of aonla in the natural habitat. In western and eastern ghats, three species *Phyllanthus emblica, Phyllanthus indofisheri* and *Phyllanthus acidus* are of common occurrence. A wild strain of aonla grows in the Himalayas up to an altitude of 1600 m. The fruits of wild aonla are relatively smaller. In the mid Himalayas almost wild aonla is distributed right from western to eastern Himalayas including Nepal (Pathak, 2003).

Ber

Out of the 50 reported species nearly 18 to 20 are native to India. A resume of species availability in different locations in India are given below:

Species	Locations
Z.apatala, Z.funiculosa, Z.incurva	North-Eastern hills
Z.mauritiana, Z.nummularia	All over the drier tracts, particularly in North-West India and UP
Z.oenoplia, Z.rugosa	Throughout India except in drier tracts, particularly in Central and Eastern India
Z.vulgaris	North-Western Himalayas
Z.rupicola	Central and Eastern India
Z.xylocarpus	MP and Peninsular region

There are more than 100 named varieties in ber and areas rich in variability have been identified in several places in UP, Rajasthan, Haryana, Gujarat, MP, Maharashtra, AP, Karnataka and Tamil Nadu. However, some popular cultivars are Umran, Gola Reshmi and Illaichi.

Custard apple

The genus *Annona* contains some 120 species originating from warm countries but a few important species became integral part of the Indian culture bearing the names of great heroes of the epic Ramayana. Some important species and their natural distribution are given below:

Species	Common name	Varieties	Location
A.squamosa	Sweetsop, Sharifa, Sitaphal	a) Green types: Balanagar, Mammoth, British Guinea, Washington-95,Barbadose seedling, Arka Sahan, (An F1 hybrid)(b) Red types: Red Sitaphal	Low and medium elevations throughout tropics
A.cherimola	Lakshmanphal	Cherimoyar	Cooler places in India
A.atemoya		Pinks mammoth, Bradley, Keller, Page, African Pride, Island Gem	Adopted to colder climate and alkali soils
A.reticulata	Ramphal or Bullock's Heart	Used mostly as a root stock	
A.glabra	Pond apple	Root stock	Flooded areas
*A.mu*ricata	Soursop, Hanuman Phal	Root stock	Mountain regions ofIndia
A.montana	Mountain soursop Used in breeding programme for quality		
A.purpurea	Soncoya	Used as resistant source for fruit cracking	
A.scleroderma	Eoshto	Used in breeding programmes for thick hard shell	

Fig

The original home of origin of fig (*Ficus carica)* is South Arabia (Egypt). There are four horticultural types in this crop viz., Smyrna, Capri, Sanpedro and Adriatic. This crop has very narrow range of diversity in India and there is a scope for introducing exotic germplasm. However, there are wild relatives found in India and some of them are given as under:

Species	Common name	Distribution
F. auriculata	Timla	Bihar,Odisha, Khasi hills, Manipur
F. benghalensis	Banyan	All over India
F. benjamina		
F. carica	Fig	Uttar Pradesh, Rajasthan, Andhra Pradesh, Maharashtra, Karnataka
F.elastica	Indian Rubber tree	Assam and Khasi hills
F.glomerata	Cluster fig	
F.hispida		Throughout India
F.krishnae	Krishna's fig	
F.lucescens		North India, MP and Western peninsula
F.palmata		N.W India and Rajasthan
F.religiosa	Peepal tree	All over India
F.rumphii		Punjab, MP, Assam
F.samicordata		Punjab, Assam, Bengal, Khasi hills and Manipur

Jamun

This genus *Syzygium* comprises about 1000 species of evergreen trees and shrubs; most of them are tropical in origin. Jamun is found in Western Ghats and very extensively in the tropical region. The diversity found is due to the high heterozygosity and seed multiplication. Some of the species are described below:

Species	Common name	Distribution
S. aqueum	Watery Rose-apple, Fruits edible	A small tree distributed in Assam and Meghalaya
S. arnottianum	Produces edible fruits	W.Ghats, The Nilgris, Palni and Anamalai hills
S. aromaticum	Clove, dried flower buds are of commercial importance	Evergreen trees cultivated in Tamil Nadu and Kerala
S. claviflorum	Fruits are acidic and edible	Andamans
S.cumini	Java plum, Jamun, Jambu	Throughout India
S.fruticosum	Wild Jamun	Avenue tree
S.jambos	Rose-apple	Many parts of India
S.mappaceum	Grown as ornamental plant	Assam, Meghalaya, Arunachal Pradesh and Tamil Nadu
S. samarangense	Wax Jumbu, fruits edible	Andamans and many parts in India
S. zeylanicum	Aromatic fruits are edible	Maharashtra,Karnataka, Odisha, Kerala and Andamans

Pomegranate

Pomegranate (*Punica granatum*)is an ancient fruit, which originated in Persia, Afghanistan and Baluchistan naturalized in Western India very early. Its wild forms are found in lower hills of Himachal Pradesh. Most of the pomegranate types cultivated in India are of seedling origin and thus providing a wide range of variability with respect to fruit shape, size, and mellowness of seed, aril colour, rind colour, sweetness and acidity of juice. Some popular varieties in different regions are furnished below:

Regions	Variety/type
Maharashtra	Ganesh, Bhagwa, Mridula, Aaraktha,G-137,P-23, P26, Muskat
Karnataka	Ganesh, Ruby, Bassein Seedless
Gujarat	Dholka
Rajasthan	Jodhpur Red, Jodhpuri White
Tamil Nadu	Yercaud, Vellodu, Kabul Red, CO-1

Apart from the above mentioned fruit species, there are several other species of fruits for which considerable diversity exists in the wild and conservation of such fruits needs to be carried out both *in situ* as well as in *ex situ*. There is also a need to work out the diversity using molecular means, so that the concept of 'core collection' can be practiced effectively.

In situ conservation of tropical fruit trees

In situ conservation of tropical fruit tree species is one of the most important aspects in the overall conservation of fruit diversity (Dinesh 2001). It is well known that many of the species of mango, when introduced to other areas do not perform well or die. It is observed that the *Mangifera andamanica, M. camptosperma* and *M. griffithi* when introduced to mild tropics could not survive (Prakash, 2001).

It is to be mentioned here that in spite of innumerable problems that are faced in the *ex situ* conservation, it is still advantageous to maintain them in the field genebank, as it keeps the biodiversity of a particular species safe when plants are destroyed in the wild. Hence, to rationalize the concept of core collection was introduced, which in a limited set represents the genetic spectrum in the whole collections (Brown, 1989). It is proposed that landraces should be preserved for future generations because they harbour a diversity of interesting traits for future breeding work and for developing new farming systems and moreover, reflect the cultural identity of certain groups of people (Altieri and Merrick 1987).

Until recently, germplasm conservation of crop landraces, as well as of their wild relatives, relied on *ex situ* methods (i.e. the conservation of biological material outside its natural habitat, (UNCED 1992), mostly in germplasm banks. More recently *in situ* (on-farm) conservation (i.e. the conservation of biological diversity in its natural habitat) has been proposed as a conservation strategy which allows evolutionary processes to continue rather than being halted as occurs to a large extent in *ex situ* conservation (Frankel *et al.,* 1995; Maxted *et al.,* 1997).

Pollen Storage as a Means to Conserve Diversity in Tropical Fruit Trees

Genetic conservation through pollen storage is desirable for a variety of horticultural plant species, since pollen is known to transmit important genetically heritable characters. Pollen is a product of genetic recombination and can provide a reliable source of nuclear genetic diversity at the haploid stage.

Fruit tree pollen is generally required to be stored for controlled crossings, either to achieve a desired breeding objective, or to overcome a constraint involved in commercial fruit production.

Gene pool conservation at the haploid stage can, therefore, be effectively accomplished through pollen, which can provide a rich source of nuclear genetic diversity. A major emphasis on research needs include pollen storage (Arora and Rao 1995) for citrus, mango, rambutan, jackfruit, durian and litchi. Pollen cryopreservation research has been recently recognized by IPGRI (IPGRI 1995). As one of the genepool components in an integrated PGR conservation programme, pollen can serve as an alternative or additional *ex situ* conservation method. Alexander and Ganeshan (1993) have extensively reviewed pollen storage research in fruit crops. Grout and Roberts (1995) have elaborately described the methodology involved in pollen cryopreservation. Hoekstra (1995) has assessed the merits and demerits of pollen storage for genetic resources conservation.

One of the main lacunae so far in the survey and collection of tropical fruit tree species has been the non-adoption of 'Geographic Information System' (GIS) tools, with the result that many of the regions with rich diversity were excluded. However, the use of Geographic Information System tools of late has helped in avoiding duplication of surveys and carrying out surveys with fair degree of accuracy. These tools can also help in isolating genotypes with exceptionally good characteristics.

GIS tools and preparation of distribution maps

Geographic Information System (GIS) is widely used in management of natural resources. Presently GIS is being widely used in mapping biodiversity by different organizations. GIS is a database management system with specific functions to handle spatial data, i.e. latitude and longitude. Many applications of GIS have been developed for commercial purposes or for specific management purposes, for example, Atlas, MapInfo for Windows, Arc/Info, etc. for commercial use, and GRID, FloraMap, DIVA, etc. for specific purpose of mapping biodiversity. For mapping biodiversity and its assessment for tropical fruit tree species, software such as FloraMap and DIVA, which were developed by the International Potato Centre (CIP) and International Centre for Tropical Agriculture (CIAT) for research purposes. GIS has two kinds of software, viz. rector-based system and raster-based system. The rector-based system stores geographic data as points, while the raster-based system stores data as grid cell. For mapping genetic diversity, rector-based system is popularly used. Using DIVA and FloraMap maps have been generated for fruit species like *Mangifera* and *Citrus*.

Biodiversity Assessment Using Molecular Techniques

With the advent of molecular biology techniques, DNA-based markers have replaced enzyme markers in germplasm identification and characterization. Because of its plasticity, ubiquity and stability, DNA is the ideal molecule for such analysis in fruit crops (Lalitha Anand, 2001). Although, molecular markers have been used to create extensive linkage maps for many annual plants (Helentjaris 1987; Bernatzky and Tanksley 1986; McCouch *et al.,* 1988), a few attempts have been made for their use in fruit crops. Genetic analysis and breeding of woody perennial fruit species can be complicated by many factors including long periods of juvenility, high ploidy levels, lack of described Mendelian markers and so on (Rowland and Levi 1994). Hence, the task of developing molecular marker-based genetic maps is both challenging and important.

References

Alexander, M.P and S. Ganeshan, 1993. Pollen Storage, (In): Advances in Horticulture, Vol I. Fruit Crops Part I (K.L. Chadha and O.P. Pareek eds).: 481-86, Malhotra Publishing House, New Delhi.

Altieri, M.A., and L.C. Merrick. 1988. Agroecology and *in situ* conservation of native crop diversity in the Third World. Pp. 361-369, (In) E.O. Wilson, (ed)., Biodiversity. Washington, D.C.: National Academy Press.

Arora, R.K and V. Rao, 1995.Proceedings of the expert consultation on Tropical Fruit Species of Asia, 17-19 May 1994, Malaysian Agricultural Research and Development Institute, Serdang, Kuala Lumpur IPGRI Office for South Asia, New Delhi 116p.

Arora, R.K. and E.R. Nayar. 1984. Wild relatives of crop plants in India. NBPGR Sci. Monogr.No. 9.90 p.

Arora., R.K., 1994, Promoting conservation and use of tropical fruit species in Asia. Proceedings of Expert Consultation on Tropical Fruit Species of Asia held at MARDI, Kaula Lumpur, Malaysia 17-19 May.

Bernatzky, R and S.D., Tanksley 1986. Towards a saturated linkage map in tomatobased on isozymes and random cDNA sequences. *Genetics*, 112: 887-898.

Brown., A.H.D.,1989. Core Collections: A practical approach to genetic resources management. *Genome,* 31: 818-824.

Dinesh, M.R., 2001. Field genebank management in tropical fruit species. Regional Training Course on "Characterization, evaluation and conservation of tropical fruits genetic resources" held at Bangalore from May 15-26th, 2001.

Frankel, O.H., A.D.H. Brown and J.J. Burdon. 1995. The conservation of plant biodiversity. Cambridge University Press

Grout, B.W.W and A. V. Roberts, 1995. Storage of free pollen, pollen embryos and zygotic embryos of seed by cryopreservation and freeze drying. In: Genetic Preservation of Plant Cells *In vitro* B. Grout (Ed) Pp 63-74, Springer Verlag, Berlin.

Haq, N. 1994. Identification of priority species of tropical fruits for research and development [Final Report IPGRI Project No. 93/11 (AOG-APO), ICUC in collaboration with IPGRI-APO]. International Centre for Underutilized Crops, Department of Biology, Southampton University, UK.

Harlan, J.R. 1975. Crops and man. American Society of Agronomy, Madison, Wisconsin. 295 p

Helentjaris, T., 1987. A genetic linkage map of maize based on RFLP's. *Trends Genet.* 3: 217-221.

Hoekstra, F.A., 1995. Collecting pollen for genetic resources conservation. In: Collecting Plant Genetic Diversity. Technical Guidelines (L. Guarino, V. Ramanatha Rao and R. Reid (Eds) Pp 527-550, CAB International, Wallingford, UK.

Hore, D.K. 1998. Fruit germplasm diversity and its development in north-eastern India. *Proc. Nat. Conf. Sci. & Tech.* Pp. 101-106.

IPGRI, 1995. Annual Report, IPGRI, Rome

Kostermans, A.I.G.H and J.M. Bompard., 1993. The mangoes: their botany, nomenclature horticulture and utilization, Academic Press, London.

Krishnamurthi, S and V.S. Seshadri., 1958. Origin and evaluation of cultivated bananas. *Ind. J. Hort.*15:135-45.

Lalitha Anand., 2001. Assessing genetic diversity using molecular techniques in tropical fruit species. (In) Proc of Regional Training Course on Characterization, evaluation and conservation of Tropical fruits genetic resources held at Bangalore 15-26th May.

Maxted, N., B.V. Ford-Lloyd and J.G. Hawkes. 1997. Complementary conservation strategies. Pp. 15-39 (In) Plant Genetic Conservation (N. Maxted, B.V. Ford-Lloyd and J.G. Hawkes, eds). Chapman and Hall, London.

McCouch, S.R., G. Kochert, Z.H. Yu., G.S. Khush, W.R. Coffman and S.D. Tanksley.,1988. Molecular mapping of rice chromosomes. *Theor. Appl. Genet.* 76: 815-829.

Mukherjee, S.K., 1949. A monograph on the genus *Mangifera* L. Lloydia (12):73-136

Mukherjee, S.K., 1985. Systematic and eco-geographic studies on crop gene pools 1. *Mangifera* L. IBPGR, Rome.

Naik, K.C., C.Bhujanga Rao and Raman V.S., 1958. Problems in crop improvement in mango. *Indian J. Hort.*, 15: 159-167.

Pandey, G., B.D. Sharma, D.K. Hore and N.V. Rao. 1993. Indigenous minor fruit genetic resource and their marketing status in north-eastern India. *J. Hill Res.* 6(1):1-4.

Paroda, R.S and R.K. Arora., 1991 Plant Genetic Resources Conservation and Management Concepts and Approaches. Published by the International Board for Plant Genetic Resources, Regional Office for South and Southeast Asia, c/o NBPGR, Pusa Campus, New Delhi 110 012, India

Pathak, R.K. 2003. Twenty five years of research and development of aonla. Lead paper presented in National Seminar on Production and Utilization in Aonla during August 2003 at Salem, Tamil Nadu, organized by Aonla Grower Association of India.

Pathak, R.K., 2003. Status Report on Genetic Resources of Indian Gooseberry-Aonla (*Emblica officinalis* Gaertn.) in South and Southeast Asia. APO Publication

Prakash. G.S., 2001. *In situ* conservation of tropical fruit species. Regional Training Course on "Characterization, evaluation and conservation of tropical fruits genetic resources" held at Bangalore from May 15-26th, 2001.

Purseglove, J.W. 1968. Tropical Crops: Dicotyledons. Essex: Longman Group and Co. Ltd. Burnt Mill, United Kingdom.

Rai, Mathura. 1995, Genetic Diversity In Sapota (*Manilkara achras* MILL.) in India. IPGRI Newsletter For Asia, the Pacific and Oceania, No.17: 12-13.

Rawat, R.B.S. and R.C. Uniyal 2003. Amla- An important species for Indian system of medicine. Paper presented in National Seminar on Production and Utilization in Aonla during August 2003 at Salem, Tamilnadu, organized by Aonla Grower Association of India.

Rowland, L.J. and A. Levi., 1994. RAPD-based genetic linkage map of blueberry derived from a cross between diploid species (*Vaccinium danowi* and *V. elliotti). Theor. Appl. Genet.*, 87:863-68.

Singh, H.P., 1996. Perspective of banana production and utilization in India. In Banana-Improvement, Production, and Utilization (H.P. Singh and K.L. Chadha eds), Proceedings of the conference on 'challenges from banana production and utilization in 21st century held in Trichy, India 24-25 September 1996.

UNCED. 1992. Convention on Biological Diversity. United Nations Conference on Environment and Development, Geneva. http://www.biodiv.org/convention/default.shtml).

Verheij, E.W.M. and R.E., Coronel., 1991. Plant Resources of South –East Asia No.2. Edible Fruits and Nuts, Pudoc, Wageningen.

Yadav, D.S., M. Mishra and A. Nath. 2001. Horticultural Research: An Overview. Pp 93-124

Yadav, I.S. and H.P. Singh, 1985. Evaluation of different ecological groups of mango cultivars for flowering and fruiting under subtropics. *Prog. Hort.* 17(3): 168-75.

Yadav, I.S. and S. Rajan, 1993. Genetic resources of mango. (In) Advances in Horticulture Vol. 1. Part 1 (Eds K.L. Chadha and O.P.Pareek). Malhotra Publishing House, New Delhi, pp. 77-93.

3

Fruit Breeding-Challenges, Approaches and Prospects

M.R. Dinesh and G.L. Veena

Fruit crops, majority of them are generally perennial in nature and owing to the outcrossing during their evolutionary stages, they are highly heterozygous. Genetic improvement is extremely difficult due to the complex genetic nature. The germplasm collection, conservation and evaluation take long time and area. Although the concept of core collection is very much valid, there is always a question mark when it comes to highly heterozygous crops as to whether the sample is good enough to represent the population. Due to the propagation by seeds, the large variability present needs to be evaluated for various characteristics. The genetical studies carried out in many of these crops are not conclusive and the progeny performance cannot be predicted. The biometrical methods to estimate the parameters viz., combining ability, heritability using the methodologies such as diallel have proven to be futile excercises to a very great extent. Parental selection, which is so vital for a successful breeding programme is very difficult as mainly selection has to be based upon the phenotype. Breeding for particular traits like biotic or abiotic stress has been difficult as resistance is not found within the spp and interspecific / intergeneric hybridization is difficult because of incompatibility barriers for eg the 'PRSV' breeding programme in papaya. In highly heterozygous crop like mango, there exists strong genotype x environmental interaction, which alters the performance of improved varieties.

Problems vis-à-vis advantages

1. Highly heterozygous
2. Long juvenile phase
3. Taxonomy is complex
4. Chromosomal aberrations
5. Nucellar embryony
6. Cross and self incompatibility.

Fruit crops most of them being perennial are highly heterozygous. This comes in the way of improvement as understanding of the genetics and inheritance pattern of the quantitative traits become extremely tedious. Hence, the selection of parents in a breeding programme becomes difficult. Parents when they are selected based on the phenotype and used in the breeding programme, progeny performance or the manifestation of hybrid vigour becomes unpredictable. The progenies also are influenced to a great extent by the environment. However, one of the advantages of the heterozygosity is that in the F_1 generation itself segregation will be maximum and the desirable trait in the progenies can be fixed by vegetative propagation.Perennial crop breeding takes long time. The seedling progenies take more than eight to nine years to fruit thereby making evaluation of progenies a long duration programme. The absence of molecular markers and pre-selection indices in most of the crops is hampering the breeding programme. Several attempts have been made in crops like mango by vegetative propagation to induce early flowering through grafting (Kulkarni, 1991; Kulkarni, 1986 and Kulkarni, 1988 a)

In most of the fruit crops, taxonomy is complex and the nomenclature ambiguity in crops like mango due to the large unevaluated variability has hindered improvement. The chromosomal aberrations in the case of seed propagated crops generally get deleted. But in vegetatively propagated crops these aberrations get fixed ultimately resulting in undesirable recombinants. The presence of nucellar embryony in crops like citrus and mango interferes in the identification of zygotic seedlings. Hence, there is an urgent need to develop markers to identify them at the nursery stage. In some of the crops cross and self incompatibility come in the way of selecting parents for the breeding programme. Raising a large population, which is the basic necessity for selection, is extremely difficult and the large area needed for the evaluation is another hindrance in the breeding programme.

Breeding approaches

The breeding approaches aimed at the improvement are almost same pertaining to fruit crops like other crops. However, the production of inbreds are totally ruled out barring a few seed propagated crops like papaya, in which also only 90% homozygosity can be achieved through selfing. Here in this chapter, only introduction and selections are discussed as hybridization and mutation are discussed separately under each of the crops. Many of the fruit crops have its base in south-east nations (Table 1). Apart from mango, some of the crops like banana, pomegranate, Citrus, etc have their primary centre of origin as south east-asia.

Table 1. Ploidy and Places of origin

Sl. No	Crop	Ploidy number	Place of origin
1.	Mango	2n=40	Indo-Burma
2.	Banana	2n=22,33,44	South East Asia
3.	Citrus	2n=18	North western China
4.	Grape	2n=38	Black sea to Caspian sea
5.	Papaya	2n=18	South Mexican and Central American Central
6.	Guava	2n=22	South Mexican and Central American Central
7.	Sapota	2n=26	Mexico
8.	Pomegranate	2n=16,18	Near-Eastern Center: Includes interior of Asia Minor, all of Transcaucasia, Iran, and the highlands of Turkmenistan.
9.	Fig	2n=26	West Asia
10.	Avocado	2n=24	Mexico and West Indies
11.	Aonla	2n=28	Tropical Asia; Indo-china
12.	Custard Apple	2n=14	Tropical America
13.	Pineapple	2n=50,75,100	Brazil
14.	Jamun	2n=40	East India- Malaya

Introduction

It is one of the most important methods of creating variability in a crop. Introduction can be within a particular region or from outside essentially from outside the country. One of the important aspects of such an introduction is the genotype x environmental interaction. If a variety has to perform in the place of introduction then it should be the least sensitive to the soil and climatic changes. A pointer to this is the introduction of the mango variety Dashehari in the southern regions of the country. Although the quality of the fruit has not changed, the size of the fruit has come down. However, even in such cases as well as the introduction from outside, care should be taken not to introduce new pests and diseases. Even if an introduction fails in its purpose, these can always be used as prebreeding lines, which would help in the development of improved recombinants. Fruit crops like papaya and guava, which have been introduced from Latin American countries, have been able to establish well in India. Many of the mango varieties viz., introduced from India to the American continent have performed well in these regions. Hence, introduction can play an important role in the improvement of crops.

Some of the introduced varieties

Sl. No.	Fruit Crops	Cultivars	Introduced from
1.	Banana	Lady Finger	Australia
		Grand Naine	France
2.	Citrus	Torocco Blood	USA
		Sunramon	Peru
		Vainiglia Sanuigno	USA
		Tha-Thai-1	Thailand
	Mandarin	Kinnow, Pixie, Wilking, Sunburst, Shamouti, Satsuma	USA
	Sweet Orannge	Blood red, Washington Navel Orange	USA
	Acidlime	Star Ruby	USA
	Lemon	Lisbon, Meyer	USSR
3.	Mango	Tommy Atkins, Zilet, Haden, Sensation, Julie	USA
		Carabao	Phillippines
4.	Papaya	Solo Sunrise, Sunset, Wilder	USA
5.	Guava	Beaumont G-35, Indonesia Seedless	Australia
		Verdie	USA
6.	Grape	Thompson Seedless, Perlette, Beauty seedless	USA
		Kishmish Beli, Kishmish Charni, Dogridge	USA
7.	Fig	Genoa White, Conadria	USA
8.	Datepalm	Khalas, Tayer, Mejnaj	Saudi Arabia
9.	Pomegranate	Wonderful	USA

Clonal selection

Clonal selection is one of the important tools in the breeding programme that helps in creating variability in woody perennials especially in crops like mango. The exact basis for clonal variants is still not known, but spontaneous mutants may be one of the reasons for such variants. In crops like mango, innumerable varieties have added to the confusion of nomenclature ambiguity by way of clonal variation. However, clonal selections have been reported in grapes, mango etc. In clarifying the nomenclature ambiguity biotechnological tools come in handy.Tissue culture can also be used for rapid multiplication of the selections made. Selections made out of clones can also be used as prebreeding lines as they may have some desirable traits other than that of the original mother clone.

One of the main drawbacks of the clonal seclections is the susceptibility to diseases and pests. If the original mother clone is susceptible, then the selections made also will be susceptible.

Open pollinated selection

As mentioned above, the number of progenies raised is one of the most important steps in the breeding programme. In some of the perennial crops like mango, due to fruit drop fruit recovery from a cross is very low. In many of the crops, flowers are delicate. In such cases there is a need to develop alternate strategies to raise large number of progenies. One of the methods would be to raise open pollinated progenies from a desired variety wherein trait improvement is sought after or in an otherwise ideal variety, which suffers from one or two defects. In these progenies, female parent is known but the male parent is not known. Validation of these half-sib progenies for their parentage can be carried out by using molecular markers once a hybrid progeny is selected.

The evaluation of hybrid progenies can be carried out by planting them in high density so that the large number of progenies can be accommodated.

Hybridization

The success of a hybridization programme depends on the selection of parents and the number of recombinant progenies selected out of a particular cross. In the case of many of the fruit crops, the flower structure is complex and the crossing is difficult. Intervarietal hybridization is quite common in almost all the fruit crops. However, interspecific or intergeneric hybridization is not common except barring a few of the temperate crops like strawberry. In most instances compatibility mechanism between the genera or species has not been worked out or there is incompatibility mechanism operating. In some of the fruit crops like mango, use of wild spp holds promise (Iyer, 1991). Intergeneric hybridization has been used in papaya for development of tolerant lines to 'PRSV (Dinesh *et al.,* 2007 and 2013). Interspecific hybridization has been adopted in developing rootstocks resistant to diseases viz., wilt resistance in guava (Edward, 1960). Interspecific hybridization is very widely practiced in some of the fruit crops like Citrus, wherein intervarietal hybridization work is very limited as the desired traits are available across the species and there is no barrier. In the development of better quality fruits, it is used in custard apple breeding programme (Jalikop, 2011). Use of widely related specis is also observed since long in banana, although, viable hybrids have not been generated.

Ploidy manipulation

Ploidy manipulation also can be considered as part of mutation breeding by some. It has been employed in many of the crop improvement programmes. Ploidy manipulation can help fruit breeders especially production of haploids and doubled haploids as they can show the way in shortening the time needed to produce homozygous lines compared to conventional breeding and also in the production of viable of triploids. However, one of the basic needs for undertaking ploidy manipulation is that the crop should have lower chromosome number. Polyploids are broadly classified based on their chromosomal composition into euploids or aneuploids. It is generally observed that euploids constitute the majority of polyploids. Euploids are polyploids with multiples of the complete set of chromosomes. Euploids are further classified into either autopolyploids or allopolyploids. Comai (2005) refers to tetraploidy as the most common of the euploids. He also observed that unlike diploids which may lose heterosis with each consecutive generation due to segregation, alloploidy and autoploidy impose pairing of homologous chromosomes, preventing intergenomic recombination. Acquaah (2007) called the tendency for a doubled set of chromosomes to pair independently of the doubled set of chromosomes of the other species as preferential or selective pairing. Because of this phenomenon, heterosis is not lost with the multiplication of successive generation in these polyploids unlike diploids.In several of the fruit crops it is attempted, especially in Citrus and banana where lot of work has been carried out. One of the limitations is that seeds will have to be taken and these are already heterozygous. In some crops, another hindrance noticed is the conversion of tetraploids into diploid like behaviour, which is termed as 'diploidization' (Ozkan and Feldman, 2009). Oritz *et al.*,(1997), through ploidy manipulations by utilizing the local triploid plantain landraces and diploid or cultivated wild bananas in interspecific hybridization in triploid-diploid crosses, followed by *in vitro* seed germination through embryo culture, and further rapid micropropagation of selected genotypes developed a tetraploid plantain type.

In Citrus, it was observed that occurrence of duplication in the female gametes varies between less than 1% to more than 20% (Soost, 1987; Iwamasa and Nito, 1988). Esen *et al.,* (1979) have suggested that this happens due to the abortion of the second meiotic division in the megaspore. This hypothesis was confirmed by Luro *et al.,* (2000, 2004) in Clementine through molecular marker analysis showing that less than 50% of maternal heterozygosity is transmitted by the 2n ovules. Chen *et al.,* (2007) proposed that 2n gametes of sweet orange result from first division restitution. Luro *et al.,* (2000) observed the formation of triploid hybrids by fertilization of a haploid ovule by diploid pollen. Grosser *et al.,* (2010), reported nine new somatic hybrids and five autotetraploids produced by protoplast fusion, with focus on the zipperskin trait. Viloria *et al.,* (2004), reported that kumquat-derived resistance can be transferred to triploid lime-like hybrids when using canker resistant 'Lakeland limequat' as a parent. Assani *et al.,* (2003), reported the production of 41 haploid (n=x=11) plants from anther culture of banana [Musa balbisiana (BB)]. In papaya, haploid plantlets and pollen-derived embryos were obtained from anthers cultured at the uninucleate stage (Litz and Conover, 1978; Tsay and Su, 1985). Oiyama and Okudai (1986) combined colchicine treatments and the shoot tip grafting technique to produce tetraploid plants in Citrus. Aleza *et al.,* (2009) developed a technique for producing tetraploids that can be used as female parent for non apomictic citrus, which will be useful for apomictic cultivars also in producing large number of triploids.

Haploids and double haploids

In fruit crops, high heterozygosity, long generation cycle with long juvenile period, large size, and often self-incompatibility do not allow the haploidization easily through the conventional methods (Germanà, 2006). *In vitro* androgenesis is single-step to produce haploid or double haploid plant from heterozygous origin (Srivastava and Chaturvedi, 2008).

Compared to the conventional breeding approach, it is in this context that haploids and double haploids can make a significant contribution in shortening the time taken for fruit breeding. Using DH technology, completely homozygous plants can be established in one generation thus saving several generations of selfing comparing to conventional methods. Development of homozygous lines through gametic embryogenesis results in single step improvement of the crop. In a conventional breeding programme, a pure line is developed after several generations of selfing. However, in fruit crops, there is no way of obtaining haploids through conventional methods. Hence, fixation by doubling the chromosome complement presents a possibility using haploids. Haploids along with marker assisted selection can help in shortening the time taken for fruit improvement.

In many of the important fruit crops like mango, citrus, banana, grapes, papaya, sapota, pomegranate, custard apple, apples, peaches, pears, plums, apricot, kiwi fruit etc, it can be a very handy tool in the improvement. In most of the fruit crops, essentially there are two parts in breeding. One is the scion breeding and another is rootstock breeding. The main objectives in scion breeding are:

a. to obtain new varieties with precocity, higher yield and regular bearing with amenability for high density planting.

b. better shelf life of fruits

c. soft seededness with better pulp quality and attractive skin colour.

With regard to the rootstock breeding, the main objective is to develop a genotype resistant or tolerant to biotic and abiotic stresses. In fruit breeding, which is based on either conventional (hybridization, selection, mutation) or biotechnological methods employing embryo culture, regeneration from protoplasts, somatic hybridization, *in vitro* mutant selection, genetic transformation, production and doubling of haploids would help in understanding the inheritance of traits. This integrated approach with both biotechnological tools and conventional can help in obtaining good results in a short time.

Some studies in fruit crops

One of the fruit crops, in which ploidy manipulation has been attempted, is Citrus. Citrus and its related genera viz., *Poncirus, Fortunella*, etc., are diploids with a monoploid number of chromosome i.e., n=x=9 (Frost, 1925). However, triploid and tetraploid forms of Citrus have also been observed to exist in nature. Karasawa (1971) successfully obtained haploid seedlings of *Citrus natsudaidai* by utilizing gamma rays. Esen and Soost (1972) obtained one haploid embryo in an immature seed from a diploid (*Clementine mandarin*) x diploid (Pearl tangelo) cross. Oiyama and Kobayashi (1993) generated three haploid plants from *in vivo* crosses of two monoembryonic diploids Clementine and Lee and a triploid hybrid of "Kawano natsudaidai" (*Citrus natsudaidai* Hayata). Gemana` and Chiancone, (2001) regenerated haploid plantlets through gynogenesis in *Citrus clementine* Hort. ex Tan, cv. Nules. It was induced by *in vitro* pollination with pollen from a triploid plant. Hidaka and Omura (1989) described cytologically in *C. aurantium* and *P. trifoliata* that the nucleus divides without cell division and multinucleate pollen grain is initially formed, which later gives rise to a multicellular structure, that develops into a proembryo and finally into an embryo. Ramirez *et al.,* (2003) observed different cellular types in these embryos after the exine breakdown. Masaki Yahata *et al.,* (2005), developed a double haploid from a haploid Pummelo using colchicine treatment of axillary buds and he noticed that the obtained plant was similar to that of a diploid plant.

The grapevine, which is a diploid (2n=2x=38), the production of haploids would help especially in seedless grape development. There are a number of spontaneous mutants obtained in grapes over a period of time. Zou and Li (1981) obtained haploid in grapes. Gresshoff and Doy (1974) reported production of haploid callus line. Regeneration of plants has been obtained by Rajasekaran and Mullins (1979), Bouquet *et al.,* (1982), Hirabayashi and Akihama (1982), Mauro *et al.,* (1986) and Cersosimo *et al.,* (1990).

In banana, Kerbellec (1996) reported successful haploid plant regeneration in banana [*Musa acuminate* (AA)]. Assani *et al.,* (2003) reported the production of 41 haploid (n=x=11) plants from anther culture of banana [*Musa balbisiana* (BB)], 18 from the genotype *Pisang klutuk*, 12 from *Pisang batu*, 7 from *Pisang klutuk* wulung and 4 from *Tani*.

Another fruit crop in which work has been carried out to produce haploids was in apple. Apple is a diploid with 2n=2x=34. Several workers (Fei and Xue, 1981; Xue and Niu, 1984; Zarsky *et al.,* 1986; Zhang *et al.,* 1987; Ho¨ fer and Hanke, 1990, 1994; Verdoodt *et al.,* 1998; Ho¨ fer, 2003) have reported induction of embryogenesis and regeneration of pollen-derived plants from anther culture in this species. In the cultivars, 'Alkmene', 'Remo', 'Rene' and 'Realka'; Ho¨ fer (2004), reported improvement of the induction phase, which allowed to obtain induction of androgenic embryos. In the case of pears, haploids and double haploids were obtained by *in situ* parthenogenesis induced by irradiated pollen or by seedling selection from three pear cultivars: 'Doyenne′ du Comice', 'Williams' and 'Harrow Sweet' (Bouvier *et al.,* 1993). Bouvier *et al.,* (2002) reported doubled haploids by either spontaneous chromosome doubling or by oryzalin treatment. Kadota *et al.,* (2002), obtained two embryos by pear anther culture, cv. Le Lectier, but their origin was not established and plant regeneration was not obtained.

The commercial strawberry (*Fragaria x ananassa* Duch.) is grown in the most arable regions of the world. The cultivated strawberry is highly heterozygous with 2n = 8x = 56 chromosomes (Hancoock *et al.,* 1996). The production of haploid plants (n = 4x = 28) through anther culture is widely used for breeding purposes, as an alternative to the numerous cycles of inbreeding or backcrossing usually needed to obtain pure lines in conventional breeding (Debnath *et al.,* 2007). Haploid plant production in strawberry has been attempted through anther culture by several workers (Niemirowicz-Szczytt *et al.,* 1983; Owen and Miller, 1996), or through interspecific hybridization of diploids with two stages of colchicine doubling (Evans, 1977). Truong Xuan Nguyen *et al.,* (2012) in a study attempted to develop an anther culture system to produce haploid plant of day-neutral strawberry *(Fragaria x ananassa* Duch) cv. 'Albion' for use in breeding program. The cytological analysis of 33 regenerants showed that 6.0% were haploid (4x = 28 chromosomes), but 3.0% hexaploid (6x = 42), 69.7% octaploid (8x = 56), 6.1% doubled octaploid (16x = 112) and 15.2% aneuploid plants were also determined.

Induction of maternal haploids using irradiated pollen for pollination is one of the methods. Pollen germination on the stigma stimulates embryo development and growth of the pollen tube within the style, although irradiated pollen is unable to fertilize the egg cell. It has been used successfully in several species

(Jana Murovec and Borut Bohanec, 2012). In most plant species, *in vitro* embryo rescue is necessary to recover haploid plants. Pandey *et al.,* (1990), Chalak and Legave, (1997) and Aleza *et al.,* (2009), reported collection of mature seeds in kiwifruit. Froelicher *et al.,* (2007), reported the same in mandarin. In apple, irradiated pollen was used to induce haploids (Zhang & Lespinasse, 1991; Hofer and Lespinasse, 1996; De Witte and Keulemans, 1994), in blackberry by Naess *et al.,* (1998), Bouvier *et al.,* (1993) in pear and Höfer and Grafe (2003) in sweet cherry.

Conclusion

The use of doubled haploids in cross-pollinated species improves selection efficiency and can be used at any or each cycle of recurrent selection. Jana Murovec and Borut Bohanec, (2012) opined that in cross-pollinated species, numerous deleterious recessive alleles are not expressed in heterozygous states, they are gradually fixed during self-pollination, causing inbreeding depression and difficulties in producing homozygous lines during conventional breeding. Doubled haploid technology helps to overcome these problems through the rapid fixation of genes in one generation and early elimination of deleterious alleles from populations. The recovered recombination products thus represent more viable combinations of genes. The complete homozygosity enables true breeding and stable field performance over generations of progeny, although the complete lack of heterozygosity and heterogeneity in varieties is thought to be more vulnerable to environmental changes and altered cropping systems.

Haploidy, doubled haploidy and gametic embryogenesis in fruit crop breeding can bring about the shortening of the breeding period along with the understanding of the inheritance pattern of some of the most important traits. A better understanding of the gametic embryogenesis process could make haploid production a powerful fruit crop breeding tool in the future. There is a need to improve the induction rate (the frequency with which gametes form embryos) and to increase the survival rate. Through haploid breeding the prediction of progeny performance be it disease resistance or any other trait in question can be addressed, which is of great importance in perennial crop breeding.

References

Acquaah G., 2007. Principles of plant genetics and breeding. Wiley-Blackwell, Malden.

Aleza, P., Juarez, J., Hernandez, M., Pina, J.A., Ollitrault, P. and Navarro, L. 2009. Recovery and characterization of a *Citrus clementina* Hort. ex Tan. 'Clemenules' haploid plant selected to establish the reference whole *Citrus* genome sequence. *BMC PlantBiology*, 9: 110, ISSN 1471-2229.

Assani, A., Bakry, F., Kerbellec, F., Haicour, R., Wenzel, G. and Foroughi-Wehr, B. 2003, Production of haploids from anther culture of banana [*Musa balbisiana* (BB)]. *Plant Cell Rep.*, 21: 511–516.

Bouquet, A.B., Piganeau,B. and Lamaison, A.M. 1982, Influence du ge´notype sur la production de cals, d'embryons et de plantes entie'res par cultures d'anthe'res in vitro dans le genre Vitis. C.R. *Acad. Sci. Paris,* 259: 569–574.

Bouvier, L., Zhang, Y.X. and Lespinasse, Y. 1993, Two methods of haploidization in pear, *Pyrus communis* L.: greenhouse seedling selection, in situ parthenogenesis induced by irradiated pollen. *Theor. Appl. Genet.*, 87: 229–232.

Bouvier L, Gue´ rif P, Djulbic M, Durel C-E, Chevreau E and Lespinasse Y 2002. Chromosome doubling of pear haploid plants and homozygosity assessment using isozyme and microsatellite markers. *Euphytica* ,123: 255–262.

Cersosimo, A., Altamura, M.M., Crespan, M. and Paludetti, G. 1990, Embryogenesis, organogenesis and plant regeneration from anther culture in Vitis. *Acta Hort.*, 280: 307–314.

Chalak ,L. and Legave, J.M. 1997, Effects of pollination by irradiated pollen in Hayward kiwifruit and spontaneous doubling of induced parthenogenetic trihaploids. *Sci. Hort.*, 68: 83 93.

Chen, C., Lyon, M.T., O'Malley, D., Federici, C.T., Gmitter, J., Grosser, J.W., Chaparro, J.X., Roose, M.L. and Gmitter Jr., F.G. 2007. Origin and Frequency of 2n Gametes in Citrus sinensis × Poncirus trifoliata and their Reciprocal Crosses, Plant, Plant Science.

Comai L., 2005. The advantages and disadvantages of being polyploid. *Nature Reviews Genetics,* 6:836-846.

Debnath, S.C., A. Jaime and T.D. Silva. 2007. Strawberry culture in vitro: Applications in genetic transformation and biotechnology. *Fruit, Vegetable and Cereal Science and Biotechnology*, 1: 1 -12.

De Witte, K. and Keulemans, J., 1994. Restrictions of the efficiency of haploid plant production in apple cultivar Idared, through parthenogenesis in situ. *Euphytica*, 77, No. 1-2: 141-146, ISSN 0014-2336.

Dinesh M.R.., G.L. Veena , C. Vasugi , M. Krishna Reddy , and K.V. Ravishankar., 2013.Intergeneric hybridization in papaya for 'PRSV tolerance. *Scientia Horticulturae,* 161: 357–360.

Dinesh, M.R., A. Rekha, K.V. Ravishankar, K.S. Praveen and L.C. Santosh, 2007. Breaking intergeneric crossing barrier in papaya using sucrose treatment. *Scientia Horticulturae,* 114: 33–36.

Edward, J.C. 1960. Wilt of Guava. *Allahabad Farmers*. 34: 289-293.

Esen, A. and Soost, R.K. 1972, Unexpected triploids in citrus: their origin, identification, and possible use. *J. Hered.*, 62: 329–333.

Esen, A., R.K. Soost and G. Geraci. 1979. Genetic evidence for the origin of diploid megagametophytes in citrus. *J. Hered.*, 70: 5-8.

Evans, W. D., 1977. The use of synthetic octoploids in strawberry breeding. *Euphytica*, 26,(2):497-503.

Froelicher, Y., Bassene, J.B., Jedidi-Neji, E., Dambier, D., Morillon, R., Bernardini, G., Costantino, G. and Ollitrault, P. 2007, Induced parthenogenesis in mandarin for haploid production: induction procedures and genetic analysis of plantlets. *Plant Cell Reports*, 26(7): 937-944.

Frost , H.B. 1925, The chromosomes of Citrus. *J. Washington Acad. Sci.,* 15:1–3.

Fei, K.W. and Xue, G.R., 1981, Induction of haploid plantlets by anther culture in vitro in apple. *Cv.* "Delicious". *Sci. Agric. Sin.*, 4: 41–44.

Germana' MA and Chiancone B 2001. Gynogenetic haploids of Citrus after in vitro pollination with triploid pollen grains. *Plant Cell Tiss. Org. Cult.*, 66: 59–66.

Germanà, M.A. 2006. Doubled haploid production in fruit crops. Plant Cell, *Tissue and Organ Culture,* 86: 131–146.

Gresshoff, P.M. and Doy, C.H. 1974, Derivation of a haploid cell line from *Vitis vinifera* and the importance of the stage of meiotic development of anthers for haploid culture of this and other genera. *Z. Pflanzenphysiol*, 73: 132–141.

Grosser, J.W., H.J. An, M. Calovic, D. Lee, C. Chen, M. Vasconcellos, and F.G. Gmitter Jr..2010. Production of New Allotetraploid and Autotetraploid Citrus Breeding Parents: Focus onZipperskin Mandarins. *HortScience,* 45(8):1-4.

Hancook, J.F., D.H. Scott, and F.J. Lawrence. 1996, Strawberries. In: *J. Janick and J.N. Moore (Eds). Fruit Breeding:* Vine and Small Fruit. Vol. II: 419 -470. John Wiley and Sons, Inc., New York.

Hidaka, T. and Omura, M. 1989, Origin and development of embryos from microspores in anther culture of Citrus. *Jpn. J. Breed.*, 39: 169–178.

Hirabayashi, T. and Akihama, H. 1982, In vitro embryogenesis and plant regeneration from the anther-derived callus of *Vitis.* In: *Fujiwara A (ed) Plant Tissue Culture,* (pp. 547–548). Maruzen Co., Ltd, Tokyo.

Höfer M and Hanke V., 1990. Induction of androgenesis in vitro in apple and sweet cherry. *Acta Hort.*, 280: 333–336

Höfer., M., 2004. In vitro androgenesis in apple – improvement of the induction phase. *Plant Cell Rep.*, 22: 365–370

Höfer M and Hanke V., 1994. Antherenkultur bei Apfel und Kirsche: Einfluß des Pollenentwicklungsstadiums – KorrelativeBeziehungen zu morphologischen Blu¨ tenmerkmalen.*Gartenbauwissenschaft,* 59: 225–228

Höfer, M. and Grafe, Ch. 2003. Induction of doubled haploids in sweet cherry (*Prunus avium* L.). *Euphytica*, 130(2):191-197.

Höfer, M. and Lespinasse, Y. 1996. Haploidy in apple. In: *In Vitro Haploid Production in Higher Plants Vol. 3: Important Selected Plants*, Jain, S.M., Sopory, S.K. and Veilleux, R.E., pp 261-276, Kluwer Academic Publishers, ISBN: 9780792335795, Dordrecht.

Iwamasa, M., N. Nito and J. T. Ling., 1988. Intra- and intergeneric hybridization in the orange subfamily, Aurantioideae . International Citrus Congress (6th : 1988 : Tel Aviv, Israel) : Margraf. 1: 123-130.

Iyer, C.P.A., 1991. Recent advances in varietal improvement in mango. *Acta Horticulturae,* 291:109-132.

Jalikop, S.H. 2011. Breeding of Pomegranate and Annonaceous fruits. Proc. IInd IS on Pomegranate and Minor, including Mediterranean fruits. *Acta Hort.*, 890: 191-196.

Jana Murovec and Borut Bohanec, 2012, Haploids and Doubled Haploids in Plant Breeding. www.intechopen.com

Kadota, M., Han D-S and Niimi, Y., 2002, Plant regeneration from anther-derived embryos of apple and pear. *Hort Science,* 37(6): 962–965.

Karasawa, K. 1971, On the occurrence of haploid seedlings in *Citrus natsudaidai* Hayata. *Bull. Sakushingakuin Junior Col. Woman*, 1: 1–2.

Kerbellec, F. 1996, Establissement d_une technique d_androgene‘ se pour l_ame´lioration ge´ne´tique du bananier (Musa ssp.). *Ph.D. Thesis*, Ecole Nationale Agronomique de Rennes, France.

Kulkarni, V.J., 1991, Physiology of flowering in mango studied by grafting. *Acta. Hort.*, 291: 95-104.

Kulkarni, V.J., 1986, Graft induced off season flowering and fruiting in mango (*Mangifera indica* L.) *J. Hort. Sci.*, 1:141-145.

Kulkarni, V.J., 1988 a. Further studies on graft induced off season flowering and fruiting in mango (*Mangifera indica* L.) *J. Hort. Sci.*, 63: 361-367.

Litz. R.E and Conover. R.A., 1978. Recent advances in papaya tissue culture. *Proc. Fla State Hort. Soc.*, 91: 180–182.

Luro F, Maddy F, Jacquemond C, Froelicher Y, Morillon R, Rist D, and Ollitrault P, 2004. Identification and evaluation of diplogyny in clementine (Citrus clementina) for use in breeding. In: XI Eucarpia Symposium on fruit breeding and genetics. *Acta Hortic,* 663:841–847.

Luro, F., Maddy, F., Ollitrault, P. and Rist, D. 2000. Identification of 2n gamete parental origin and mode of nuclear restitution of spontaneous triploid Citrus hybrids. In: *Proceedings of the Ninth InternationalCitrus Congress*. Orlando, Florida, pp. 168–169.

Masaki Yahata, Hisato Kunitake, Tsutomu Yabuya, Yukiko kashihara and Haruki Komatsu, 2005, Production of a double haploid from a haploid Pummelo using colchicine treatment of axillary buds. *J. Amer. Soc. Hort. Sci.*, 130(6): 899-903.

Mauro, M.C., Nef, G. and Fallot, J. 1986, Stimulation of somatic embryogenesis and plant regeneration from anther culture of *Vitis vinifera* cv. Cabernet Sauvignon. *Plant Cell Rep.*, 5: 377-380.

Naess, S.K., Swartz, J.H. and Bauchan, G.R. 1998. Ploidy reduction in blackberry. *Euphytica.*, 99:57-73.

Niemirowicz-Szczytt K. Zakrzewska Z. Malepszy S. and Kubicki B. 1983. Characters of plants obtained from *Fragaria* X *ananassa* anther culture. *Acta Hortic.,* 131:231–237.

Oiyama I, and Okudai N 1986. Production of Colchicine-induced Autotetraploid Plants through micrografting in monoembryonic Citrus Cultivars. *Jpn J Breed.,* 36:371–376

Oiyama, I. and Kobayashi. 1993, Haploids obtained from diploid triploid cross of Citrus. *J. Jpn. Soc..Hort. Sci.*, 62:89-93.

Ortiz, R. and D.Vuylsteke, 1997. Developing new plantain varieties for Africa. *Plant Varieties & Seeds,* 10(1): 39-57

Ozkan H., Feldman M., 2009. Rapid cytological diploidization in newly formed allopolyploids of the wheat (Aegilops-Triticum) group. *Genome,* 52:926-934.

Owen, H.R. and A.R. Miller. 1996. Haploid plant regeneration from anther cultures of three North American cultivars of strawberry (*Fragaria x ananassa Duch*) . *Plant Cell Reports* 15: 905-909.

Pandey, K.K., Przywara, L. and Sanders, P.M. 1990, Induced parthenogenesis in kiwifruit (*Actinidia deliciosa*) through the use of lethally irradiated pollen. *Euphytica*, 51: (1): 1-9.

Rajasekaran, K. and Mullins, M.G. 1979, Embryos and plantlets from cultured anthers of hybrid grapevines. *J. Exp. Bot.*, 30: 399–407.

Ramirez, C., Chiancone, B., Testillano P. S., Garcia-Fojeda, B., Germana,‘ M.A. and Risueno, M.C. 2003, First embryogenic stages of Citrus microspore-derived embryos. *Acta Biol. Cracov. Bot.,* 45: 53–58.

Soost, R.K., 1987. Breeding Citrus-Genetics and nucellar embryony(In) improving vegetatively propagated crops (Abbott, A.J and Atkin, R.K-eds) Academic Press, New York, pp 83-110.

Srivastava, P. and R. Chaturvedi. 2008. *In vitro* androgenesis in tree species: an update and prospect for further research. *Biotechnol Adv.,* 26:482-491.

Tsay, H.S. and C. Y. Su. 1985. Anther culture of papaya (*Carica papaya* L). *Plant Cell Rep.*, 4(1):28-30. (SCI).

Truong Xuan Nguyen, Ye-Su Song1 and Sung Min Park, 2012, Haploid plant production through anther culture in Day-neutral strawberry *(Fragaria x ananassa* duch) cv. Albion. *J. ISSAAS,* 18(1):173-184.

Verdoodt, L., Van Haute, A., Goderis, I.J., De Witte, K., Keulemans, J. and Broothaerts, W. 1998, Use of the multi-allelic self-incompatibility gene in apple to assess homozygosity in shoots obtained through haploid induction. *Theor. Appl. Genet.*, 96: 294–300.

Viloria, Z., D.L. Drouillard, J.H. Graham, and J.W. Grosser. 2004. Screening triploid hybrids of’Lakeland’ limequat for resistance to citrus canker. *Plant Disease.*, 88:1056-1060.

Xue GR and Niu JZ 1984. A study on the induction of apple pollen plants. *Acta Hort. Sin.*, 11:161–164.

Zhang, Y. X., Boccon-Gibod, J. and Lespinasse,Y. 1987, Obtention d'embryons de pommier apre' s culture d'anthe' res. C.R. *Acad. Sci. Paris Ser.*, III 305: 443–448.

Zhang, Y.X., Lespinasse,Y. and Chevreau E., 1991, Induction of haploidy in fruit trees. *Acta Hort.,* 280: 293–304.

Zou, C.J. and Li, P.F. 1981, Induction of pollen plants of grape (Vitis vinifera L.). *Acta Bot. Sin.*, 1: 79–81.

Zarsky, V., Tupy, J. and Vagera, J. 1986, *In-vitro*-culture of apple anther. (In) Moet-Henessy *Conference Fruit Tree Biotechnology*, Moet Henessy, Paris, October, p. 75.

4

Apomixis and Fruit Breeding

G.L. Veena and M.R. Dinesh

Apomixis was defined by Winkler as replacement of the normal sexual reproduction by asexual reproduction, without fertilization. This definition notably does not mention meiosis. Thus "normal asexual reproduction" of plants, such as propagation from cuttings or leaves, has never been considered apomixis, but replacement of the seed by a plantlet, or replacement of the flower by bulbils are types of apomixis. Apomixis deviates from sexual reproduction in several aspects: modification or complete omission of meiosis, formation of unreduced megaspores, parthenogenetic development of embryo, and autonomous, or pseudogamous, endosperm development (Ramulu *et al.,* 1998). Apomixis produces embryos or seed progeny that are exact genetic replicas of the mother plant. The main advantage of apomixis over sexual reproduction is the possibility to select individual plants with superior characteristics and to propagate them clonally through seeds. Apomictically produced offsprings are genetically identical to the parent plant. In flowering plants, the term "apomixis" is commonly used in a restricted sense to mean agamospermy, i.e. asexual reproduction through seeds.

However, one of the most important aspects is that the female gamete either reduced or unreduced develops into a seed. Apomixis is of great use in fruit crops as the seed can be stored and transported more easily. Apomixis has not been used in the breeding programme extensively. Despite many years of breeding research, apomixis has not yet been exploited in commercial breeding programmes, and its genetic regulation is poorly understood.Hence, there is a need to use this so that it can be used in the breeding programme for the improvement of fruit crops. In the production of seed generated from apomixis the basic mechanisms involved are diplospory, apospory and adventive embryony. Diplospory and apospory result in the formation of a megagametophytic structure without meiotic reduction and the embryo develops from a cell in unreduced megagametophyte. Diplospory and apospory are, therefore, commonly referred to as gametophytic apomictic processes. By contrast, adventive embryony is initiated late in ovule development and usually occurs in mature ovules. Embryos are initiated directly from individual cells in the tissues of the ovule that are external to the megagametophyte. Therefore, adventive embryony has been described as sporophytic apomixes (Nogler, 1984; Asker and Jerling, 1992). Most plants with gametophytic apomixis are polyploids and genera with adventive embryony are generally diploids (Asker and Jerling, 1992). Many agronomic advantages of apomixis viz., rapid generation and multiplication of superior forms through seed from novel and underused germplasm; the reduction in cost and time of breeding; the avoidance of complications associated with sexual reproduction, such as pollinators and cross-compatibility; and the avoidance of viral transfer in plants that are typically propagated vegetatively, such as potatoes (Savidan, 2000a, 2000b). Apospory is probably controlled by a single dominant gene in *Panicum maximum* (Savidan 1982), *Ranunculus aricomus* (Nogler, 1984), *Cencchrus ciliaris* (Sherwood *et al.,*1994) and *Brachiaria* (Pessino *et al.,* 1997). Genetic control of displospory in *Taraxacum* resides on a single chromosome and probably a single locus, as suggested by Mogie (1988). Results provided by Leblanc *et al.,* (1995) with restriction fragment length polymorphism (RFLP) analysis in maize-Tripsacum hybrids support the hypothesis of monogenic control. However, nothing is known yet about the genetic control of nucellar embryony.

Apomixis in fruit crops—some studies

Although most of the fruit crops are not seed propagated, apomixis can be a handy tool in fixing the character. Apomixis is of great importance to agriculture primarily because of its potential to facilitate multiplication of F1 hybrid seeds. Hybrid seeds are preferred for use in modern agriculture mainly because they provide a yield advantage. Hybrid seed is usually produced from two inbred parental lines and outperforms the yield and vigour of either of the parental lines. Currently, production of hybrid seed is expensive because the parental lines need to be maintained as pure lines and during normal sexual development the hybrid vigour present in the F_1 seed breaks down in subsequent generations because of assortment and recombination, thus necessitating purchase of hybrid seed each season. If apomixes were to be available as a controlled tool where it could be switched on and off as required in plant breeding, hybrid vigour could be fixed, enabling indefnite multiplication of the hybrid of uniform quality without decrease in the yield advantage. Carl Ramage *et al.,* (2004) in a study of mango steen (*Garcinia mangostana*), used the novel molecular marker technique RandomlyAmplified DNA Fingerprinting (RAF) to survey genetic relationships between 37 accessions of the tropical fruit *G. mangostana* (mangosteen) and 11 accessions from eight other *Garcinia* species. Although mangosteen is believed to reproduce exclusively through apomixis, and the results showed that considerable genetic diversity existed within *G. mangostana* and between other Garcinia species. Among the 37 *G. mangostana* accessions examined, nine different genotypes were identified which clustered into three distinct groups based on correspondence analysis (reciprocal averaging), for 26 (70%) of the accessions no marker variation was detected over 530 loci screened. A further eight (22%) accessions exhibited very low levels of variation (0.2–1%) suggesting at least one well conserved mangosteen genotype. The remaining three accessions (8%) showed extensive variation (22–31%) compared with the majority of accessions. The three mangosteen groups were 63–70% dissimilar to the other *Garcinia* species investigated.

Many members of the genus *Citrus* and some closely related genera belonging to Rutaceae reproduce apomictically by nucellar embryony (Frost 1943). Since nucellar embryos develop asexually by ordinary mitotic division of cells of the nucellus and the male gamete does not contribute to their formation, nucellar seedlings are identical to the seed parent. The presence of extra (nucellar) embryos in addition to that from sexual origin leads to polyembryony. Polyembryony is the development of two or more embryos in one seed. Hence, a polyembryonic variety is commonly considered apomictic (yielding apomictic, nucellar, seedlings) and a monoembryonic variety, sexual or zygotic. However, it is well known that extra embryos are occasionally sexual, produced either from the fission of one fecundated egg or from two or more functional embryo sacs in a single ovule (Bacchi 1943; Cameron and Garber 1968).The propagation of citrus rootstocks depends on the production of clonal plants from nucellar seedlings. Uniformity of the rootstock genotype is essential for reliable plant performance following budding and orchard establishment. Most rootstock cultivars presently grown are polyembryonic, producing seeds that contain both nucellar and zygotic embryos. The proportion of zygotic embryos developing into seedlings varies with both genotype and environment (Cameron 1979), and uniformity is seldom complete. Off-type seedlings are generally rogued in the nursery, but the efficiency of such roguing is not known for many rootstocks. Using enzyme analysis, Anderson *et al.,* (1991) determined that rouging based primarily on size and growth habit of seedlings is effective in removing some, but not all, zygotic seedlings. They found that zygotic rootstocks escaped the roguing in two of the three groves studied and, in some instances, an apparent graft-incompatibility developed in young trees. In any case, zygotic seedlings may affect the performance of trees budded on them (Weber 1932; Roose and

Traugh 1988). Therefore, one of the most important traits to be considered within a breeding program for citrus rootstocks is apomictic reproduction by seed, *i.e.* that the new genotype yields the least number of zygotic seedlings. This makes apomixis one of the most important traits in breeding programs for citrus rootstocks.

Genetical basis

The genetical basis for apomixes was first given by the inheritance studies carried out by Gregor Mendel on *Hieracium.* However, later studies carried out in various species have shown that the inheritance of gametophytic apomixis is associated with the transfer of either a single locus or a small number of loci. Simple inheritance pattern has been noticed for apomixis by the transfer of a single dominant factor. Simple dominant inheritance has also been reported for apospory by Nogler, (1984b) and Bicknell *et al.,*(2000). Among the diplosporous apomicts, independent inheritance of diplospory and parthenogenesis have been observed (Noyes, 2000;Noyes and Rieseberg, 2000), whereas Voigt and Burson (1983) reported the simple dominant inheritance. It is widely generalized that there is either one locus, or only a small number of loci, involved in the inheritance of apomixis in native systems. Some of the lacunae observed in these studies are that the inheritance of apomixis was recorded after crosses between closely related sexual and apomictic species. Very often in these studies, the expression of apomixis restricts the use of the apomict to the pollen parent. Reciprocal crosses are seldom reported, and the influence of maternal effects remains largely untested. Additionally, particular care must be exercised in scoring progeny during these studies. Apomixis is a complex trait, which it should be measured strictly through the production of genetically identical seedlings, coupled with an embryological study to confirm the reproductive mode of each progeny plant and an assessment of ploidy in all individuals (Nogler, 1984a). It is now believed that number of controlling loci has been underestimated and rather than a single locus, it appears that as many as three unlinked dominant factors may need to be inherited to ensure the transfer of autonomous aposporous apomixes (Bicknell and Koultunow, 2004).The genetic control of apomixis was studied in a 50-tree progeny derived from the cross *C. volkameriana*×*P. trifoliata* using 69 molecular markers and bulked segregant analysis. The proportion of nucellar seedlings was estimated by isoenzymatic analysis of 25 seedlings per tree for 2 consecutive years. The type of embryony (polyembryonic versus monoembryonic seeds) was also determined for fruit yielding trees. Separate genetic maps for each parental species were developed. The integration and comparison of these maps could be accomplished using common multiallelic segregant loci. Differences in gene synteny between the two species-specific genetic maps were shown. Important distortions in the segregation of markers at several genomic regions, some of them also involving differences in the C-methylation pattern, have been observed, especially for the pollen parent. Analysis of quantitative trait loci (QTLs) revealed presence of six genomic positions (two in *P. trifoliata* and four in *C.volkameriana*) contributing individually up to 24% of the total variation for apomixis. Within the same species, QTLs with positive and negative allele effects were present, even in the same linkage group. One of the markers associated to apomixis (Apo2) is also associated to embryony type. Therefore, the genetic control of apomictic reproduction found in citrus (nucellar embryony) is quite complex compared to what has been reported for gametophytic apomixis. Molecular markers linked to QTLs governing apomixis will be useful to assist selection of future apomictic rootstocks for citrus varieties.

Okoro *et al.* (2011) in banana, each mother and ratoon were assayed over 2 flowering cycles, resulting in a total of 16 inflorescences analyzed. Pollination barrier procedures performed involving bagging of the inflorescences, isolating them from natural pollinators. Scoring for apomixis was based on seed-set in isolated (bagged) inflorescences without pollination. Parthenocarpy was determined by seedless pulp development in open pollinated or as artificial pollination by Calcutta 4, using standard *Musa* crossing procedures. Floral and fruit developmental traits indicated that Borneo was monoecious, comprising traits of apomixis and vegetative parthenocarpy. Borneo × Calcutta 4 cross, a P1-lacking, P2 and P3 modelled *Musa*, suggests that Borneo may be genetically modelled to comprise P1 but lack P2 and/or P3. Such modifier genes appear to correlate with rates of genome by environment dependent seed development regimes and if not synchronized with fruit development regimes, results in dehiscence, death of developing fruit, and non-parthenocarpic expression.

References

Anderson CM, Castle WS and Moore GA 1991 Isozymic identification of zygotic seedlings in Swingle citrumelo *Citrus paradisi×Poncirus trifoliata* nursery and field populations. *J. Am. Soc. Hortic. Sci.,* 116:322–326.

Asker, S. and Jerling, L., Apomixis in Plants, CRC Press, Boca Raton, 1992.

Bacchi, O. 1943 Cytological observations in Citrus. III. Megasporogenesis, fertilization and polyembryony. *Bot Gaz,* 105: 221–225.

Bicknell, R.A., Borst, N.K., and Koltunow, A.M. 2000. Monogenic inheritance of apomixis in two *Hieracium* species with distinct developmental mechanisms. *Heredity,* 84:228–237.

Bicknell, R. A. and Koltunow, A. M., 2004. Understanding apomixis: recent advances and remaining conundrums. *Plant Cell*, 16: S228–S245.

Cameron., J.W, and Gaber., M.J., 1968. Identical-Twin Hybrids of Citrus×*Poncirus* from strictly sexual seed parents. *Am. J. Bot.,* 55:199–205.

Cameron., J.W., 1979. Sexual and nucellar embryony in F_1 hybrids and advanced crosses of Citrus with *Poncirus*. *J. Am. Soc. Hortic. Sci.,* 104:408–410.

Carl Ramage., M., L Sando, C.P. Peace, B.J. Carroll and Drew, R.A 2004. Genetic diversity revealed in the apomictic fruit species *Garcinia mangostana* L.(mangosteen). *Euphytica*, 136:1-10.

Frost HB 1943. Seed reproduction: development of gametes and embryos. In: Webber HJ, Batchelor LD (eds) The Citrus industry, vol 1. University of California Press, Berkeley, Calif., pp 767–815.

Leblanc O, Grimanelli D, González-de León and Savidan Y., 1995. Detection of the apomictic mode of reproduction in maize Tripsacum hybrids using maize RFLP markers. *Theor Appl Genet,* 90:1198–1203.

Mogie M 1988. A model for the evolution and control of generative apomixis. Biol. J. Linn. Soc. 35:127-154.

Nogler, G. A., 1984. Gametophytic apomixis. In Embryology of Angiosperms, Springer-Verlag, Berlin, pp. 475–518.

Nogler, G.A. 1984a. Gametophytic apomixis. In Embryology of Angiosperms, B.M. Johri, ed (Berlin: Springer-Verlag), pp. 475–518.

Nogler, G.A. 1984b. Genetics of apospory in apomictic *Ranunculus auricomus. V. Conclusion. Bot. Helv.,* 94: 411–422.

Noyes, R.D., and Rieseberg, L.H. 2000. Two independent loci control agamospermy (apomixis) in the triploid flowering plant *Erigeron annuus.Genetics,* 155: 379–390.

Noyes, R.D. 2000. Diplospory and parthenogenesis in sexual × agamospermous (apomictic) *Erigeron* (Asteraceae) hybrids. *Int. J. Plant Sci.,* 161:1–12.

Okoro, P, Shaibu, A. A., Ude, G. and Olukolu, B. A., Ingelbrecht, I, Tenkouano, A, Ogburia, M. N., Moonan, F., and Dimkpa, C. 2011. Genetic evidence of developmental components of parthenocarpy in Apomictic *Musa* species. *Journal of Plant Breeding and Crop Science,* 3(8):138–145.

Pessino SC, Ortiz JP, Leblanc O, do Valle CB, Evans C and Haywar MD 1997. Identification of a maize linkage group related to apomixis in Brachiaria. *Theor. Appl. Genet.,* 94:439–444

Savidan Y 1982. Nature et hérédité de l'apomixie chez *Panicum maximum* Jacq. PhD thesis. University of Paris XI, Paris, France.

Savidan, Y. 2000a. Apomixis, the way of cloning seeds. *Biofutur,* 2000, 38–43.

Savidan, Y. 2000b. Apomixis: Genetics and breeding. *Plant Breed. Rev.,* 18:13–86.

Sherwood RT, Berg CC, Young BA 1994. Inheritance of apospory in buffalo grass. *Crop Sci.,* 34:1490–1494.

Ramulu K.S, Dijkhuis, R. Pereira A, Angenent, G.C., van Lookeren Campagne MM and Dons H 1998. EMS and transposon mutagenesis for the isolation of apomictic mutants in plants. (In) Jain SM, Brar DS, and Ahloowalia BS (eds) Somaclonal variation and induced mutation in crop improvement. Kluwer,

Roose ML and Traugh S.N. 1988. Identification and performance of citrus trees on nucellar and zygotic rootstocks. *J. Am. Soc. Hortic. Sci.,* 113:100–105.

Voigt, P.W. and Burson, B.L. 1983. Breeding of apomictic *Eragrostis curvula.* (In) Proceedings of the 14th International Grasslands Congress (1981), Lexington, KY, J.A. Smith and Hayes, V.W. (eds) (Boulder, CO: Westview Press), pp. 160–163.

Weber, H.J. 1932. Variations in citrus seedlings and their relation to rootstock selection. *Hilgardia,* 7:1–79.

5

Mutation Breeding in Fruit Crops

M.R. Dinesh, G.L. Veena and Bharathi Nirujogi

Mutations are sudden and heritable changes in the genetic material. They can result in deletion, inversion and translocation of certain portions in the chromosomes, which can result in the changed expression of a trait due to the interaction of alleles within or without the locus, sometimes the effects revert back or can be heritable.

Most of the fruit crops are highly heterozygous. However, heterozygosity is one of the advantages when it comes to mutation. In highly heterozygous crops wherein there is a need for improvement in a single trait, mutations have been of great use. One of main limitations is use of vegetative material, which due to the unlikelyhood of uniform penetration and distribution of mutagenic agent can result in chimera. Although hybridization and selection have made a fair dent in the improvement of fruit crops, mutation is one of the methods that can be relied upon to develop pre-breeding lines and to bring about change in the quantitavie traits. Another very important fact is that there are crops like grapes wherein a lot of spontaneous mutants have become established varieties *viz*., Sonaka and Saritha to quote a few. Some of the existing varieites wherein one or two traits need to be improved, mutation would greatly help in this in comparatively short time. In some of the crops like sapota, wherein the conventional breeding would take long time, mutation can be resorted to both for inducing variability in the germplasm as well as for trait improvement. It is also to be noted here that mutation breeding can be a good supplement for conventional breeding, more so with the emergence of tissue culture propagation, which often throws up spontaneous mutants. Mutation as a methodology can also be tried in some fruit crop spp where there is no disease resistance source within. It can also be adopted in cases where there is a need to break linkages.

Types of mutations

Like in any other crop, mutations can be spontaneous or can be induced. The sudden heritable changes brought about by the nature through radiation and temperature is known as spontaneous mutations. Several examples can be found in fruit crops that are of spontaneous in origin. Mutations can also be induced by the shocks due to the damage to any of the parts. There are several examples of mutations in grapes and citrus.

Apart from spontaneous mutations, mutations can be induced through radiations which is known as physical mutagens and through chemicals which is known as chemical mutagens. In the case of seed propagated crops like papaya, chemical mutagens can be utilized to bring about changes. Some of the chemical mutagens which are used are ethyl methane sulphonate, diethyl sulphate, nitroso compounds, nucleoside analogues, azide, colchicine and N-nitroso-N-ethyl urea. Mutation can be detected by its effect on the phenotype. As a result only those mutations, which have substantial effect on the phenotype, can only be detected. Work carried out in papaya with induced mutation using gamma irradiation has resulted in the development of a dwarf dioecious variety Pusa Nanha (Mansha Ram and Srivastava, 1984). Radiation and other chemical mutagens such as E.M.S (Ethyl Methane Sulphonate) and D.E.S. (Diethyl Sulphate) induce a much intensive mutation than other chemical mutagens in most of the crops (Hildering *et al.,* 1965). However, EMS has proved to be one of the most effective chemicals.

Plant materials for treatment

Any material used for propagation can be used for mutation treatment. Seeds, pollengrains, buds/cuttings or complete plant can be used for mutagenesis. It depends on whether the crop is sexually or asexually propagated and type of mutagen to be used. Pollen grains are also infrequently used because, they are difficult to collect in large quantities, hand pollination(with treated pollens) is rather difficult and survival percentage of pollens is relatively low. Pollen grains are the only plant part which can be treated with ultra violet rays. The plant material for mutagenic treatment is a generally desirable cultivar that needs improvement in one particular characterstic. Great chances of inducing a mutation are offered by those traits that are in heterozygous condition or those determined by polygenes. Polyploids tolerate more extensive structural changes than diplopids and respond well to mutagenic treatment. Haploids are a favored material for mutation studies as both recessive and dominant mutation can be identified, and subsequent chromosome doubling results in a homozygous genotype. A mutation originates in a single cell. In order that a mutation can manifest itself in a whole plant, the target material for mutagenic treatment must be meristematic tissues. Although nearly all the parts of a plant can be induced to develop meristematic tissue, shoot tip meristems in leaf buds or in seeds are typically targets for mutation.

Procedure for development of mutants in sexually propagated fruit Crops

In sexually propagated crops, seeds are commonly used because seed can tolerate extreme environmental conditions. From treated seeds M_1 generation is developed. In M_1 generation large number of plants are grown with wider spacing. Dominant mutations are recorded if any (generally mutations are recessive and do not express in M_1). Chlorophyll sectors and fertility are recorded. M_1plants are selfed and their seed is harvested separately. In M_2seed obtained from M_1 is sown in wide spacing, selected mutants are selfed, Oligogenic mutations are detected in M_2 and are harvested separately. M_3 progeny is raised from selected M_2 and evaluated for homozygosity. Selected homozygous M_3 progenies are bulked together to conduct yield trials in M_4. M_4 progenies are raised in replicated trials using local check for comparison.From M_5-M_9 generations, selected lines are tested in multiplication coordinated trials.

Procedure for development of mutants in vegetatively propagated crops

In vegetatively propagated crops, mutations are expressed in the form of chimeras. The chimera refers to genetically different tissue in an individual. The individual has one type of tissue in one part and another type of tissue in another part. In apical buds, axillary buds and adventitious buds, there are two functional layers, outer layer and inner layer. When the changes occur in entire inner or outer layer, it is also known as Periclinal Chimera and when only a part of inner or outer layer is altered, it is called Sectorial Chimera. Chimera is a plant or plant part composed of genetically different layers. The most common example is a "variegated" plant where different regions of the leaf are yellow or white due to lack of chlorophyll synthesis, i.e. these are chlorophyll mutants. However, there are many kinds of chimeras. Thornless blackberries are chimeras where the L-I epidermis lacks the ability to produce thorns. Some fruits have sweet and sour regions of flesh, which may be a chimera.

Chimeras

In mericlinal Chimera, a mutation occurs in one layer and along the side of the apex. Due to its position, the cell division products of those mutated cells occur as a layer only one side of the plant. In other words, only a section of one of the layers is mutated. Mericlinal chimeras are not stable.

In periclinal chimera, a mutation occurs in one (or more) layer at the top of the apex. Due to its position, the cell division products of the mutated cells spread and cover the entire layer of the apex. In other words, the entire layer is mutated.

Sectorial Chimera: A mutation occurs in multiple layers at the top of the apex. Due to its position, the cell division products of the mutated cells give rise to a section of mutated cells. In other words, an entire section of the plant is mutated. They are stable to very stable, and comprise the most common type of chimeras in horticulture.

Handling of mutagenic generations

Treatment of seeds and vegetative propagules commonly produce chimeras. Shoot tip meristem has three layers, L1-give rise to epidermis; L2-Part of leaf mesophyll and gametes; L3-Yield the rest of plant. Periclinal chimera is when the whole of L1, L2or L3layer is affected. Sectorial chimera is when only a part of L1, L2or L3layer is affected.

Depending on the severity of the mutagenic treatment, affected cells may show mitotic delay and chromosome damage or lose viability and function. Moderate exposures result in a decreased rate of synthesis of DNA, RNA, ATP and IAA. Over doses are detrimental to plants, result in expression of undesirable traits and inhibit the synthesis DNA, RNA and proteins.

More number of seeds per fruit is a limiting factor in the export of kinnow on a large scale.Benefiting from the induced mutations for selectivity, especially in the vegetatively propagated fruitcrops like citrus, induced mutation for seedlessness in Kinnow with gamma irradiation of dormant bud was attempted by Shahid Akbar Khalil *et al.,* (2011). In their study dormant bud irradiation-cum-grafting technique was carried out using *Citrus jambhiri* rootstock for propagation of the scion. A sparse seeded (5 ± 3 seeds/fruit) mutant was evolved from an exposure dose of 20 Gy. The mutant was put to the conventional propagation up to mV5, for confirmation of the continuity of the mutation. The seedlessness in Kinnow, in essence, is caused by the abortion of both ovules and seeds. The sparse seeded mutation was found to be a solid and stable one. The quality parameters in fruit and juice of the mutant were comparable with its parent. Sutarto *et al.,* (2009) found that bud woods of mandarin *cv*. So E and Garut and pummelo *cv*. Nambangan when irradiated with gamma rays @ 20Gy gave seedless fruits, with yellow skin colour, green pulp colour and thick endocarp. Lucie Fernandez *et al.,* (2006) identified in grapevine (*Vitis vinifera*) a mutant and named as fleshless berry (flb) that reduced by 20 times the weight of the pericarp at ripening without any effect on fertility or seed size and number. The flb mutation strongly impaired division and differentiation of the most vacuolated cells in the inner mesocarp. The timing of ripening was not altered by the mutation although the accumulation of malic acid in the green stage was noticeably reduced while sucrose content (instead of hexoses) increased during ripening. The mutation segregated as a single dominant locus.

Hafiz *et al.,* (2005) treated the stones of the mango var. Dashehari with the gamma irradiation, they observed that CO 60 @ 2.5 Kr gave the highest seed germination, higher average for height of seedlings and lateral branching. However, primary leaves appeared serrated in the seedlings that emerged from the treated seeds, which turned normal in the subsequent growth of flushes. In the papaya variety Dai Loan Tim, Nhat and Chau (2010) reported that LD50 was observed to be 30Gy in the seeds when irradiated with gamma rays ranging from 10 to 60Gy. Four mutant plants, which were particularly outstanding in vigour with medium dwarf stature, bearing the first flower at a height of 50-60 cm from the ground and improved fruit quality when compared to control were selected from a population of one thousand M1 progenies.

Hensz (1971) developed the seedless 'Star Ruby'grapefruit through irradiation of seed of the seedy 'Hudson' by thermal neutrons. Hearn (1984) produced seedless strains of 'Pineapple' orange and 'Duncan' and 'Foster' grapefruit from γ-ray irradiation of seeds. In the case of lemon, seedless strains were developed from 'Eureka' lemon and 'Monreal' clementine through γ-irradiation of budsticks by Starrantino *et al.,*(1988). Chen *et al.,* (1991) were able to produce seedless strains of 'Jin Cheng' orange through γ-irradiation of seeds, however, chromosomal aberrations were observed in these strains.

Induction of mutation *in vitro*

Several attempts were made in fruit crops to create variability by inducing mutation *in vitro.* Yulin *et al.,* (2008) exposed shoot tip tissues of *in vitro* grown banana (*Musa* spp. AAA Group) to ethyl methane sulphonate (EMS) solution with different concentrations at different times for observing resistance to bacterial wilt. The observed variability viz., dwarf types, narrow-leaf types, distortion types, wave types, and mosaic type. After inoculation in both indoor and outdoor environments, there were 3.1 and 6.3 per cent plants that survived respectively, and these plants were initially considered to be the resistant mutants. It was also found that there were higher proportions of resistant mutants from the dwarf and narrow leaf type mutation than from others.In mango, Litz (2009), irradiated at 100 Gy embryogenic cultures of monoembryonic and polyembryonic mango on semi solid medium, two weeks after irradiation, cultures were inoculated with *Colletotricum gloesporoides*. The somatic embryos were recovered and their shoots were used by *ex vitro* grafting. Somsri *et al.,* (2009), showed successful induction of mutation *in vitro* through gamma irradiation using both chronic and acute procedures resulting in improvement of quality characteristics viz., seedlessness of fruit in tangerine (*C.reticulata*) and pummelo (*C. grandis*). Zheng *et al*, (2009), conducted a study on inductionof mutation in Jujube (*Zizyphus jujube*) using tissue culture combined with $^{60}CO\gamma$ irradiation, which improved the germination. Mutant characters such as narrow leaf, yellow leaf, multiple twigs and the genotypes with different blooming time and different quality of the fruits were observed.Zamir *et al*, (2009), studied *in vitro*-mutagenesis followed by micropropagation via axillary bud proliferation in shoot tips of guava (*Psidium guajava*) cultivar Safeda. In the shoot tips irradiated with 90 Gy gamma rays, shoot proliferation was observed after 7 weeks of culture initiation. The bud woods of guava cv. Safeda were more radiosensitive than seeds.

In some of the fruit crops like grapes, spontaneous mutants are frequently observed. Mutation breeding does hold promise in heterozygous perennials as the first effect in the M_1 generation is the branching habit and dwarf statured plants. This will help in developing varieties for high density planting. Although there is every chance of mutants reverting back, raising large population and fixing the mutagenic changes through vegetative propagation would help in developing improved recombinants.

References

Chen, S., Gao, F. and Zhang, J., 1991 Studies on the seedless character of citrus induced by irradiation. *Mutation Breeding Newsletter,* 39:8–9.

Hafiz, I.A., Anwar, N., Abbasi, N.A. and Asi, A.A., 2005, Effect of various doses of gamma radiation on the seed germination and seedling growth of mango. *Sarhad J. Agric.*, 21(4): 563-567.

Hearn, C.J., 1984. Development of seedless orange and grapefruit cultivars through seed irradiation. *J. Am. Soc. Hort. Sci.,* 109: 270-273.

Hensz , R.A., 1971. Star Ruby, a new deep-red-fleshed grapefruit variety with distinct tree characteristics. *J. Rio Grande Valley Hort. Soc.*, 25:54-58.

Hildeing, G.J. and Verkerk, K., Chimeric. 1965. Structure of the tomato plant after seed treatment with EMS and X-ray, in use of induce mutations in plant breeding. Food and Agriculture Organisation. Rome. 317.

Litz, R.E., 2009, Recovery of mango plants with anthracnose resistance following mutation induction and selection in vitro with culture filtrate *of Colletotrichum gloesporoides* Penz. *IAEA proceedings.* 1615: 7-16.

Lucie Fernandez, Charles Romieu, Annick Moing, Alain Bouquet, Mickael Maucourt, Mark R. Thomas, and Laurent Torregrosa., 2006. The Grapevine fleshless berry Mutation. A Unique Genotype to Investigate Differences between Fleshy and Non fleshy Fruit. *Plant Physiology*, 140:537–547.

Mansha Ram and Srivastava, S. 1984. A note on the occurrence of mutants in papaya. In:Proceedings of National Seminar on Papaya and Papain Production: 28. Coimbatore.

Nhat hang, N.T. and Chaun, M., 2010, Radiation induced mutation for improving papaya variety in Vietnam. *Acta Hort.,* 851: 93-98.

Shahid Akbar Khalil, Abdus Sattar and Roshan Zamir., 2011. Development of sparse-seeded mutant Kinnow (*Citrus reticulata* Blanco) through budwood irradiation. *African Journal of Biotechnology,* 10(65):14562-14565.

Somsri, S., Jompook, P., Kanhom, P., Thayamanont, P., Meecharoen, S. and Kumcha, U., 2009, Development of seedless fruits mutants in citrus including tangerine (*C.reticulata*) and Pummelo (*C. grandis*) through induced mutations and biotechnology. *IAEA Proceedings.* 1615: 33-40.

Starrantino, A., Mannino, P. and Russo, F. 1988. The genetic stability of three seedless clonal selections obtained by gamma rays of the seedy 'Monreal' clementine. In: *Proceeding of the Sixth International Citrus Congress.* Margraf, Tel Aviv, Israel, 1:175–182.

Sutarto, I., Agisimanto, D. and Supriyanto, A., 2009, Induced plant mutations in the genomics era. *Food and Agriculture Organization of the United Nations, Rome, Italy,* 306-308.5(1): 81-84.

Yulin, H.U., Xie jiang hui. Liang guolu., Chen jia ying and wang zhongmeng, 2008, Induction and selection of banana mutants resistant to fusarium wilt disease. *J. Fruit Sci.*, 25(2):188-192.

Zamir, R., ALI, N., Tariq shah, S., Mohammad, T. and Ahmad, J., 2009, Guava (*Psidium guajava* L) improvement using invivo and invitro induced mutagenesis. *IAEA proceedings,* 1615: 101-112.

Zheng, H.R., Liu, Z.C., Ye, Z.W., Su, M.S., and Jin, Y.F., 2009, Induction of mutation in Jujube (*Zizyphus jujube* Mill) using tissue culture combined with ^{60} CO γ ray irradiation. *IAEA Proceedings,* 1615: 113-122.

6

Role of Insects in Fruit Crop Pollination

P. V. Rami Reddy and K. Sreedevi

Introduction

Pollination is an essential ecosystem service for the maintenance and conservation of biodiversity on the earth. Many food crops, barring cereals, are entomophilous in nature and rely on insects for pollination. The pollinators in turn are benefited by obtaining floral resources such as nectar, pollen or both. This mutualism has evolved over centuries and been helping both natural terrestrial ecosystems as well as man-made agro-ecosystems. Pollinators such as bees, birds and bats contribute to 35 per cent of the world's crop production, increasing outputs of 87 of the leading food crops worldwide (FAO, 2009). The total economic value of crop pollination worldwide has been estimated at €156 billion annually (Gallai *et al.,* 2009). The area covered by pollinator-dependent crops has increased by more than 300 per cent during the past 50 years (Aizen and Harder 2009). Pollinators are found in diverse groups of the animal kingdom, including birds, bats, reptiles, insects, etc. Among them, insects particularly honey bees, dominate in providing pollination services to several plants. There are about 19,000 described species of bees in the world (Linsley, 1958) and, with the exception of one species, *Apis mellifera* L., the domestic honey bee, all of them are grouped under the general term 'wild bees'. India is endowed with the greatest biodiversity as far as honey bee species are concerned and is home to five species *viz*., *A. cerana*, *A. florea*, *A. dorsata*, *A. andreniformis* and *A. laboriosa* besides hosting *A. melllifera,* an introduced one. In addition, several other pollinators including stingless bees, carpenter bees, bumble bees, megachilids, halictids, sphecids, andrenids, syrphids, etc. occur in several agro-ecosystems (Thakur, 2012). Studies on crop pollination by insects are becoming increasingly critical because of a perceived global decline in pollinator stocks, with great economical and conservation consequences (Ghazoul, 2005a, 2005b; Steff an-Dewenter *et al.,* 2005). In the recent years, there has been a concern about declines in both wild and domesticated pollinators, especially honey bees (Reddy *et al.,* 2012a). A large proportion of horticultural crops is potentially vulnerable to declines in honey bee and other pollinator insects (Table 1). One of the essential requirements to adequately evaluate the importance of animal pollination for food production and the impact of pollinator losses is to determine the pollination needs of different cultivated plants. However, detailed studies of crop pollination requirements are incomplete in many crops (Klein *et al.*, 2007).

Table 1: Economic impacts of insect pollination on major food crops and their rate of vulnerability to pollinator loss – World scenario (Gallai *et al.,* 2009)

Crop group	Total Production Economic Value in 10^9€ (EV)	Insect Pollination Economic Value in 10^9€ (IPEV)	Rate of Vulnerability (%) (IPEV/EV)
Fruits	219	50.6	23.1
Vegetables	418	50.9	12.2
Nuts	13	4.2	31.0
Oilseeds	240	39.0	16.3
Roots and Tubers	98	0.0	0.0
Spices	7	0.2	2.7
Cereals	312	0.0	0.0
Pulses	24	1.0	4.3

Pollination is not a limiting factor in wind pollinated or completely self pollinated crops but it is a matter of concern in crops that depend on external agents like insects and deserves a special attention. Majority of fruit crops are cross-pollinated and depend on insects for pollination. We made an attempt in this chapter to systematically analyse the pollinator fauna associated with and the pollination requirements of important entomophilic fruit crops in the light of the available information. Proven and potential measures to enhance or augment the pollinators are also discussed.

Mango (*Mangifera indica*)

Floral biology

The mango inflorescence is a branched terminal panicle, 4 to 24 inches long, consisting of a few hundred to several thousand individual flowers, requiring up to a month for all to open. The number of panicles may range from 200 to 3,000 per tree with 500 to 10,000 flowers per panicle i. e.100,000 to 30 million per tree. There are perfect and staminate flowers on the same panicle. The proportion of perfect to staminate flowers may vary from 1 to 35 per cent (Cobley, 1956; Mukherjee, 1953; Singh, 1978) depending on the cultivar, the early or late emergence of the panicle and its position on the tree (Singh, 1960). In the perfect or hermaphrodite flower, a nectar secreting fleshy disk surrounds the ovary. The stamen is on the outer margin of this disk. The pistil and stamen are the same length; therefore, pollinating insects that feed on either nectar or pollen are likely to transfer pollen from the anther to the stigma (Juliano and Cuevas, 1932; Sturrock, 1966). The flower opens early in the morning and the stigma is immediately receptive. Maximum pollen shedding is from about 8 a.m. to noon. This delayed pollen shedding can result in inadequate stigma fertilization (Spencer and Kinnard, 1956). When the flowers open, they secrete nectar in considerable quantity, which attracts a large number of insects (Mukherjee, 1953; Free and Williams, 1976).

Pollination requirements

Mango trees produce an abundance of flowers but relatively a few mature fruits. The scanty productiveness of many Indian mangoes has been attributed by several workers to defective pollination. Popenoe (1917) stated that the mango is self fertile but cross-pollination increases the fruit set and in his later study (Popenoe,1920) pointed out that the mango flowers have none of the characteristics of a wind-pollinated flower, and he considered the mango to be an insect-pollinated plant. Singh (1961) reported that over 65 percent of the perfect flowers were never pollinated which is a strong indication that wind is not an effective pollinating agent. Singh *et al.* (1962) reported that crossed flowers set fruit whereas selfed ones did not, indicating a degree of self sterility. Within the cultivar there is a definite need for transfer of pollen from anther to stigma by an outside agent. Naik and Rao (1943) obtained no parthenocarpic fruit set of more than 100,000 flowers studied. Ruehle and Ledin (1955) considered that the lack of efficient pollination might be one of the reasons for the low yields of some Florida cultivars. The studies indicate that the need for cross-pollination between mango cultivars is not critical, at least for most cultivars, but there is need for pollinating insects to transfer the pollen from anthers to stigma within the cultivar to obtain satisfactory fruit set. Studies in India by Sharma and Singh (1970) revealed that four mango cultivars were self-unfruitful and self-pollination gave only 0 to 2% fruit set compared to 6 to 23% for cross-pollination. The majority of the fruits from self-pollinated flowers that had set initially had fallen within four weeks, and all had fallen before they were half grown. Bhatia *et al.*, (1995) and Singh (1997) recorded zero fruit set in bagged panicles while Reddy *et al.* (2011) reported 60-80 per cent reduction in fruit set of different varieties when panicles were covered thus confirming the essential role played by insect pollinators. Similar results were also reported from other parts of the world. The cultivar studied by Phoon (1984) in Malaysia also appears to benefit from cross-pollination where inter cultivar pollination gave 82% set compared to 33% from self-pollination. In South Africa, de Wet and Robbertse (1986) found evidence of some self incompatibility, and that pollen germination and pollen tube growth was faster after cross-pollination.

Pollinator diversity

Several agents have been given credit as pollinators of mango. Wagle (1929) showed that there was some selfing and some wind pollination, but insects (bees, ants, and flies) played an important part. Surveys conducted by Singh (1997) of insect species visiting mango flowers indicated that highest number (17) of species were from order Diptera (flies) followed by Coleoptera (beetles and weevils), Hymenoptera (ants, bees, wasps, sawflies), Heteroptera (bugs), and Lepidoptera (butterflies, moths). The number of pollinators/panicle/minute increased as the average temperature increased to 19.5° C but decreased with further increase in temperature and with rains. Verghese and Tandon (1990) studied the pollination behaviour of *Apis florea* on mango and found it to be an important pollinator. Reddy *et al.* (2012b) recorded four species *viz. Apis florea* (Hymenoptera:Apidae), *Chrysomya megacephala* and *Stomorhina discolor* (Diptera:Calliphoridae), *Eristalinus arvorum* (Dipteral:Syrphidae) as the dominant foragers significantly contributing to mango pollination in Bengaluru, while stingless bee, *Trigona iridipennis* as the most abundant in the *konkan* region in India. Pollinator abundance seems to be influenced by varietal differences. In terms of total pollinator activity, 'Ratna' attracted maximum numbers (3.24/10 panicles) while it was lowest in 'Sindhu' (0.63). Dipterans and bees showed varied preferences to varieties as was evident in case of 'Alphonso' which recorded the highest number of bees (2.6) while dipteran activity was lowest (0.02). Across the varieties, *A. florea* was the most dominant forager followed by *E. arvorum* (Reddy *et al.,* 2013b). In Israel, forty-six distinct species or types were found, most belonged to the orders Diptera (26), Hymenoptera (12) and Coleoptera (6). Of them two blow flies (*Chrysomya albiceps* and *Lucilia sericata*), the honey bee (*Apis mellifera*) and the housefly (*Musca domestica*) played a significant role in mango pollination in most orchards (Dag and Gazit, 2000). In Taiwan, one hundred and twenty-six individual insects belonging to 39 species in 23 families and five orders were recorded as a visitor or pollinator on mango (Sung *et al.,* 2006). In South Africa, honey bees were observed collecting extra-floral nectar from the bases of mango panicles which according to Du Toit and Swart (1993) would ensure sufficient numbers of pollinators by helping to sustain honey bee colonies.

Pollinator enhancement

Placing colonies of honey bees within the mango orchard would result in increased floral visitation and possibly more stabilized fruit set. As the mango flowers tend to open in large numbers at a time of year when many other flowers are also available, so visitation in commercial orchards is likely to be far lesser than that is necessary for maximum floral visitation. Hence four to six colonies per acre may be necessary to obtain maximum fruit set. Studies from Israel also report the positive impact of the introduction of honey bee and the bumblebee (*Bombus terrestris*) on mango yield (Dag and Gazit, 2000). Besides honey bees, the activity of other pollinators, especially that of *C. megacephala* can be enhanced by placing simple fish meal based traps. These traps were found to attract adult flies which breed on fish based diet. These traps should be erected just before flowering so that population can be enhanced by the time full blossom occurs (Reddy *et al.,* 2013a).

Guava (*Psidium guajava* L.)

Floral biology

The hermaphrodite flowers of guava are 2.5-3.0mm in diameter with four or five white petals. They are solitary and borne in cymes of two or three flowers. The single greenish-yellow style projects above the numerous stamens and produce an abundance of pollen. The ovary has four or five locules with numerous ovules. Purseglove (1968) reported that *P. guajava* is visited by bees and other insects and although it is self fertile about 35% cross pollination occurs. Kumar *et al.* (1996b) studied the floral biology of guava and emphasized the role of cross pollination.

Pollination requirements

Guava is a highly cross pollinated crop and pollination by honey bees results in higher fruit yield. The different modes of pollination in guava showed complete flower drop in self-pollination indicating high degree of self incompatibility. In India, the percent fruit set, fruit weight, number of seeds per fruit and test weight were significantly higher in hand pollination and also in guava orchard provided with five hives of *A. cerana* compared to orchard without bee hives (Kuberappa *et al.*, 2002). Partial self- incompatibility was found in all of the species of *Psidium* studied cross-pollination by insects resulted in higher yields. In Brazil, flowers excluded from flower visitors set only 15.09% while it was 39.5% in open pollinated ones (Alves and Freitas, 2007).

Pollinator diversity

Honey bees are the main pollinators of guava (Morton, 1987). Reddy and Jacob (1998) studied the foraging behavior of honey bees on guava and found that the peak activity of all species was between 9 and 11 AM. Among the three species studied, *Apis dorsata* was the dominant forager followed by *A. mellifera* and *A. cerana*. Similar observations were reported by Kuberappa *et al.*, (2002). They recorded seven species of insects (*A. dorsata, A. mellifera, A. cerana indica, A. florea, Trigona iridipennis, Megachile lanata, Chalicodoma lanatum* and *Xylocopa fenestrata*) foraging on guava in Bangalore, India. The honey bees accounted for 92.86% of total pollinators. Anitha *et al.* (2012) observed maximum activity of honey bees on guava between 10.00AM and 12.00noon. They tested certain bee attractants for their efficacy and found that a commercial formulation *viz*., Bee Scent @1.5ml/L was most effective in attracting highest number of bees. Plants sprayed with attractants had fruits of higher weight and size compared to control.

Litchi (*Litchi chinensis* Sonn.)

Floral biology

Das *et al.,* (2002) reviewed the flowering phenology of litchi. Litchi inflorescence is determinate and composed of several multiple branched panicles on the present season wood. Litchi produces three basic types of flowers i. e. male with no pistil (M_1), female with rudimentary stamen (F) and male with incomplete pistil (M_2) on the same panicle in succession (Costes, 1988; Stern *et al.,* 1993). As litchi flowers are functionally unisexual, fruit set can occur only if pollen from male flowers (M1 and M2) is transferred to the stigma and hence they need an external agent (Stern and Gazit, 1996). Peak releases of pollen occur in the morning. Female flowers remain receptive for three days from anthesis (Chaturvedi and Saxena, 1965). Adequate pollination is ensured by the fact that the F flowering wave occurs between two pollen-releasing waves (M1 and M2), and by the abundance of pollinating agents (Butcher, 1957; DeGrandi-Hoffman, 1987).

Pollination requirement

Litchi flowers are entomophilic and pollination is usually performed by insects, mainly the honey bee (Free, 1993; King *et al.*, 1989; McGregor, 1976). Pollination of the litchi by *Apis mellifera* was studied by Stern and Gazit (1996). Pollination was low during first male phase (M) and reached a high only when pseudohermaphrodite (M2) flowering started. The density of bees found on inflorescence was high during female flowering. Pollen density collected on bees was low during M1 flowering and increased to high values during M2 flowering. They also found a highly significant positive correlation between bee density and sugar concentration in the nectar. Du Toit and Swart (1995) concluded that insect pollination is necessary to ensure optimum fruit set in litchi.

Pollinator diversity

In India 16 species of bees, flies, wasps and other insects were recorded visiting litchi flowers by Singh and Chopra (1998). Most common were the honey bees *viz.*, *A. cerana, A. dorsata, A. mellifera* and *A. florea* and syrphid flies *viz.*, *Melanostoma univittatum* and *Episyrphus balteatus*. Pollination by these insects resulted in 387% increase in fruit set and 50% increase in retention. In northern states of India and also in Thailand, *A. mellifera* was credited as the most important pollinator of litchi (Abrol, 1999; Kumar *et al.,* 1996; Wongsiri and Chen, 1995). Besides, stingless bee, *Trigona iridipennis* is also reported to contribute to litchi pollination (Kumar *et al.*, 1996). In Thailand, *A. cerana* is reported to be the preferred species for small scale pollination of litchi (Wongsiri and Chen, 1995). It was suggested that colonies should be provided at a minimal rate of one per acre (Du Toit, 1990). In India, Pandey and Yadava (1970) reported that only 0.03 to 0.10 percent of flowers caged to exclude insects set fruit, whereas 0.7 to 11.2 percent of flowers exposed to insect pollination set fruit. They also reported that *Apis* spp. and *Melipona* spp. comprised 98 to 99 percent of the total visitors. Choudary *et al.,* (2003) studied the effect of bee pollination on the morphophysiological characters of litchi fruit. The highest volume of fruits (1860 ml), fruit size (3.15×2.87 cm), fruit weight (19.96 g), pulp weight (12.47 g), skin weight (3.05 g), seed weight (4.34 g), and fruit:seed ratio (1:0.22) were observed in bee-pollinated litchi trees. About 80% reduction in fruit set was recorded when flowers were bagged to exclude pollinators in India (Kumar *et al.,* 1996a) and South Africa (Du Toit, 1994).

Citrus (*Citrus* spp.)

Pollination requirement

The extensively cultivated genus *Citrus*, which includes orange (*Citrus sinensis*), mandarin (*C. reticulata*), grapefruit (*C. paradisi* Macf.), and lemon (*C. limon*), exhibits tremendous variation in pollination requirements (Roubik, 1995). It is considered that there are few problems in citrus with pollination. Navel oranges produce seedless fruit without pollination. Most commercial seeded types are self-fruitful but require pollination. For instance, 'Clementine' tangerine and some sweet oranges are completely or partially self-incompatible (McGregor,

1976; Sanford, 1992) and thus require insects for pollination. This genus also exhibits large variation in the incidence of parthenocarpy (McGregor, 1976), and thus, the sexual process is not always necessary for fruit development. On the other hand, many cultivars of *Citrus* are known to be self-incompatible and therefore an appropriate supply of cross pollen and pollinating insects are needed for high fruit yields (Sanford, 1992).

Pollinator diversity

Some studies state that cross-pollination is not necessary for grapefruit (Roubik, 1995), whereas others demonstrate that the cross pollination mediated by insects increases fruit production (Burger, 1985). Chacoff and Aizen (2007) recorded 50 species of insects visiting grapefruit flowers. The most frequent visitor was *Apis mellifera* followed by stingless bees(*Trigona argentina* and *Tetragonisca angustula*) and solitary bees (*Xylocopa eximia* and *Megachile* sp.) They found that hand- and open-pollinated flowers set about six times more fruit than emasculated and bagged (insect excluded) flowers. In addition, cross-pollen performed better in terms of grain germination and tube growth than self pollen. The fruit set and mean fruit weight were 35.30% greater in uncovered sweet orange compared to the uncovered flowers. Some compounds like Bee-Here, eugenol, geraniol, citral, and lemon grass extract, mainly diluted in water, were effective in attracting honey bees (Malerbo-Souza *et al.,*2004). In India, Phartiyal *et al.* (2012) recorded a total of 12 insect visitors on flowers of sweet orange in *Terai* region. Among the pollinators, the Hymenoptera constituted major group of insects including *A. dorsata*, *A.cerana indica*, *Trigona* spp., *Campsomeriella megachalis* F., and *C. collaris.* Among the Dipterans, syrphid flies were the most frequent visitors including *Syrphus corollae* Fab., *Episyrphus balteatus* De Geer, *Spherophoria* spp., and *Melanostoma* spp. The percent fruit setting was higher (8.4 - 12%) in panicles under open pollination as compared to completely bagged panicles (1.2 - 8.0%).

Custard Apple (*Annona* spp.)

Floral biology

Annonaceous flowers are protogynously dichogamous, opening as females with receptive stigmas and closed anthers, and later losing stigmal receptivity as the flowers turn into pollen-shedding males (Gottsberger, 1970). This prevents deposition of pollen on to the stigmas in the same flower leading to temporal separation of female and male function within the flower and limits its potential to self-pollinate without external factors. The flowers attract their pollinators through strong odours. Some species have thermogenic flowers that attain a temperature higher than that of surrounding air, presumably to intensify volatilization of the odoriferous chemicals in the petals (Gottsberger, 1970). Annonaceous flowers produce no nectar, but rewards for pollinators include fleshy edible petals. The flowers are also used as mating sites by some pollinators (Gottsberger, 1988).

Pollinator diversity

Inadequate pollination is a major factor limiting production of commercial Annona fruits in many locations (Gazit *et al.,* 1982). The majority of *Annonaceae* are pollinated by beetles Gottsberger (1970), although some are pollinated by thrips, true bugs (Farre *et al.,* 1997), and even cockroaches (Nagamitsu and Inoue, 1997). Reiss (1971) was the first to document pollination by nitidulid beetles, in cultivated atemoyas and cherimoyas and Villatla (1988) documented scarab beetle pollination of soursop. Today, natural pollination of *Annona* is believed to be largely brought about by beetles, though some 'self-fertility' is mentioned in much of the literature and awaits elucidation. The number of beetles per flower affects the likelihood of fruit set, and also the quality of the fruit in some cases. All studies provide for increased fruit set as numbers of visiting beetles increase (Gazit *et al.*, 1982; George *et al.*, 1989; Nadel and Pena, 1994; Lopez and Uquillas, 1997). The species complex of beetles and the performance of each species vary geographically. The four major pollinators in Israel *viz.*, *Carpophilus humeralis, C. heemipterus, C. mutilates* and *Haptoncus luteolus* are equally effective as pollinators. In Florida, USA, out of nine species of nitidulids, *C. mutilatus* was the most important pollinator in terms of efficacy and abundance (Nadel and Pena, 1994). The number of species visiting flowers in Ecuador, Colombia and the Caribbean region was similar to that found in Florida (Pena and Bennet, 1995).

Mixed results were reported from the trials to increase fruit set in atemoya orchards by augmenting nitidulid populations. In Israel, one-time additions of decaying apples as an attractant failed to increase the number fo beetles in the flowers or fruit set (Galon *et al.*, 1982). In Australia, success was achieved by adding pineapples in an atemoya orchard throughout the flowering season. (George *et al.*, 1992). Nitidulid pollination in Annonas can also be improved by using chemical lures. The effect of nitidulid-pheromone bait station on sugar apple and atemoya fruit set was determined in southern Florida (Pena *et al.*, 1999). There are no reports from India on pollinator studies in *Annona* spp.

Passion Fruit (*Passiflora edulis* Sims.)

Floral biology

Passion fruit has hermaphrodite, solitary flowers, located in the leaf axils. There are usually five stamens. The ovary is borne over the androgynophore. There are three styles united at their base, and at the top of style there are three bifurcated stigmas (Vanderplank, 1996). Flowers are large, attractive, colourful and fragrant and produce plentiful nectar and pollen that facilitates insects for cross-pollination. Passion fruits are protandrous as anther dehisces before stigma becomes receptive and stigma remains receptive from the time of flower opening to closing (Cox, 1957). The flowers of passion fruit are fragrant when open and nectar is secreted in a groove at the base of the gynophores, and the pollen is heavy and sticky. These features, and the position of the anthers and the functional position of the stigmas, indicate that flowers of passion fruit are adapted to pollination by insects rather than by wind (Akamine *et al.,* 1954; Nishida, 1963; Free, 1993). When the flowers first open, the stamens hang down, and the anthers dehisce on the undersides, exposing the pollen, the style remain erect, and there is no stickiness on the stigmatic surfaces. Eventually, the erect styles curve downwards and outwards, and when the process is completed, they are more likely to be touched by insects collecting nectar and pollen. However, for the first hour of flowering, a visiting insect is likely to receive pollen on its body but is unlikely to deposit it on the stigma. Under natural conditions, the stigma remains receptive only on the day of flower opening, and the pollen losses it viability after 24 h (Vasconcellos, 1991). In India, Kishore *et al.* (2010) studied the floral biology of purple, giant, yellow and *P. foetida* in the Sikkim region. Purple, giant and *P. foetida* had major bloom during March-April, July-August and September-October, respectively while in yellow, it was mainly during May-June and September-October. They recorded the maximum stigma receptivity on the day of anthesis in all the passion fruits. Completely curved style was more common in all passion fruits that gave the maximum fruit set.

Pollinator diversity

In Hawaii, the principal insects visiting passion fruit flowers were honey bee, *Apis mellifera* L. (Apidae) and carpenter bee, *Xylocopa varipuncta* Patton (Anthophoridae) (Akamine *e tal.,* 1954). Considering the body size, the carpenter bee is more effective as pollinator than honey bee (Ruggiero *et al.*, 1976). Approximately, there are 700 species of bees in the genus *Xylocopa* and are found throughout the tropical regions of the world (Hurd, 1978). Nishida (1963) reported insect species within Diptera, Hymenoptera, Coleoptera, Thysanoptera and Orthoptera visiting flowers of passion fruit in Hawaii, with *X. varipuncta* and hover fly, *Eristalis arvorum* (Syrphidae), being the most abundant species. The insects most commonly visiting passion flowers in El Salvador are *Bombus* spp., *Trigona* spp., and *Xylocopa* spp. (Free, 1993). In Brazil, the most common species visiting passion fruit flowers are *Xylocopa* spp., *Epicharis* spp. and *Apis mellifera scutellata* (Nishida, 1963; Ruggiero *et al.,* 1976). Akamine and Girolami (1957) noticed that hover fly *E. arvorum* and long-horned grass-hopper *Conocephalus saltator* (Tettigoniidae) feed on pollen of passion flowers, but considered them important from pollination point of view. In Malaysia, Mardan *et al.* (1991) observed that carpenter bee, *Patinopoda latipes* was the most important pollinator of passion fruit while three species of *Xylocopa viz., X. mordax* Smith, *X. scutellata* Moure, and *X. (Megaxylocopa) fimbriata* Fabricius, were the most important pollinators in Venezuela. Free (1993) suggested that honey bees are also are important because they forage in the lowlands and in the humid uplands, and may be easily manipulated. Cox (1957) reported that honey bees are abundant and effective pollinators of passion fruit in Australia. In Florida, USA, where honey bees are the sole pollinators of *P. edulis*, up to 25% of its flowers produce fruit (Hardin, 1986). In India, the most common pollinating bee in purple, giant and yellow passion fruit was *Apis mellifera*, while *A. cerena* was in *P. foetida*. The maximum number of bees observed between 7-8 hrs in purple and giant and between 13-14 hrs in yellow (Kishore *et al.,* 2010)..

Pollinator enhancement

Different ways were suggested to facilitate pollination by *Xylocopa* spp. (Nishida, 1959; Cobert and Willmer, 1980). Since the carpenter bee builds its nest in wood or plant stems, placing wooden posts throughout the passion fruit plantation would enhance their numbers. Camillo and Garófalo (1982) tested nine types of woods and found that eucalyptus was the most preferred wood by *Xylocopa* bees. Hoffmann (1997) recommended the use of posts of 20 cm diameter and 60-70 cm length with two holes of 1.5 cm diameter. Camillo (1996) increased the pollination of passion fruit by introducing *Xylocopa* spp. into plantations.

Apple (*Malus sylvestris* Mill.)

Floral biology

The apple flower produces both nectar and pollen in abundance. The average blossoming period for apples is about nine days. Cool weather lengthens and warm or dry windy weather shortens this period (Morris, 1921). Bee activity on apples during the day is usually greatest about 9 a.m. (Brittain 1933). Although numerous blossoms appear on the apple tree, a set of only 5 percent will produce a fair apple crop (Brittain, 1935). The need for an appropriate agent to transfer pollen from one self- incompatible cultivar to another was established by Waite (1899). Wind has been suggested and disproved at various times as a possible agent in the transfer of apple pollen (Free, 1966).

Pollinator diversity

Various wild bees have been mentioned as important pollinators of apples, including the genera *Andrena, Bombus, Halictus,* and *Osmia* (Free, 1964). Honey bees are reported as the main pollinators of apple and other temperate fruits in India. The proportion of the different honey bee species in the Punjab was: *Apis dorsata* (62%), *A.florea* (9%), and *A. mellifera* (29%) (Mann and Singh, 1981). In other parts of India *A. cerana* is the most important pollinator (Kumar *et al,* 1985; Verma and Chauhan, 1985, Abrol, 1988). *Eristalis* spp. are often important and abundant pollinators of apple and pear (Verma and Chauham, 1985) and sometimes outnumber honey bees. Solomon and Kendall (1970) introduced *E. tenax* to cages containing an apple tree and appropriate pollenizer branches, so there were always three to eight active flies per tree. The fruit set was greater than on trees caged without pollinating insects. Verma and Dulta (1986) observed that some *A. mellifera* nectar-gatherers (6-11 %) collected pollen incidentally as they pushed against the anthers while approaching the nectaries of apple flowers, *A. cerana* nectar-gatherers, probably because of their smaller size collected nectar without contacting the anthers, and no *A. cerana* forager collected both nectar and pollen. Because honey bees collecting pollen necessarily contact the anthers and stigmas, and usually work faster than nectar-gatherers, they are generally regarded as more efficient pollinators.

Other crops

Though not widely studied, there are reports that bees play a significant role in effecting the pollination in fruit crops like aonla, avocado, jackfruit, Macadamia, rambutan, strawberry and tamarind. In papaya, besides wind, sphinx moths and honey bees are given credit as pollinators (Stambaugh, 1960). In case of fig, small wasps of the genus *Blastophaga* are the only insects involved in pollination. The wasps develop in special gall flowers wild fig known as Capri. They emerge from these and accumulate pollen as they exit the fruit. They then enter the "eye" of a Smyrna fig and attempt to oviposit, unsuccessfully, in the long styles of the female blossoms. This activity effects pollination. In commercial plantings the Capri figs containing wasps will be collected and hung on the branches of the Smyrna variety. However certain newer varieties of edible figs are available that do not require this procedure (Free, 1993).

Apis florea

Apis cerana

Apis dorsata

Chrysomya megacephala

Eristalinus arvorum

An unidentified fly

Fig 1. Pollinator diversity on mango

Fig. 2. *Apisdorsata* foraging on guava flower

Fig 3. Honeybees (*Apisdorsata* and *A. cerana*) foraging on passion fruit

Conclusion

Conservation and management of native insect pollinators is very important for sustaining agricultural productivity in the long run. Today the modern beekeeping suffers from a magnitude of problems, including parasitic mites, honey bee diseases, inability of honey bees to work at low temperature and adverse climatic conditions. These difficulties threaten the honey bees general utility as an agricultural pollinator. Therefore, conservation of biodiversity of honey bees and wild pollinators is important to realize the potential yields of several cross-pollinated crops (Thakur, 2012). Efforts must be made to identify, assess and develop techniques/methodology to rear and manage them for enhancing crop pollination. Even though both the need and the potential exist, the practice of rearing and managing natural pollinators for crop pollination has not been given due attention. Another area that deserves to be seriously looked into is the enhancement of non-*Apis* pollinators like stingless bees, syrphid and calliphorid flies.

The following measures will be of great help to conserve and sustain the pollinator populations in orchards (FAO, 2009).

- Providing sufficient flora for off-season sustenance of pollinators
- Protecting and conserving nest sites of natural pollinators
- Ensuring connectivity of natural habitats in farming areas, so that bees can more easily disperse and make needed range shifts in response to changing climates.
- Providing more non-crop flowering resources in fields, such as cover crops, strip crops or hedgerows.
- Avoiding insecticide applications during blossom period. If essential, should use selective and relatively safer chemicals. Dust formulations are more harmful than liquid formulations.
- Spreading the awareness among growers on the importance of pollinator conservation

References

Abrol, D.P. 1988. Pollination studies in almond. *(Prunus amygdalus* L.). *Current Science* 57: 397-398.

Abrol, D.P. 1999. Pollination importance and value of bees. *Insect Environment*, 5:140-141.

Aizen, M. A. and Harder, L.D. 2009. The global stock of domesticated honey bees is growing slower than agricultural demand for pollination. *Biology,* 19: 915-918.

Akamine, E.K. and Girolami, D.G. 1957. Problems in fruit set in yellow passion fruit. *Hawaii Farm Science,* 5: 3–5.

Akamine, E.K., Hamilton, R.A., Nishida, T., Sherman, G.D. and Storey, W.B. (1954) *Passion Fruit Culture*. University of Hawaii, Extension Circular 345, 23 pp.

Alves, José Everton and Freitas, B. 2007. Pollination requirements of guava. *Cienc. Rural*, .37 (5):1281-1286.

Anitha, M., Sivaram,V. and Jayaramappa, K. V. 2012. Influence of bee attractants on pollination and yield parameters in Guava (*Psidium guajava* L.). *The International Journal of Plant Reproductive Biology*, 4(1):41–46.

Bhatia, R., Gupta, D., Chandel, J. S. and Sharma, N. K. 1995. *Indian Journal of Agricultural Sciences,* 65(12): 907-912 .

Brittain, W. H. 1933. Apple pollination studies in the annapolis valley N.S., canada, 1928-1932. *Canada Dept. Agr. Bul.*, 162: 198 pp.

Brittain, W. H. 1935. Studies in bee activity during apple bloom. *Jour. Econ. Ent.* 28: 553-559.

Butcher, F.G. 1957. Pollinating insects on lychee blossoms. *Proc. Fla. Stat. Hort. Soc.*, 70:326–328.

Camillo, E. 1996. Use of species of *Xylocopa* (Hymenoptera, Anthophoridae) in the pollination of passion fruit. In: Meeting on bees, 2, 1996 Ribeirão Preto.. Proceedings Ribeirao Preto: University of Sao Paulo, Faculty of Philosophy, Sciences and Letters of Ribeirão Preto, p. 141-146.

Camillo, E., Garófalo, C.A. .1982. On the bionomics of *Xylocopa frontalis* (Oliver) and *Xylocopa grisescens* (Lepeletier) in Southern Brazil. I—Nest construction and biological cycle. *Rev. Bras. Biol.* 42:571–582.

Chacoff, N. P. and Aizen, M. A. 2007. Pollination Requirements of Pigmented Grapefruit (*Citrus paradisi* Macf.) from Northwestern Argentina. *Crop Science* 47:1143-1150.

Chaturvedi, R.B. and G.K. Saxena. 1965. Studies on the blossom biology of litchi *(Litchi chinensis* Sonn.). *Allahabad Farmer*, 39:10-13.

Chaudary, D.K., Singh, B., Singh, A.K, Singh, P.P. 2003 , Influence of bee pollination of morphophysiological characters on litchi fruit. *J. Res.*, 15(1),121-3

Cobert and P.G. Willmer. 1980. Pollination of the yellow passion fruit. *J. Agril. Sci.*, 95: 655-666.

Cobley, L. S. 1956. *An Introduction to the Botany of Tropical Crops.* London: Longmans.

Costes, E. 1988. Analyze architecturate et modelisation du litchi *(Litchi chinensis* Sonn.). PhD diss., Univ. of Montpelier, France.

Cox, J.E. 1957. Flowering and pollination of passion fruit. *Agri. Gaz. N.S Wales.*,68: 573-576.

Dag, A. and Gazit, S. Eisenstein D. El-Batsri R. and Degani C. 1999.*Scientia Horticulturae,* 82(3-4): 325-329.

Dag, A. and Gazit, S. 2000 . Mango pollinators in Israel. *Jour. Appl. Hort,* 2(1):39-43.

Das, A.K.B., Sureja, A.K. and Jindal, P. C. 2002. Flowering and fruit set in lychee: A review. *Agri. Rev*, 23(1): 59-64.

De Grandi-Hoffman, G. 1987. The honey bee pollination component of horticultural crop production systems. *Hort. Rev.* 9:237–260.

De Wet, E. and Robbertse, P.J. 1986. Studies on pollination problems in Mango cultivars. *Proceedings of a Symposium,* Plant Cell Biology Research Centre, University of Melbourne, May 12-13.

Du Toit, A. P. and Swart, D. J. 1995. Pollination and fruit set in Bengal litchiduring 1993 at Politsi: Second Report : *Year Book of South African Litchi Growers Association*, 7:31-32.

Du Toit, A.P. and Swart, D.J. 1993. *Yearbook South African Mango Growers' Association* 13: 129-130.

Du Toit, A. 1994. Pollination of avocados, mangoes and litchis. *Inligtingsbulletin Institut vir Tropiese en Subtropiese Gewasse*, 262: 7-8.

Du Toit, A.P. 1990. The importance of certain insects as pollinators of sunûower (*Helianthus annuus* L.). *South African of Plant and Soil,* 7: 159-162.

FAO, 2009. Global Action on Pollination Services for Sustainable Agriculture. Food and Agriculture Organization of the United Nations. Viale delle Terme di Caracalla, 00153 Rome, Italy.

Farre, M., Perez, J.M., Hermoso, M. De los A. and Orgata R.J.M. 1997. Estudio sobre polinizacion de chirimoyo (*Annona cherimola* Mill) en Espana. *Memorias Congreso International de Annonaceas. Chapingo, Mexico*, pp. 43-55.

Free, J.B. 1964. Comparison of the importance of insect and wind pollination of apple trees. *Nature*, 201(4920): 726-727.

Free, J.B. 1993. Insect Pollination of Crops. Academic Press, London, UK.

Free, J. B. and Williams, I. H. 1976. Insect pollination of *Anacardium occidentale* L., *Mangifera indica* L., *Blighia sapida* Koenig.

Gallai, N., Salles, J. M., Settele, J. and Vaissi, B.E. 2009. Economic valuation of the vulnerability of world agriculture confronted with pollinator decline. *Ecological Economics*, 68:810-821.

Galon, I., Gazit, S. and Podoler, H. 1982. Improvement of natural fruit set of Annona by increasing the population of pollinating inscets. *Alon Halontea,* 9:611-614.

Gazit, S., Galon, I. and Podoler, H. 1982. The role of nitidulid beetles in natural pollination of *Annona* in Israel. *Journal of the American Society Horticultural Sciences,* 107, 840-852.

George, A.P., Nissen, R.J. and Campbell, J.A. 1992. Pollination and selection in *Annona* species (cherimoya, atemoya and sugar apple). *Acta Horticulture,* 321:178-185.

George, A.P., Nissen. R.J., Ironside, D.A. and Anderson, P. 1989. Effects of nitidulid beetles on pollination and fruit set of *Annona* spp. hybrid. *Scientia Horticulturae* 39:289-299.

Ghazoul, J. 2005. Buzziness as usual? Questioning the global pollination crisis. *Trends Ecol. Evol.* 20:367.

Gottsberger, G. 1970. Beitrage zur Biologie von Annonaceen-Bluten. *Osterreichische Botanische Zeitschrift,* 118:237-279.

Gottsberger, G. 1988. The reproductive biology of primitive angiosperms. *Taxon,* 37:630-643.

Hardin, L. C. 1986. Floral biology and breeding systems of the yellow passion-fruit, *Passiflora edulis* f. *flavicarpa*. *Proc. Interamerican Soc. Tropical Hort.*, 30, 35-44.

Hoffmann, M., 1997. Polinização do maracujá amarelo *Passiûora edulisf.* ûavicarpa Deg. In: Manica, I. (Ed.), Maracujá: temas selecionados (1). Cinco Continentes, Porto Alegre, pp. 58–70.

Hurd, P.D. 1978. An Annotated Catalog of the Carpenter Bees (Genus *Xylocopa Latreille*) *of* the Western Hemisphere (Hymenoptera: Aphididae). *Journal of Economic Entomology* 79: 1599-1602.

Juliano, J. B. and Cuevas, N.L. 1932. Floral morphology of the mango (*Mangifera indica* L.) with special reference to the Pico variety. *Phillipine Agr.,* 21: 449-472.

King, J., E.M. Exley, and V. Vithange. 1989. Insect pollination for yield increases in lychee, p. 142–145. In: D.J. Batten (ed.). *Proc. 4th Austral. Conf. Tree Nut Crops*. Exotic Fruit Growers Assn., Lismore, Austral.

Kishore, K., Pathak, K.A., Shukla, R. and Bharali, R. 2010. Studies on floral biology of passion fruit (*Passiflora* Spp.). *Pak. J. Bot.*, 42(1):21-29

Klein, A.M., B.E. Vaissiere, J.H. Cane, I. Steffan-Dewenter, S.A. Cunningham, C. Kremen, and T. Tscharntke. 2007. Importance of pollinators in changing landscapes for world crops. *Proceedings of the Royal Society B-Biological Sciences*, 274: 303-313.

Kuberappa, G.C., Eswarappa, G., Vishweshwaraiah, C.H. and Sajjanar, S.M. .2002. Activity and abundance of insects visiting guava (*Psidium guajava* L.) flowers. *Pest Management and Economic Zoology,* 11:59–63.

Kumar, J., Mishra, R.C., Gupta, J.K. and Dogra, G.S. 1985. Pollination requirements of some peach varieties. *Indian Bee J.,* 47:3-6.

Kumar, R., Chaudhary, O.P. and Lenin, J.K. 1996a. Studies on pollination of litchi in Dehradun, India. *Indian Bee Journal*, 58: 9-12.

Kumar, R., Lenin J.K. and Chandran, K. 1996b. Honey bee and pollination of *Psidium guajava* L. *Indian Bee Journal*, 56 (4): 206-207.

Linsley, E.G. 1958. The ecology of solitary bees. *Hilgardia*, 27:543-599.

Lopez, E. and Uquillas, C. 1997. *Carpophilus hemipterus* (Coleoptera: Nitidilidae) as cherimoya (*Annona cherimola* Mill.) pollination agent under controlled conditions. *Acta Entomologica* Chilena 21, 89-99.

Malerbo-Souza D.T., Nogueira-Couto, R.H., Couto, L. A.2004. Honey bee attractants and pollination in sweet orange, *Citrus sinensis* (L.) Osbeck, var. Pera-Rio. 2004. *J. Venom. Anim. Toxins incl. Trop. Dis*., 10 (2): 144-153.

Mann, G.S. and Singh, G. 1981. Activity and abundance of flower visiting insects of almond *(Prunus amygdalus* Batsch) at Ludhiana (Punjab). *J. Bombay Nat. Hist. Soc.* 78:617-618.

Mardan, M., Yatim I.M. and Khalid, M.R. 1991. Nest biology and foraging activity of carpenter bee on passion fruit. In: *The 6th International Symposium on Pollination.* Tilburg, pp. 127–132.

Mc Gregor, S.E. 1976. Citrus. p. 149–157. *In* Insect pollination of cultivated crop plants. Agric. Handbk. 496. U.S. Gov. Print. Office, Washington, DC.

Morris, O. M. 1921. Studies in apple pollination. *Wash. Agr. Expt. Sta. Bul.* 163, 32 pp.

Morton, J. 1987. Guava. In: J.F. Morton. Fruits of warm climates. Julia F. Morton, Maimi, FL.

Mukherjee, S.K. 1953. The mango, its botany, cultivation, uses and future improvement. *Econ. Bot.* 7(2): 130-162.

Nadel, H. and Pena, J.E. 1994. Identity, behavior and efficacy of nitidulid beetles (Coleoptera: Nitidulidae) pollinating commercial *Annona* species in Florida. *Environmental Entomology* 23, 878-886.

Nagamitsu, T. and Inoue, T. 1997. Cockroach pollination and breeding system of Uvaria elmeri (Annonaceae) in a lowland mixed- dipterocarp forest in Sarawak. *American Journal of Botany* 84, 208-213.

Naik, K.C., and Rao, M.M. 1943. Studies on the blossom biology and pollination in mangoes (*Mangifera indica* L.). *Indian Jour. Hort.* 1(2): 107-119.

Nishida, T. 1963. *Ecology of the Pollinators of the Passion Fruit.* Hawaii Agricultural Experiment Station, Honolulu, *Technical Bulletin,* 55:356-363.

Pandey, R.S. and Yadava, R.P.S. 1970. Pollination of litchi (*Litchi chinensis* Sonn.) bloom by insects with special reference to honey bees. *J. Apic. Res.* 9, 103-105.

Pena, J.E. and Bennett, F.D. 1995. Arthropods associated with Annona spp. in the neotropics. *Florida Entomologist,* 78:329-349.

Pena, J.E., Castineiras, A., Bartelt, R. and Duncan, R. 1999. Effect pf pheromone bait stations for sap beetles (Coleoptera: Nitidulidea) on Annona fruit set. *Florida Entomologist,* 82:475-480.

Phartiyal, T., Srivastava, P., Khan, M.S. and Srivastava, R.M. 2012. Abundance of insect pollinators in citrus crop and their impact on fruit setting under tarai Agroclimatic condition *J. Ent. Res.,* 36(3):211-213.

Phoon, A.C.G. 1984. Insect pollination of some Malaysian fruit trees with special reference to the honey bee *Apis cerana.* Ph.D. Thesis, University Pertanian Malaysia.

Popenoe, W. 1917. The pollination of the mango. U.S. Dept. Agr. Bul. 542, 20pp.

Popenoe, W. 1920. Manual of tropical and subtropical fruits. 474 pp. The Macmillan Co., New York.

Purseglove, J.W. 1968. *Tropical Crops. Dicotyledons.* London: Longmans.

Reddy, P.V.R., Verghese, A. and Varun rajan, V. 2012a. Potential impact of climate change on honey bees (*Apis* spp.) and their pollination services. *Pest Management in Horticultural Ecosystems,* 18 (2): 121-127.

Reddy, P.V.R., Verghese, A., Varun Rajan, V. ,Rashmi, T. and Kavitha. 2012b. Diversity and foraging ecology of pollinators in mango (*Mangifera indica*): An Indian perspective. International Congress of Entomology, Daegue, South Korea 19-24 August 2012.

Reddy, P.V.R., Dinesh, M.R., Verghese, A. and Vasugi, C. 2011. Insect Pollinators: A key component to increase fruit set in mango. Presented in the Global conference on Augmenting Mango Productivity and Utilization. 21-24 June 2011, Lucknow.

Reddy, P.V.R., Varun Rajan, V. and Verghese, A. 2013a. A semi-synthetic diet based trap to attract and monitor the calliphorid pollinator, *Chrysomya megacephala* (F.).100th Indian Science Congress, Kolkata, 3-7 January 2013.

Reddy, P.V.R., Varun Rajan, V., Vasugi, C. and Verghese, A., 2013b. Pollinator activity in relation to varietal differences in mango (*Mangifera indica* L.) International Conference on Insect Science, Bangalore, 14-17 February 2013.

Reddy, P.V.R. and Jacob T.K. 1998. Foraging behaviour of honey bees in guava. *Insect Environ.*, 4: 96.

Reiss, A. 1971. Pollination and fruit set of *Annona cherimola* and *Annona squamosa*. M.Sc. Thesis, The Hebrew University of Jersalem, Israel. Report (1970), 101-102.

Roubik, D.W. 1995. Pollination of cultivated plants in the tropics. FAO *Agric. Serv. Bull.*118, FAO, Rome.

Ruehle, G. D. and Ledin, R. B. 1955. Mango growing in Florida. *Fla. Agr. Expt. Sta. Bul.*, 574:90 pp.

Ruggiero, C., Banzatto, D.A. and Lam-Sanchez, A. 1976. Studies on natural and controlled pollination in yellow passion fruit (Passiflora edulis f. flavicarpa Deg.). *Acta Horticulturae* 57:121–124.

Sanford, M.T. 1992. Pollination of Citrus by honey bees. Rf-AA092. Florida Crop Ext. Serv., Inst of food and Agri. Sci., Univ. of Florida.

Sanford, M.T. 1992. Pollination of citrus by honey bees. RFAA092. Florida Coop. Ext. Serv., Inst. of Food and Agric. Sci., Univ. of Florida.

Sharma, D. K. and Singh, R. N. 1970. Self-incompatability in mango (*Mangifera indica* L.).

Singh, G. 1997. Pollination, pollinators and fruit setting in mango. Proceedings of the 5th International Mango Symposium Vol 1. *Acta Horticulturae* No. 455. 116-123.

Singh, L. B. 1960. *The Mango*. London: Leonard Hill.

Singh, R. and Chopra, S.K. 1998. Flower visitors of litchi and their role in pollination. *Pest Management and Economic Zoology*, 6: 1-5.

Singh, R.N. 1978. *Mango*. New Delhi: Indian Council of Agricultural Research.

Singh, R. N., majumdar, P. K., and Sharma, D. K. 1962. Self-incompatibility in mango (*Mangifera indica* L.) var. Dashehari. *Cur. Sci.,* 31(5): 209.

Singh, S.N. 1961. Studies on the morphology and viability of the pollen grains of mango. *Hort. Adv.*, 5: 121-144.

Solomon, M.E. and Kendall, D.A. 1970. Pollination by the Syrphid fly, *Eristalis tenax. Long Ashton Research Station. Annual.*

Spencer, J.L. and kennard W.C. 1956. Limited stigmatic receptivity may contribute to low fruit set in the mango (*Mangifera indica* l.). *Amer. Soc. Hort. Sci. Proc.,* 67: 287- 289.

Stambaugh, S. V. 1960. Forty years of papaya development. *Fla. State Hort. Soc. Proc.,* 73:311-314.

Steffan-Dewenter, I., Potts, S.G. and L. Packer. 2005. Pollinator diversity and crop pollination services are at risk. *Trends Ecol. Evol.* 20:651–652.

Stern, R. A. and Gazit, S. 1996. Lychee Pollination by the Honey bee. *J. Amer. Soc. Hort. Sci.,* 120(1):152–157.

Stern, R. A., 1. Adato, M. Goren, D. Eisenstein, and S. Gazit. 1993. Effect of autumnal water stress on litchi flowering and yield in Israel. *Scientia Hort.*, 54:295–302.

Sturrock, T.T. 1966; the mango inflorescence. *Fla. State Hort. Soc. Proc.*, 79: 366-369.

Sung, I.H., Ming-Ying Lin, Chin-Hsing Chang, Ann-Shiou Cheng, and Wen-Shyong Chen. 2006. Pollinators and their behaviors on mango flowers in Southern Taiwan. *Formosan Entomol.* 26: 161-170.

Thakur, M. 2012. Bees as Pollinators – Biodiversity and Conservation. *International Research Journal of Agricultural Science and Soil Science,* 2(1) : 001-007.

Vanderplank, J. 1996. Passion flowers. Cassel Publishers Limited, London

Vasconcellos, M.A.S. 1991. Biologia floral do maracujá doce (*Passiflora alata* Dryand) nas condições de Botucatu – SP. MSc thesis, UNESP, Jaboticabal, SP, Brazil, 98 pp.

Verghese, A. and Tandon, P. L. 1990. Pollination behavior of *Apis florea* on mango. In:Social Insects – an Indian Pesrpective. IUSSI, Bangalore, India 60-63 pp.

Verma, L. R. and Chauhan, P. 1985. Distribution, abundance and diversity of insect pollinators in apple orchard of Shimla hills. *Indian J. Ecol.* 12, 286-292.

Verma, L. R. and Dulta, P. C. 1986. Foraging behaviour of *Apis cerana indica* and *Apis mellifera* in pollinating apple flowers. *J.Apic. Res.* 25, 197-201.

Villalta, R. 1988. Estudio de la biologia floral e identification de agentes polinizadores de la guanabana (*Annona muricata* L.) en la University, Costa Rica. Thesis, National University, Costa Rica, 61 pp.

Wagle, P. V. 1929. A preliminary study of the pollination of the Alphonso mango. *Agr. Jour. India*, 24(14): 259-263.

Waite. 1899. Pollination of pomaceous fruits. *U.S. Dept. Agr. Year Book*,1898: 167-180.

Wongsiri, S. and Chen, P. 1995. The effects of agricultural development on honey bees in Thailand. *Bee World*, 76:3-5.

7

Molecular Markers in Fruit Breeding

Kanupriya and K.V. Ravishankar

Introduction

Fresh and processed products derived from fruit crops provide vital contributions to human nutrition, health, and well-being, and collectively their production constitutes the economic backbone of many rural communities worldwide. Rapid development and distribution of new cultivars with improved characteristics which meet the industry, market (both domestic and export), and consumer preferences is the need of the day. Conventional methods of perennial fruit crop improvement have almost exclusively relied on traditional breeding techniques. These techniques are expensive, time consuming and space requiring, especially for tree species due to the need of a large population of seedlings from which only a few promising ones are selected. The juvenile period of tree crops may last three to five years before onset of fruits so that 20 years or more are common before the successful introduction of a gene of interest (Baird *et al.,* 1996). Therefore in this context, molecular tools recently made available for breeders have become tools of choice because they permit to reduce cost by fastening the selection process of woody fruit plants. Further, traits of importance like fruit quality, resistance to diseases, high yield etc. are polygenic in nature for which classical breeding is generally not suitable for detection and selection of traits. A quantitative trait locus (QTL) analysis with help of markers and linkage maps may be more appropriate for detection of such traits of interest. Also, molecular markers are being used for maintenance of core collections, for the assessment of genetic diversity between genotypes and fingerprinting.

Genetic diversity and fingerprinting

The assessment of genetic diversity of fruit crops is an important step in breeding since a diverse range of plant materials including landraces, modern cultivars, crop wild relatives and other wild species are available as the starting material. Genetic diversity existing in the plant genetic resource collections can provide plant breeders and researchers with possibilities to develop new and more productive crops, preferably resistant to various pests and diseases and adapted to changing environments. It is also felt that conservation of germplasm itself is of little importance without characterization and further utilization of the stored plant materials. To intensify these activities DNA based fingerprinting has been implemented in the characterization of germplasm. A large number of molecular markers have been developed and applied for analysis of genetic diversity and relatedness, and all markers have their strengths and weaknesses (Schlötterer, 2004; Weising *et al.,* 2005). It has been suggested that the markers should be chosen based on the objectives of the study as well as mating system of the studied species and available financial support. In general, two main groups of molecular markers have been utilized for PGR characterization:

i. markers for identification of accessions, detection of the structure of genetic resources collection and relatedness of available plant material, and

ii. gene specific markers, subsequently applicable for Marker Assisted Selection (MAS).

Commonly used markers

Random amplified polymorphic DNA (RAPD) markers

These markers were first described by Williams *et al.,* (1990). The method uses short arbitrary primers, usually 10 bp (decamers), to amplify anonymous PCR fragments in noncoding as well as coding regions of template DNA. The PCR products are separated by electrophoresis in an agarose gel. The gel is stained with ethidium bromide and the amplification products are visualized under UV light. A PCR product is generated when primers anneal in an inverted way and at an amplifiable distance from one another (150–3000 bp). The number of amplification products is related to the number and direction of the genome sequences complementary to the primer. RAPD polymorphism is caused by a nucleotide substitution within the target site, or an insertion or deletion of a DNA-fragment within the amplified region. These mutations result in an absence of a fragment or changes in fragment size (Weising *et al.,* 2005). RAPD is a relatively cheap and technically simple method. It is especially useful for analysing large number of samples belonging to species for which there is no genome sequence information. The main drawbacks of RAPD markers are possible problems with reproducibility (Jones *et al.,* 1997), co-migration of non-homologous fragments (Rieseberg, 1996), competitive priming (Halldén *et al.,* 1996) and their dominant inheritance, i.e., heterozygotes cannot be distinguished from dominant homozygotes (Williams *et al.,* 1990). The problem of reproducibility within the same laboratory can be avoided by careful optimisation and adherence to a well-defined protocol, by screening large numbers of primers for reproducibility and by scoring only strong, clear and consistently appearing bands (Koller *et al.,* 1993; Mulcahy *et al.,* 1993; Belaj *et al.,* 2003). Comigration of non-homologous fragments can be a problem at higher taxonomic levels, but is largely negligible at the intraspecific level or when the studied species are closely related (Rieseberg, 1986). RAPD markers have, despite the above-mentioned limitations, been widely used in plant genetics and breeding, e.g., to estimate genetic diversity, detect phylogenetic relationships, and create genetic maps (Weising *et al.,2*005).

Simple sequence repeats or microsatellites (SSR)

Microsatellites are short (1–6 bp) tandemly repeated sequences of DNA (Litt & Luty, 1989; Tautz, 1989). In microsatellite analysis, two primers are designed complementary to the sequences flanking a specific microsatellite DNA sequence. The sequence is amplified by PCR, fragments are separated by electrophoresis and allele sizes are determined. The allelelic polymorphism of microsatellite loci is usually caused by a variable number of integer repeat units. Microsatellites are abundant in plant genomes, co-dominantly inherited, multiallelic and highly polymorphic. Microsatellites have therefore become a marker system of choice for identification of genotypes, population genetics, paternity testing. Although SSR markers generally show highly reproducible results within and between laboratories, some discrepancies in allele sizing between laboratories have been reported, possibly caused by the many factors that can influence the migration of PCR fragments through the gel (Galli *et al.,* 2005; Amos *et al.,* 2007). The main drawback of microsatellite analysis is the high expenses at the initial stages of primer development, when sequences for primer design must be isolated and cloned. However, some cross-species transferability of primer pairs (Yamamoto *et al.,* 2001, 2004) or software-designed primers based on publicly accessible sequence information from gene-bank accessions (Tonguc & Griffiths, 2004) can decrease the initial costs considerably.

Sequence-characterized amplified regions (SCARs)

SCARs were originally developed from RAPD-derived markers (Paran & Michelmore, 1993), but have later also been developed from AFLP-derived markers (Mienie *et al.,* 2002). A DNA fragment in an agarose gel, tightly linked to the character of interest, is cloned and sequenced, and approx. 24 bp long primers, complementary to the ends of the original fragment, are designed. These primers can then be used in amplification of a single marker for the character of interest. SCAR markers demand stringent PCR conditions and thus have no problems with reproducibility. The initial stages of development of SCAR markers are laborious, but once developed they are user-friendly since the PCR-fragments can be detected in an agarose gel with ethidium bromide. Apple cultivars have been characterized with SCAR markers to detect alleles that control apple skin colour (Cheng *et al.,* 1996), incompatibility (S-alleles) (Broothaerts, 2003) and resistance against apple scab caused by the fungus *Venturia inaequalis* (Hemmat *et al.,* 2002, Mattisson & Nybom, 2005; Melounova *et al.,* 2005).

Single nucleotide polymorphism (SNPs)

Single base differences between homologous DNA fragments are referred to as Single Nucleotide Polymorphism (SNPs). These markers represent most frequent polymorphism and are becoming more popular than SSRs as genetic markers in linkage analysis because these are more abundant, and are suitable for automatic allele calling (Novelli *et al.,* 2004; Selmer *et al.,* 2009). In temperate fruit crops like the apple, Illumina Infinium SNP assay containing 8,000 SNPs is being used as a crucial tool for application of a novel selection strategy called genomic selection (GS) in breeding programmes. Availability of this assay has resulted in dramatic reduction in genotyping costs for large scale genotyping of apple seedlings (Chagne *et al.,* 2012). SNPs have been developed only for a few tropical fruit species like banana, where 20 SNP markers are available with NCBI. At present, the cost of developing and testing SNPs is still higher than that for developing SSRs. Nonetheless, SNP discovery is actively being pursued in crops like avocado (Arias *et al.,* 2012), mango and longan under a USDA sponsored project on genomics and single nucleotide marker discovery in tropical and subtropical fruit crops. In fruit trees where reference genome sequence exists, like papaya, clementia and sweet orange and banana, genotyping- by- resequencing using next generation sequencing will allow rapid and efficient identification of SNPs and other types of polymorphism that are related to structural variations (such as copy number variations CNVs, insertion-deletion polymorphisms IDPs and presence-absent variants PAVs between genomes). These can be used for genome-wide and candidate gene-based association mapping studies to identify QTL-trait associations (Khan and Korban, 2012). Such studies have been made in cereal crops like maize, where extensive structural variation (including hundreds of CNVs and thousands of PAVs) has been discovered and implicated in important biological phenomena like heterosis (Springer *et al.,* 2009).

A number of approaches have been developed to reduce the complexity of the genomes for cost efficient and simplified discovery of SNP markers using NGS, like, RNA-Seq, complexity reduction of polymorphic sequences (CRoPS), restriction-site-associated DNA sequencing (RAD-Seq), and genotyping-by-sequencing (GBS) (Dawey *et al.,* 2011). Among these, RAD-Seq and GBS are proving to be particularly efficient in rapid and mass discovery of SNPs in highly heterozygous species. This is expected to help develop of ultrahigh density maps and association mapping. Together with NGS device, these techniques help discover SNPs by sequencing a large set of restriction fragments. Fragments from one individual are ligated to a modified adapter containing a barcode. Fragments from many individuals can, therefore, be pooled together and sequenced on a single lane. The resulting reads can be separated bioinformatically by identifying the barcode. By sequencing a family or other populations of interest this way, and comparing the tags to phenotypes of individuals, many biologically relevant SNPs and genetic loci can be identified in a single experiment. If each genotype of a mapping population has been characterized phenotypically for traits of interest, and has an individual barcode, then RAD and GBS markers can be used routinely for linkage mapping and QTL analysis (Rowe *et al.,* 2011).

Approaches for QTL identification

The identification of QTLs is in most cases based on a technique called linkage mapping (or family mapping) for which a segregating population is generated from two parents which usually considerably differ for one or several trait(s) of interest. Molecular markers revealing a polymorphism between the parents and segregating in the population are screened to construct a genetic linkage map based on the recombination frequency obtained between every marker. Two markers are linked if the recombination frequency is lower than 0.5, and this recombination frequency is then translated into a genetic distance unit (centimorgans, cM). On the other side, phenotypic data are scored in the population and generally on the parents. When a significant association is identified between the marker and the studied trait and after subsequent validation in different backgrounds or environments, the marker can be further used in a Marker-Assisted Selection (MAS) program to introduce the trait of interest in commercial cultivars.

Genomics in fruit crops

Banana

Cultivated bananas belong to the family Musaceae of the order Zingiberales. It is the fourth most important cultivated crop in the world. It has two genus

Musa and *Ensete*. Two important species of the genus *Musa* are *M. acuminata(2n=22; referred as A genome)* and *M. balbisiana (2n=22; referred as B genome)*. Being one of the oldest and highly evolved fruit it has a number of cultivated types selected by human's involving diploid AA, AB genomic group and triploid AAA, AAB & ABB genomic groups as per the classifications made by Simmonds (1962). All available markers like RAPD, AFLP, ISSR, MSAP, SSR, STMS, mini & micro satellite markers, chloroplast & mitochondrial specific markers have been used to study various aspects in banana like phylogenetic relationships, diversity, evolution, parentage analysis, differentiation of A & B genomes and to develop trait specific markers in *Musa* (Careel *et al.,* 2002; Rekha *et al.,* 2001, Ravishankar *et al.,* 2012, Creste *et al.,* 2006, Miller *et al.*, 2010). Of late SNP'S are found to be more useful markers as they have an impact on protein function and thus on phenotype. The studies have revealed the close relationship among the cultivars and species. It helped to differentiate the cultivars, species and different genomic groups. The chloroplast and mitochondrial specific markers helped the breeders to understand the evolution and involvement of male and female parents among natural hybrids. Several SCAR markers and RGA's were identified in relation to the important diseases like Fusarium wilt and leaf Sigatoka (Uma *et al.,* 2011). The first saturated map proposed as a "reference *Musa* map" was prepared from 489 markers (167 SSRs and 322 DArTs) covering 1197 cM containing 11 linkage groups between two genetically distant accessions of *M.acuminata*, 'Borneo' and 'Pisang Lilin' (P. Lilin) . The RFLP and SSR markers were employed to develop BSV (Banana Streak Virus) related map and RAPD & AFLP were used to develop wilt resistance related map in order to map these traits. Recently, whole 523-megabase genome of a double-haploid *Musa acuminata* genotype of banana has been sequenced (D'Hont, *et al.*, 2012). This will prove to be an important stepping stone for improvement of banana which has been a challenge for breeders. It will help in identifying genes responsible for important agronomic characters, such as fruit quality, disease and pest resistance as they affect the banana production and consumer acceptance.

Mango

Mango (*Mangifera indica* L.) is popularly known as the 'king of fruits' is believed to have originated in eastern India (Knight, 1980). The genus *Mangifera* belongs to the order Sapindales in the family Anacardiaceae, which is a family of mainly tropical species. It consists of 73 genera, 69 species and mostly restricted to tropical Asia. The mango genome consists of 4.39×10^8 bp. (Arumuganathan and Earle, 1991) and has 20 chromosomes, most of them are small. Molecular studies in mango are a very few and restricted to RAPD, AFLP and SSR markers. Initially RAPD markers were used to study genetic diversity of mango

(Schnell *et al.,* 1995; Ravishankar *et al.,* 2000; Ravishankar *et al,* 2004; Kumar *et al.,* 2001). AFLP markers were found suitable for cultivar identification, estimation of genetic relationship and mapping of QTLs in mango. Kashkush *et al.,* (2001) constructed a genetic linkage map of mango consisting of 13 linkage groups and covering 161. 5 cM defined by 34 AFLP markers. The information generated by AFLP analysis regarding genetic relatedness and diversity existing in the mango gene pool are useful for breeding of improved mango varieties (Yamanaka *et al.,* 2006). Fang *et al.,* (1999) reported fingerprinting of two mango cultivars viz., Keitt and Tommy employing AFLP. A study using 31 F1 progenies from crosses between "Alphonso" and "Palmer" led to the construction of maps for each cultivar that were useful for analysing correlations of traits like fruit size, shape and colour (Phumichai *et al.,* 2000). The phylogenetic relationships among 14 *Mangifera* L. species including three economically important species has been analyzed by comparing 217 amplified fragment length polymorphism (AFLP) markers (Eiadthong *et al*., 2000).

SSR markers are widely used as a versatile tool in plant breeding programs as well as in evolutionary studies because of their ability for showing diversity among cultivars (Viruel *et al*., 2005, Schnell *et al.,* 2006; Ravishankar *et al.,* 2011). SSR, also known as microsatellites, are an efficient type of molecular marker based on tandem repeats of short (2-6 nucleotides) DNA sequence (Charters *et al.,* 1996) and target highly variable and numerous loci. Among the 64 Florida cultivars evaluated in the parentage analysis by Schnell *et al.,* (2006), the genetic background was found to be based on as few as four Indian cultivars and the polyembryonic cultivar 'Turpentine'. Two Indian cultivars, 'Mulgoba'and 'Sandersha', are in the background of most Florida types with 'Amini', 'Bombay','Cambodiana', 'Long', 'Julie', 'Turpentine', and 'Nam doc Mai' making lesser contributions. Microsatellites were successfully used for genetic diversity analysis of the indigenous type 'Appemidi' mango. The analysis of the 211 bands detected by the 14 SSR markers showed the unambiguous discrimination of the 43 mango genotypes. The dendrogram resulted in the grouping of accessions into two major clusters *viz.,* cluster I with highly acidic types and cluster II with less acidic and high TSS group (Vasugi *et al.,* 2012). The work related to SNP in mango is of very recent origin. Hence the available study is very scanty. Ishtiaq and Azim (2012) used SNPs for mango cultivar identification and phylogenetics. They characterized rpl20-rps12 and atp-rbcL intergenic spacers of chloroplast DNA from 19 mango cultivars to analyze suitability of these regions for assessing intraspecific variation.

More genomic resources and markers should be developed in mango since the use of molecular markers in mango breeding would be greatly assisting in solving a number of difficult challenges for breeders such as the development of complex family structures for recombination mapping and for recurrent selection.

Papaya

Papaya (*Carica papaya* L., 2n=2x=18, 372Mbp) belongs to family Caricaceae, and until recently, this family was thought to comprise about 31 species in four genera. However, a recent taxonomic revision proposed that some species formerly assigned to *Carica* were more appropriately classified in the genus *Vasconcella* (Badillo, 2002). The species in this genus are called wild papayas and they are distantly related to *Carica papaya*. They possess several desirable traits like resistance to pathogens, cold tolerance and high sugar content of fruits which can be used for improvement of cultivated papaya if the pre and post zygotic barriers are overcome (Drew, 2002). Papaya is a tropical fruit tree, first domesticated in the eastern lowlands of central Americas (Nakasone and Paull, 1998) and now cultivated around the tropical and warmer subtropical areas of the world (Villegas, 1997).

Several microsatellite markers have been developed for papaya (Sharon *et al.*, 1992; Parasnis *et al.*, 1999; Santos *et al.*, 2003; Collard and Mackill, 2008; Eustice *et al.*, 2008; Olivera *et al.*, 2008, 2010b). Reliable and highly polymorphic markers have been used for the construction of genetic maps (Chen *et al.*, 2007), sexual differentiation (Parasnis *et al.*, 1999; Santos *et al.*, 2003) and to assess genetic diversity (Perez *et al.*, 2007; Oliveral *et al.*, 2010b).

The developed SSR markers are being put to several uses like identification of plants presenting high levels of homozygosity in segregating progenies for the development of new papaya inbred lines, which are needed in order to avoid F1 segregation. The crossing between these inbreds will allow development of hybrids with superior vigour over and beyond those of the parents (Olivera *et al.*, 2010a). SSR markers are also being used for genetic characterization of papaya plants derived from first backcross generation (Ramos *et al.*, 2010). In papaya, fruit flesh color is controlled by a single gene as segregating populations show a 3:1 inheritance of yellow: red fruit flesh color. Using a candidate gene and map based cloning approach a chromoplast- specific lycopeneâ cyclase CpCYC-b gene was identified. A 2bp insertion found in CpCYC-b coding region has been identified to result in a frame shift mutation and a premature stop codon. Using this information, a simple PCR- based screening test has been developed based on the CPFC1 marker to identify individuals in a segregating population with desired flesh color (Blas *et al.*, 2010).

The sex inheritance in papaya is determined by a dominant M allele for males, a different dominant Mh allele for hermaphrodites and a recessive m allele for females. The MhMh, MhM, and MM genotypes are lethal, thus leading to a 2:1 (hermaphrodite: female) segregation ratio after hermaphrodite plants (Mhm) have self pollinated (Ming *et al.,* 2007); thus, all hermaphrodite progeny is not possible. Since female plants have less commercial value than hermaphrodites, identification of hermaphrodite at an early stage is desirable. Several male-hermaphrodite specific markers have been developed by RAPD or AFLP, which were converted to SCAR markers e.g., T12 and W11 (Deputy *et al.,* 2002), Napf (Parasnis *et al.,* 2000) and PSDM (Urasaki *et al.,* 2002) to select for hermaphrodites from female papaya seedlings. Recently, loop mediated isothermal amplification (LAMP) reaction has been developed for quick identification of sex in the papaya field (Hsu *et al.,* 2012). LAMP is a novel technique which requires a special primer set design which recognises a total of six regions of the targeted DNA sequence (Notomi *et al.,* 2000). The advantage is the visual judgement based on turbidity or fluorescence of the reaction mixture without the need of specific equipments (Hsu *et al.,* 2011).

The first genetic map of papaya consisted of three morphological markers: sex form, flower colour and stem colour (Hofmayer, 1939). When RAPD markers were developed, the second genetic map of papaya became available with 62 RAPD markers (Sondur *et al.,* 1996). A high density linkage map was constructed with 1498 AFLP markers, papaya ringspot coat protein marker, morphological sex type and fruit flesh color markers using 54 F2 plants derived from cross Kapoho x SunUp. These markers were mapped to 12 major linkage groups covering a total length of 3294.2 cM. This map revealed severe suppression of recombination around the sex determination locus with 225 markers cosegregating with sex types (Ma *et al.,* 2003). A total of 707 markers, including 706 microsatellite loci and the morphological marker fruit flesh color, were mapped into nine major and three minor linkage groups. The resulting map spanned 1069.9 cM with an average distance of 1.5 cM between adjacent markers (Chen *et al.,* 2007).

Quantitative trait locus (QTL) mapping has been done to identify the multiple genes controlling the papaya fruit size and shape. A large variation is present in the germplasm for these traits. The fruit weight may vary from 0.2 to 10 kg, while the shape of the fruit is sex linked ranging from sperical to ovate, cylindrical or pyriform. An F2 mapping population produced from a cross between the Thai variety Khaek Dum, bearing 1.2 kg, red fleshed fruit was crossed with Hawaii Solo type bearing 0.2 kg, yellow fleshed fruits. Fourteen QTL with phenotypic effects ranging from 5 to 23% were identified across six linkage groups (Blas *et al.,* 2009). The 1-LOD interval surrounding each QTL was searched for candidate genes. Five candidate genes (without QTL association) that showed homology to one of three previously identified tomato loci (ovate, sun or fw 2.2) affecting the tomato fruit size and shape were identified in papaya genome (Blas *et al*., 2009).

Papaya was selected for genome sequencing since it is considered to be an excellent model crop to study the evolution of sex chromosomes and for tropical fruit tree genomics. At 372 Mbp papaya genome is 3 times the size of *Arabidopsis*, with which it shared a common ancestor about 72 million years ago (Wikstrom *et al.,* 2001). The transgenic female SunUp variety was chosen for sequencing using whole genome shotgun (WGS) with Sanger sequencers (Ming *et al.,* 2008). Some interesting features emerged from the genome sequencing like papaya has about 25% fewer genes than *Arabidopsis*, with reduction across most gene families and biosynthetic pathways.

Papaya has great nutritional, industrial and medicinal value. The whole genome sequencing of papaya along with integrated genetic and physical maps, EST database, metabolomic and proteomic studies and other genomic resources will provide a breakthrough in papaya improvement and the exploration of its nutritional and medicinal applications.

Guava

Guava (*Psidium guajava* L. 2n=2x=22; 496Mbp) belongs to genus *Psidium* of the family Myrtaceae. This genus has 9 species (USDA, Natural resource conservation service) and all are native to the tropical regions of northern South America (Risterucci *et al.,* 2005). They are now spread throughout the tropics and subtropics of Asia, Africa, subtropical regions of North America and Australia. Mexico, Brazil, India and Thailand are the largest producer countries of guava. The fruits which are a rich source of dietary fibre, Vitamins A and C and folic acid are eaten fresh or processed into juices and jams. The leaves and bark have pharmacological actions and have been used in folk medicine for aliments of gastrointestinal tract, gum problems, fever and cough (Kamath *et al.,* 2008).

The development of genomic resources in guava was started by Valdes-Infante *et al*., (2003) by using the amplified fragment length polymorphism (AFLP) markers to characterize the guava accessions of Cuba. They were also the first workers to develop the linkage map in the population 'Enana Cubana roja' x' N6'. This linkage map was later enhanced by Rodriguez *et al.,* (2007) and contained 220 AFLP markers. Several QTLs for vegetative traits and fruit characteristics have been identified and integrated into this map to make it more informative (Rodriguez *et al.,* 2007). The development of SSR markers was started by Resterucci *et al*., (2005) using (GA)n and (GT)n microsatellite-enriched library and 23 nuclear simple sequence repeat (SSR) loci were characterized in the guava species (*Psidium guajava* L.). These SSR markers have been put to use by several workers for assessing the genetic diversity of guava in different countries. The guava accession of Mexico has been analysed by Teyer *et al*., (2008) using 6 SSR loci and 5 AFLP primer combinations in an automatic sequencer. In this analysis, a total of 57 individual accessions from 10 different regions of Mexico were included. The AFLP- and SSR-based dendrogram grouped the guava accessions into 2 main groups with at least 5 different subclusters. In another study, microsatellite markers were used to characterize the diversity of guava germplasm in the United States (Vijji *et al.,* 2010). Seven PCR primers amplifying fourteen previously developed microsatellite loci were used to identify and characterize 13 guava accessions. The diversity in nine Indian pink and white flesh cultivars of guava were characterized using 23 microsatellite markers by Kanupriya *et al.,* 2011. The pink cultivars and white cultivars separated into two distinct groups. The EU project (2005-2009) on "Improvement of guava: Linkage mapping and QTL analysis as a basis for marker assisted selection" provided the necessary impetus for development of hundreds of microsatellite (SSR) markers in this crop. The Guava map project webpage contains a list of 344 SSR primer pairs. These primers are available in public domain and can be used for development of linkage maps and identification of markers and QTLs for traits of interest.

Conclusion

Compared to temperate fruit crops of Rosaceae family (peach, apple, strawberry, cherry, apricot, pear, raspberry, etc.) and annual crops the amount of genomic information in tropical and subtropical fruit crops, is very less. However, with advent of next generation sequencing technology large scale development of SSR and SNP markers can be attempted and then these can be used for breeding. Another aspect is that the mere development of these genomic resources will not assist in breeding as seen in case of Rosaceae fruit crops where significant amount of genomics research including EST libraries, genetic and physical maps, QTLs, and whole genome sequences has not yet been translated into routine practical application in breeding. This is largely attributed to significant technical, economic, logistical, informational, and sociological barriers. Until these barriers are overcome, these genomic resources can not be applied for routine applications (www.rosbreed.org).

References

Amos, W., Hoffman, J.I., Frodsham, A., Zhang, L., Best, S. and Hill, V.S. 2007. Automated binning of microsatellite alleles: problems and solutions. *Molecular Ecology Notes,* 7:10–14.

Arias De Ares, R.S., Borrone, J.W., Tondo, C.L., Kuhn, D.N. and Schnell, R.J. 2012. Genomics of Tropical Fruit Tree Crops. Book Chapter. (In): Schnell, R.J., Priyadarshan, P.M., (editors). Genomics of Tree Crops. Heidelberg, Germany: Springer. 209-39.

Arumuganathan, K. and Earle, E.D. 1991. Nuclear DNA content of some important plant species. *Plant Molecular Biology Reporter,* 9:208–218.

Badillo, V.M. 2002. Carica L. vs. Vasconcella St. Hil. (Caricaceae) con la Rehabilitacion de este Ultimo. *Ernstia,* 10:74-79.

Baird, W.V., Ballard, R.E., Rajapakse, S., and Abbott, A.G. 1996. Progress in *Prunus m*apping and application of molecular markers to germplasm improvement. *Hort. Science* 31:1099-1106.

Belaj, A., Satovic, Z., Cipriani, G., Baldoni, L., Testolin, R., Rallo, L. and Trujillo. I. 2003. Comparative study of the discriminating capacity of RAPD, AFLP and SSR markers andof their effectiveness in establishing genetic relationships in olive. *Theoretical and Applied Genetics,* 107:736–744.

Blas, A.L., Ming, R., Liu, Z., Veatch, O.J., Paull, R.E., Moore, P.H. and Yu, Q. 2010. Cloning of the papaya chromoplast-specific lycopene b-cyclase, CpCYC-b, controlling fruit flesh color reveals conserved microsynteny and a recombination hot spot. *Plant Physiol.,* 152:2013–2022.

Blas, A.L., Q. Yu, C. Chen, O. Veatch, P. H. Moore, R. E. Paull, Ming, R. 2009. Enrichment of a papaya high-density genetic map with AFLP markers. *Genome,* 52:716-725.

Broothaerts, W. 2003. New findings in apple S-genotype analysis resolve previousconfusion and request the re-numbering of some S-alleles. *Theoretical and AppliedGenetics,* 106:703-714.

Carreel, F., Gonzalez de Leon, D., Lagoda, P., Lanaud, C., Jenny, C., Horry, J.P. and Tezenas du Montcel, H. 2002. Ascertaining maternal and paternal lineage within Musa by chloroplast and mitochondrial DNA RFLP analyses. *Genome*, 45(4):679-692.

Chagné D, Crowhurst RN, Troggio M, Davey MW and Gilmore B, 2012. Genome-Wide SNP Detection, Validation, and Development of an 8K SNP Array for Apple.PLoS ONE; 7(2):e31745.

Chen, C., Q. Yu, S. Hou, Y. Li, M. Eustice, R. L. Skelton, O. Veatch, R. Herdes, L. Diebold, J. Saw, Y. Feng, L. Bynum, L. Wang, P.H. Moore, R.E. Paull, M. Alam and Ming, R. 2007. Construction of a sequence-tagged high density genetic map of papaya for comparative structural and evolutionary genomics in Brassicales. *Genetics,* 177:2481-2491.

Cheng, F.S., Weeden, N.F. and Brown, S.K. 1996. Identification of co-dominant RAPD markers tightly linked to fruit skin colour in apple. *Theoretical and Applied Genetics,* 93:222–227.

Collard, B.C.Y. and Mackill, D.J. 2008. Marker-assisted selection: an approach for precision plant breeding in the twenty-first century. *Phil. Trans. R. Soc. B.,* 363:557–572.

Davey, J.W., Hohenlohe PA, Etter PD, Boone JQ, Catchen JM and Blaxter M.L. 2011. Genome-wide genetic marker discovery and genotyping using next-generation sequencing. *Nature Rev Genet*, 12(7): 499-510.

D'Hont, A., Denoeud, F., Aury, J.M., Baurens, F.C., Carreel, F., Garsmeur, O. and Noel, B., 2012. The banana (*Musa acuminata*) genome and the evolution of monocotyledonous plants. *Nature,* 488:213–217.

Deputy, J.C., Ming, R., Ma, H., Liu, Z. and Fitch, M.M.M. 2002. Molecular markers for sex determination in papaya (Carica papaya L.). *Theoretical and Applied Genetics,* 106 (1):107–111.

Eiadthong, W., Yonemori, K. Kanzaki, S. and Sugiura, A. 2000. Amplified fragment lenght polymorphism analysis for studying genetic relationships among *Mangifera* species in Thailand. *Journal of the American Society for Horticultural Science*, 125:160-164.

Fang JG, Liu DJ, Zhang Z, Hillel J. and Lavi U.1999. The construction of the fingerprinting of two mango cultivars using AFLP. *Journal of Nanjing Agriculture University,* 22(2):25–27.

Halldén, C., Hansen, M., Nilsson, N.-O., Hjerdin, A. and Säll, T. 1996. Competition as a source of errors in RAPD analysis. *Theoretical and Applied Genetics,* 93:1185–1192.

Hemmat, M., Brown, S.K. and Weeden, N.F. 2002. Tagging and mapping scab resistance genes from R12740-7A apple. *Journal of American Society for Horticultural Science,* 127:365–370.

Hofmeyr, J. D. J.1939. Sex-linked inheritance in *Carica papaya* L. *South African Journal Science,* 36:283–285.

Hsu, T.H., Adiputra, Y.T., Ohta, H. and Gwo, J.C. 2011. Species and sex identification of Formosa landlocked salmon using loop-mediated isothermal amplification. *Molecular Ecology Resources,* 11:802–807.

Jones, C.J., Edwards, K.J., Castaglione, S., Winfield, M.O., Sala, F., van de Weil, C., Bredemeijer, G., Vosman, B., Matthes, M., Daly, A., Brettschneider, R., Bettini, P., Buiatti, M., Maestri, E., Malcevschi, A., Marmiroli, N., Aert R., Volcaert, G., Rueda, R.,Linacero, R., Vazgues, A. and Karp, A. 1997. Reproducubility testing of RAPD, AFLPand SSR markers in plants by a network of European laboratories. *Molecular Breeding,* 3:381–390.

Kamath, J.V., Rahul, N., Ashok Kumar, C.K. and Laxmi, S.M. 2008. *Psidium guajava* L.A. Review. *International Journal of Green Pharma*, 2(1):9-12.

Kanupriya., MadhaviLatha, P., Aswath, C., Reddy, L., Padmakar,B Vasugi,C and Dinesh, M.R. 2011. Cultivar Identification and Genetic Fingerprinting of Guava (*Psidium guajava*) Using Microsatellite Markers. *International Journal Fruit Science,*11:184-196.

Kashkush, K., Jinggui, F., Tomer, E., Hillel, J. and Lavi, U. 2001.Cultivar identification and genetic map of mango (*Mangifera indica)* . *Euphytica,* 122:129–136.

Koller, B., Lehmann, A., McDermott J. M. and Gessler, C. 1993. Identification of apple cultivars using RAPD markers. *Theoretical and Applied Genetics,* 85:901–904.

Kumar, N.V.H., Narayanaswamy, P., Prasad, D.T., Mukunda, G.K. and Sondur, S.N. 2001. Estimation of genetic diversity of commercial mango (*Mangifera indica* L.) cultivars using RAPD markers. *Journal of Horticultural Science and Biotechnology,* 76(5):529–533.

Khan MA and Korban SS. 2012. Association mapping in forest trees and fruit crops. *J. Expt. Bot.,* 63(11): 4045-60.

Litt, M. and Luty, J.A. 1989. A hypervariable microsatellite revealed by in vitro amplification of a dinucleotide repeat within the cardiac muscle actin gene. *American Journal of Human Genetics,* 44:397–401.

Mattisson, H. and Nybom, H. 2005: Application of DNA markers for detection of scab resistant apple cultivars and selections. *International Journal Horticulture Science,* 11: 59–63.

Mulcahy, D.L., Cresti, M., Sansavini, S., Douglas, G.C.,Linskens, H.F., Bergamini Mulcahy, G., Vignani, R. and Pancaldi, M. 1993: The use of random amplified polymorphic DNAs tofingerprint apple genotypes. *Scientia Horticulturae*, 54:89-96.

Melounova, M., Vejl, P., Sedlák, P., Blazek, J., Zoufala J., Milec, Z. & Blazkova, H. 2005. Alleles controlling apple skin colour and incompatibility in new Czech apple varieties with different degrees of resistance against *Venturia inaequalis* CKE. *Plant Soil and Environment,* 51:65–73.

Mienie, C.M.S., Fourie, H., Smit, M.A., Van Staden, J. and Botha, F.C. 2002. Identification of AFLP markers in soybean linked to resistance to *Meloidogyne javanica* and conversion tosequence characterized amplified regions (SCARs). *Plant Growth Regulation,* 37: 157–166.

Ming, R., Hou, S., Feng, Y., Yu, Q., Dionne-Laporte, A., Saw, J.H., Senin, P., Wang, W., Ly, B.V. and Lewis K.L. 2008. The draft genome of the transgenic tropical fruit tree papaya (*Carica papaya* Linnaeus). *Nature,* 452:991-996.

Ming, R., Mu, Q. and Moore, P.H. 2012. Papaya genome and genomics. (In) Schnell, R.J and Priyadarshan, P.M. (eds) Genomics of tree crops. Springer, pp 241-261.

Ming, R., Yu, Q.Y., Moore, P.H. 2007. Sex determination in papaya. *Seminars in Cell Development Biology,* 18:401–408.

Nakasone, H.Y. and Paull, R.E. 1998 Papaya. (In) Nakasone, H.Y., Paull, R.E. (eds) Tropical fruit. CABI, New York.

Novelli, V.M., Cristofani, M., Souza, A.A. and Machado, M.A. 2006. Development and characterization of polymorphic microsatellite markers for the sweet orange (*Citrus sinensis* L. Osbeck). *Genet Mol Biol.,* 29(1): 90-6.

Oliveira, E.J., Amorim, V.B.O., Matos, E.L.S., Costa, J.L., Castellen, M.S., Pádua, J.G. and Dantas, J.L.L. 2010b. Polymorphism of microsatellite markers in papaya (Carica papaya L). *Plant Molecular Biology Reporter,* 28:519-530.

Oliveira, E.J., Dantas, J.L.L., Castellen, M.S. and Machado, M.D. 2008. Identificac¸ ão de microssatélites para o mamoeiro por meio da explorac¸ ão do banco de dados de DNA. *Rev. Bras. Frutic.,* 30:625–628.

Oliveira, E.J., Santos Silva, A., Carvalho, F.M., Santos, L.F., Costa, J.L., Amorim, V.B.O. and Dantas, J.L.L. 2010a. Polymorphic microsatellite marker set for *Carica papaya* L. and its use in molecular-assisted selection. *Euphytica,* 173:279–287.

Parasnis, A., Ramakrishna,W., Chowdari, K., Gupta, V., Ranjekar, P. 1999. Microsatellite (GATA) n reveals sex-specific differences in papaya. *Theoretical and Applied Genetics,* 99:1047–1052.

Paran, I. and Michelmore, R.W. 1993. Development of reliable PCR-based markers linked to downy mildew resistance genes in lettuce. *Theoretical and Applied Genetics,* 85:985–993.

Pérez, J.O., d'Eeckenbrugge, G.C., Risterucci, A., Dambier, D. and Ollitrault, P. 2007. Papaya genetic diversity assessed with microsatellite markers in germplasm from the Caribbean region. *Acta Hortic.,* 740:93–101.

Phumichai, J.C., Babprasert, C., Chunwongse, C. and Sukonsawan, S. 2000. Molecular mapping of mango cultivars "Alphonso" and "Palmer". *Acta Horticulturae,* 509:193-207.

Ravishankar, K.V., Mani, B.H., Dinesh, M.R. and Lalitha Anand 2011. Development of new microsatellite markers from Mango (*Mangifera indica*) and cross-species amplification. *American Journal of Botany,* 98(4):e96-e98.

Ravishankar, K.V., Vidhya, L., Cyriac, A., Rekha, A., Goel, R., Singh, N.K. and Sharma, T.R. 2012. Development of SSR markers based on a survey of genomic sequences and their molecular analysis in banana (*Musa* spp.) *Journal of Horticultural Science and Biotechnology,* 87:84-88.

Ravishankar, K.V. Chandrashekara P. Sreedhar, S.A. Dinesh· M.R. Lalitha Anand and Saiprasad, G.V.S. 2004. Diverse genetic bases of Indian Polyembryonic and Monoembryonic Mango (*Mangifera indica* L) Cultivars. *Current Science,* 87:870-71.

Ravishankar, K.V., Lalitha Anand and Dinesh, M.R. 2000. Assesment of genetic relatedness among a few Indian mango varieties using RAPD marker. *The Journal of Horticultural Science and Biotechnology,* 75:198-201.

Risterucci, A.M., Duval, M.F., Rohde, W. and Billotte, N. 2005. Isolation and characterization of microsatellite loci from *Psidium guajava* L. *Molecular Ecology Notes,* 5(4):745–748.

Ritter, E., Herran, A., Valdés-Infante, J., Rodríguez-Medina, N.N., Briceño, A., Fermin, G., Sanchez-Teyer, F., O'Connor-Sanchez, A., Muth, J., Boike, J., Prüfer, D., Santos, C.A., Nunes dos Santos, I.C., Rodrigues, M.A., Risterucci, A.M., Billotte, N., Becker, D. and Rohde, W. 2010a. Comparative linkage mapping in three guava mapping populations and construction of an integrated reference map in guava. *Acta Horticulturae* (ISHS) 849:175-182.

Ritter, E., Rodríguez-Medina, N.N., Velásquez, B., Rivero, D., Rodríguez, J.A., Martínez, F. and Valdés-Infante, J. 2010b. Qtl (quantitative trait loci) analysis in guava. *Acta Horticulturae*, (ISHS) 849:193-202.

Rowe, H.C., Renaut, S. and Guggisberg, A. 2011. RAD in the realm of next generation sequencing technologies. *Mol. Ecol.*, 20(17):3499-502.

Santos, S.C., Ruggiero,C., Silva, C.L.S.P. and Lemos, E.G.M. 2003. A microsatellite library for *Carica papaya* L. cv. Sunrise 'Solo'. *Rev. Bras. Frutic.*, 25:263–267.

Schnell, R.J., Ronning, C.M. and Knight Jr. R.J.1995. Identification of cultivars and validation of genetic relationships in *Mangifera indica* L. using RAPD markers. *Theoretical Applied Genetics,* 90:269-274.

Schnell, R.J., Brown, J.S., Olano, C.T., Meerow, A.W., Campbell, R.J. and Kuhn, D.N. 2006. Mango genetic diversity analysis and pedigree inferences for Florida cultivars using microsatellite markers. *Journal of American Society of Horticultural Science,* 131(2): 214–224.

Schlötterer, C. 2004. The evolution of molecular markers – just a matter of fashion? *Nature reviews/Genetics,* 5:63–69.

Selmer, K.K., Brandal, K., Olstad, O.K., Birkenes, B., Undlien, D.E., Egeland, T. 2009. Genome-wide linkage analysis with clustered SNP markers. *J. Biomol Screen*, 14:92–96.

Sondur, S.N., Manshardt, R.M. and Stiles, J.I. 1996. A genetic linkagemap of papaya based on randomly amplified polymorphic DNA markers. *Theoretical and Applied Genetics,* 93: 547–553.

Tautz, D. 1989. Hypervariability of simple sequences as a general source for polymorphic DNA markers. *Nucleic Acids Research,* 17: 6463–6471.

Tonguc, M. and Griffiths, P.D. 2004. Genetic relationships of *Brassica* vegetables determined using database derived simple sequence repeats. *Euphytica,* 137:193–201.

Uma.S, M.S. Saraswathi, A. Rekha and Ravishankar, K.V. 2011. Banana and Plantain. In Advances in Horticultural Biotechnology- Moleular Markers and Marker Assisted Selection Vol.III: Fruit crops, Plantation crops and Spices.

Urasaki, N., Tokumoto, M., Tarora, K., Ban, Y., Kayano, T., Tanaka, H., Oku, H., Chinen, I. and Terauchi, R. 2002. A male and hermaphrodite specific RAPD marker for papaya (*Carica papaya* L.). *Theoretical and Applied Genetics,* 104:281–285.

Valdés-Infante, D., Becker, D., Rodríguez, N., Velázquez, B., González, G., Sourd, D.,Rodríguez, J., Ritter, E. and Rohde, W. 2003. Molecular characterization of Cuban accessions of guava (*Psidium guajava* L.), establishment of a first molecular linkage map and mapping of QTLs for vegetative characters. *Journal of Genetics & Breeding,* 57:349-358.

Viji, G., Harris, D.L., Yadav, A.K. and Zee, F.T. 2010. Use of microsatellite markers to characterize genetic diversity of selected accessions of guava (*Psidium guajava*) in the United States. *Acta Horticulturae,* 859:169-176.

Villegas, V.N. 1997. Edible fruits and nuts - *Carica papaya* L. (In) EWM Verheij, RECoronel, (eds) Volume 2. Wageningen University, The Netherlands

Weising, K., Nybom, H., Wolff, K. and Kahl, G. 2005. *DNA fingerprinting in plants: principles, methods and applications*. Taylor & Francis/CRC Press.

Wikstrom, N., Savolainen, V. and Chase, M.W. 2001. Evolution of the angiosperms: calibrating the family tree. *Proc R Soc London*, 268:2211-2220.

Williams, J.G.K., Kubelik, A.R., Livak, K.J., Rafalski, J.A. and Tingey, S.V. 1990. DNA polymorphisms amplified by arbitrary primers are useful as genetic markers. *Nucleic Acid Research,* 18:6531–6535.

Yamamoto, T., Kimura, T., Sawamura, Y., Kotobuki, K., Ban, Y., Hayashi, T. and Matsuta,N. 2001. SSR isolated from apple can identify polymorhism and genetic diversity in pear. *Theoretical and Applied Genetics,* 102:865–870.

Yamanaka, N., Hasran, M., Xu, D.H., Tsunematsu, H., Idris, S. and Ban, T. 2006. Genetic relationship and diversity of four Mangifera species revealed through AFLP analysis. *Biomed Life Science of Earth and Environment,* 53(5):949–954.

Yamamoto, T., Kimura, T., Soejima, J., Sanada, T., Ban, Y. and Hayashi, T. 2004. Identification of quince varieties using SSR markers developed from pear and apple. *Breeding Science,* 54: 239–244.

8

Statistical Techniques in Quantitative Genetics for Fruit Crop Improvement Research

R. Venugopalan

Quantitative genetics is often considered to be the bedrock of crop breeding research. Understanding and analyzing complex nature of genetic data based on its quantitative traits is of much importance in any perennial crop improvement research. Genetics and statistics are the two sides of a same coin to unravel complex problems in the study of inheritance and heredity of quantitative traits. Application of statistical methods in genetics probably dates back to 1880's when Galton (1886), in his historic study about inheritance of mid-parent's heights on their children, adopted the theory of regression. In the early 1990's, Fisher (1918), in his study on correlation in Mendelian inheritance, utilized the analysis of variance (ANOVA) technique. The distribution of quantitative traits in a population follows the normal distribution (also known as Gaussian distribution or bell curve), the central concept of statistical theory. Number of research experiments after these two historic studies essentially concentrated on analyzing vast genetic data of several quantitative traits and attributing the source of variability in traits to genetic variation. Furthermore, as emphasized by Narain, (2010) the classical genetic basis of continuous variation based on the infinitesimal model of Fisher and mostly using statistical methods has since undergone major modifications. Largely due to rapid advances in computational tools for data analysis, complex techniques, which better explain the reality in the vast data generated by the researchers, are being employed. Accordingly, an attempt is made not necessarily be exhaustive, to discuss about traditional data analysis techniques leading to modern aspects with special reference to perennial fruit crops.

Statistical techniques for analyzing quantitative traits

Statistical techniques are extensively used in biometrical research to help crop breeders in studying inheritance of quantitative traits. These techniques are applicable in breeding research, broadly in four different ways.

i) To assess the genetic variability present in a population under study. For this purpose, preliminary statistical measures such as range, variance, standard deviation, standard error, coefficient of variation and D^2 mahalanobis statistic are utilized.

ii) To select elite inbred lines from mixed populations using techniques such as correlation, path and discriminant analysis.

iii) To select parents and breeding procedures for in-depth study through line x tester, diallel, partial diallel etc. and finally

iv) To determine the adaptability of various genotypes in different environments (stability/adaptability analysis).

Assessing genetic variability

To assess the genetic potential of collected resources, the most commonly utilized statistical measures are average, range, variance, standard deviation, standard error, coefficient of variation and D^2 mahalanobis statistic. A simple dendrogram generated out of employing cluster analysis technique, which is essentially based on D^2 mahalanobis distance statistic, group the collected resources into homogenous clusters enabling breeders to select farthest individuals across clusters for use in future crop hybridization program.

Selection of elite/promising lines

To select elite lines from mixed populations, statistical techniques such as correlation, path-coefficient and discriminant analysis are normally utilized.

Path coefficient analysis

The plant breeder tries to partition the correlation coefficients into components of direct and indirect effects by employing the path coefficient analysis, which involves measurement of influence of one trait upon the set of other traits through standardized partial regression coefficient to increase the efficiency of selection. The path analysis was originally developed by Wright (1921). Basically this technique aims to improve a dependent character like yield when the independent characters have a significant relation in desirable direction and positive direct effect or indirect effect through other traits on the dependent characters.

Choice of parents and breeding procedures

Hybridization is the most useful method for developing varieties having high yield potential. However, care must be taken while selecting suitable parents to be used in such a program. Hence, in addition to evaluating lines based on their phenotypic superiority, it is highly essential for any breeder to assess the genetic value of the lines considered. One such famously used techniques, which uses advanced statistical methods, is diallel analysis.

a) Diallel mating design

Diallel cross refers to mating of selected parents in all possible combinations and evaluation of a set of diallel crosses is known as diallel analysis. The mating among selected parents may include reciprocal crosses also. The all-possible mating in a diallel cross are equal to p(p-1)/2 for direct or one-way crosses and double of this, i.e., p(p-1) when reciprocals are included, where P is the number of parents. It is used to evaluate several inbred lines in terms of combining ability variances and effects.

For example, if there are 5 parents, then the set of all possible crosses excluding reciprocals would be 10 and including reciprocals would be 20.

Parents (p)		1	2	3	4 5
1	*	x	x	x	x
2	+	*	x	x	x
3	+	+	*	x	x
4	+	+	+	*	x
5	+	+	+	+	*

Note: x: direct crosses +: reciprocals and *: parents.

i) Concept of combining ability

It refers to the capacity of a genotype to transmit superior performance to its crosses. The two types of combining ability are: general combining ability(GCA) and specific combining ability (SCA). GCA is defined as the average performance of a genotype in a series of hybrid combinations and the performance of a parent in a specific cross is termed as SCA. These two effects are effectively used in breeding programs to select a better parent for hybridization.

ii) Types of Diallel Cross

The diallel mating design involves three types of materials; namely direct crosses or one way crosses, reciprocals and parents. Depending upon number of crosses to be made, diallel cross is of two types.

Full Diallel: In this design all possible matings among the selected parents are made in both directions i.e., direct and reciprocals. In a full diallel, each parent is used as male and female for each mating.

Half Diallel: In this design all possible crosses among the selected parents are made in one direction only i.e., direct crosses.

iii) Differences between Full and Half Diallel

Full/Diallel

- Total number of crosses is equal to p(p-1), where p is number of parents.
- Used when reciprocal differences are significant.
- Used when parents do not have male sterility.
- Maternal effects can be estimated.
- Each parent is used as male and female in the mating.

Half/Diallel

- Half crosses are required i.e., p (p-1)/2.
- Used when reciprocal differences are insignificant.
- Can be used when male sterility or incompatibility is present.
- Maternal effects cannot be estimated.
- Each parent is used either as male or female in the mating.

In nutshell, diallel is a set of all possible matings between several genotypes and are useful to estimate the genetic components of total variance of a quantitative character. They are also used to estimate general and specific combining abilities of inbred lines involved in the crosses. We now delineate about the statistical procedures for performing diallel analysis.

iv) Statistical approaches of Diallel Analysis

Breeders engaged in crop improvement research perform diallel analysis in order to select suitable parents for use in the hybridization program and also for choice of appropriate breeding procedure for the genetic improvement of various quantitative traits in a crop species. There are two approaches of diallel analysis, *viz.*, (1) Hayman's graphical approach, and (2) Griffing's statistical approach.

v) Hayman's Graphical Approach

This approach is based on the estimation of components of variation and it is confined only to a set of inbred line, which is used in making diallel crosses. This method facilitates the estimation of several components of variation such as, genetic variance, dominance variance and environmental variance along with the covariance estimates for each of the parents (lines) under study. The relationship between variances and covariance will then be plotted in a form of graph to arrive at useful conclusions. Another approach (Griffing's), which is based on a statistical model uses the estimated GCA and SCA variance and effects of inbred lines to select better parents.

vi) Griffing's Statistical Approach

In this approach, in addition to the considerations made in the above, also regards the inbred lines having been derived from a random mating population by inbreeding. Here the treatment effect is split into general combining ability, specific combining ability and reciprocal effects. The following steps are used in this approach.

For testing the significant differences among the crosses, consider a model:

$$\text{Model : } X_{ijk} = U + g_{ij} + b_k + e_{ijk} \qquad (1)$$

X_{ijk} is the mean of the observations intersecting ijth cross in the kth block/ replication

U is the general mean

g_{ij} is the effect of ijth cross/treatment

b_k is the effect of kth block/replication

e_{ijk} is the error effect.

As the above model is of simple two-way analysis type, the following ANOVA table can be formulated.

Sources of variation	D.F	Sum of squares	Mean sum of squares	F-Statistic ratio
Replication/Block	r-1	SSR	MSR	
Crosses/Treatment	c-1	SSC	MSC	MSC/MSEE
Error	(r-1)(c-1)	SSE	MSEE	

A significant result while testing cross ms over error ms confirms about the presence of significant difference among the lines. Hence, we can proceed to perform combining ability analysis as discussed below.

Fixed effect model: $X_{ijk} = U + g_i + gj + S_{ijk} + 1/k\ \Sigma e_{ijk}$ (2)

As earlier the sum of squares due to GCA and SCA effects are obtained as:

SS due to GCA = $1/\ (p\text{-}2)\ \ \Sigma\ Xi.^2 - (4/p(p\text{-}2))\ X..^2$

SS due to SCA = $\Sigma\Sigma\ Xij^2 - (1/(p\text{-}2))\ \Sigma\ Xi.^2$

and MSS due to error = MSEE/k.

The ANOVA table for combining ability analysis:

Sources of variation	D.F	Sum of squares	Mean sum of squares	F-Statistic ratio
GCA	p-1	SSR	MSG	MSG/MSS
SCA	P(p-3)/2	SSC	MSS	MSS/MSE
Error		SSE	MSE	

Using the ANOVA table, several components of variance are estimated as below:

Var (GCA) = (MSG-MSS)/(p-2)

Var (SCA) = (MSS – MSE)

V1: Additive genetic variance = 2 Var(GCA)

V2: Dominance deviation = Var(GCA)

V3: Environmental variance = MSE

Heritability = V1/(V1+V2+V3)

Estimate of GCA effects: gi = (1/(p(p-2)) [p Xi. – 2 X..]

Estimate of SCA effects $S_{ij} = X_{ij}$- (1/(p-2)) (xi.+x.j) + 2/(p-1)(p-2) X..

The computational details of the above method are well explained in Narian *et al.,* (1979). Based on the estimates of GCA and SCA values, breeders can make some of the following important observations.

If the GCA variances are higher than SCA variances, it implies that progeny selection will be effective for the genetic improvement of such traits. On the other hand if the reverse is true then heterosis breeding will be more effective. However, if the magnitude of both the variances is equal then we may go for reciprocal recurrent selection for population improvement.

b. Line × Tester Analysis

The line × tester analysis is another biometrical technique which quiet often are used by the breeders to select better parents for hybridization program. Like in diallel, in this approach also estimates of GCA and SCA effects are utilized to decide about the better parents. In any hybridization method, the primary objective is to evaluate the potential of new inbred lines based on a system of crossing them to a common parent and compare the performance of their hybrids. The common parent is called as the testers and the hybrids produced are termed as test crosses. This design is useful in deciding the relative ability of a number of female and male inbreds to produce desirable hybrid combinations. This mating can also provide information regarding the usefulness of male and female inbreds as parents for hybridization to generate segregating populations.

In this method, the genotypes to be evaluated are selected from germplasm collection made by breeders. Some of these selected are designated as male testers and some others as female lines. Each male parent is mated to each female line. The total number of crosses to be made is equal to l x t. These L×T crosses along with the parents (L+T) are tested in a RBD with certain number of replications. For example, if there are 8 parents (5 lines 1-5 and 3 testers 6-8) then there will be 23 lines to be evaluated under RBD (4 replications). L×T analysis is based on a fixed effects model as given below:

Fixed Effects Model: $X_{ijk} = u + G_i + S_{ij} + b_k + e_{ijk}$ (3)

where X_{ijk} is an observation on cross (ixj) in k^{th} replication,

G_i is the GCA effect of ith plant,

S_{ij} is the SCA effect of cross (ixj),

b_k is replication effect

e_{ijk} environmental effect peculiar to $(ijk)^{th}$ unit,

u is the population means effect.

The ANOVA for fixed effects model is

Sources	D.F.	SS	MS
Replication	r-1	S_n	$M_\wedge$
Testers	l-1	S_m	M_1
Lines	t-1	S_f	M_2
L x T	(l-1)(t-1)	S_{mf}	M_3
Error	(r-1)(lt-1)	S_e	M_4

The GCA and SCA effects are estimated as

Gca L= $(x_i../rt)$ -u

Gca T = (x.j /rt)-u

$$S_{ij} = \frac{X_{ij}}{r} - \frac{X_i..}{rl} - \frac{x.j.}{tr} - \frac{X...}{ltr};\quad u = \frac{x...}{ltr}$$

The SEs of combining ability effects are

$SE(g_i) = \sqrt{M4/rt}$ and $\sqrt{M4/rl}$

$SE(s_{ij}) = \sqrt{M4/r}$

It is to be noted here that the MS due to lines and testers are to be tested against MS due to LxT, and the latter in turn is to be tested against the MS due to error. Thus, utility of advanced statistical techniques in breeding trails involving biometrical methods help the breeder in future hybridization program. Needless to mention that proper selection of superior parents for evolving mating designs using proper biometrical methods directly results in high success of breeding experiments.

i) *Estimation of genetic parameters*

With a view to understand the extent to which the observed variations were due to genetic factors, estimation of genetic parameters such as phenotypic variance (PV), genotypic variance (GV), phenotypic coefficient of variation (GCV), genotypic coefficient of variation (GCV), broad sense heritability (h2), genetic advance (GA) and genetic advance as per cent over mean (GAM), are essential.

ii) *Genotypic, phenotypic and environmental variances*

$$\text{Genotypic variance } (\sigma_g^2) = \frac{\text{Treatment MSS + Error MSS}}{r}$$

$$\text{Environmental variance } (\sigma_e^2) = \text{Error mean sum of squares}$$

$$\text{Phenotypic variance } (\sigma_p^2) = \sigma_g^2 + \sigma_e^2$$

iii) *Coefficient of variation*

The coefficient of variation (CV) being a standardized form of variance is useful for comparing the extent of variation between different characters with different scales. Genotypic and phenotypic coefficients of variation were estimated according to Burton and Dewane (1953) based on estimate of genotypic and phenotypic variance.

$$\text{Genotypic coefficient of variation (GCV)} = \frac{\sqrt{\sigma_g^2}}{\overline{X}} \times 100$$

$$\text{Phenotypic coefficient of variation (PCV)} = \frac{\sqrt{\sigma_p^2}}{\overline{X}} \times 100$$

Where, $\overline{X}$ = General mean of the character

σ_g^2 = Genotypic variance

σ_p^2 = Phenotypic variance

PCV and GCV were classified as per Sivasubramanian and Menon (1973) as shown below:

0-10%- low; 11-20%- moderate; >20%- high.

iv) *Heritability (h^2)*

GCV helps in the measurement of the range of genetic diversity in a character and provide means to compare the genetic variability in the quantitative characters. However, it is not possible to estimate the heritable variation with the help of GCV alone. Burton (1952) suggested that GCV together with heritability estimates would give the best picture of the amount of advance to be expected by selection. The heritable portion of variability was thus determined with the help of heritability estimates.

Heritability in broad sense was calculated as the ratio of genotypic variance to the phenotypic variance and expressed in percentage [Lush (1949) and Hanson *et al.*, (1956)].

$$\text{Heritability } (h^2) = \frac{\sigma_g^2}{\sigma_p^2}$$

Where, σ_g^2 = Genotypic variance σ_p^2 = Phenotypic variance.

As suggested by Johnson *et al.,* (1955) h^2 estimates were categorized as

0-30%- Low; 30.1-60%- Medium >60% - High

v) Genetic advance (GA)

Genetic advance for each character was estimated by using the following formula of Johnson *et al.,* (1955).

$$GA = h^2 \sigma_p \; K$$

Where,

h^2 = Heritability estimate in broad sense

σ_p = Phenotypic standard deviation of the trait

K = Standard selection differential which is 2.06 at 5 per cent selection intensity.

Genetic advance was classified as high (>20), moderate (10-20) and low (<10).

Further the Genetic advance as per cent of mean was computed by using the following formula

$$\text{GA as per cent of mean} = \frac{\text{GA}}{\text{Grand mean}}$$

Genetic advance as per cent mean was categorized as suggested by Johnson *et al.,* (1955).

Less than 10%: Low; 10-20%: Moderate; More than 20%: High.

vii) Interpretation of genetic parameters

Low estimates of GCV and PCV for certain traits, indicating low genetic variability in germplasm material studied, necessitates the need for generation of new variability for these characters. Moderate estimates of GCV, PCV for certain traits, suggest the traits are more sensitive to environmental effects and their expression depends more on other non-genetical factors. High estimates of PCV and GCV for certain traits, indicating existence of wide range of genetic variability, suggests good scope for improvement of these traits. A narrow difference between GCV and PCV estimates, if recorded, indicates improvement of these characters through selection, thereby indicating less degree of environmental influences in the expression of characters and are comparatively stable. High heritability indicates the effectiveness of selection through phenotypic performance but it does not mean a high genetic gain. However, high heritability associated with high genetic advance proves more useful for efficient improvement of a character through simple selection.

c) Statistical models for Genotype-Interaction studies

The success of any crop improvement research is solely based on the stability and adaptability of the inbred lines/ genotypes/ races/ varieties tested over varying environments/seasons/regions/periods. In such a program, it is highly essential for the breeders to ensure proper identification of superior varieties (strains) for mass propagation, which is not only stable but also adaptable over varying environments. Any breeding efforts to evolve high yielding strains will go a long way in identifying a best variety for further breeding programs and commercial exploitation in farmers/users field, and also for taking care of a situation wherein yield thresholds have been attained.

In such studies, the breeders' main interest will be to estimate the average response of the varieties and also to test the consistency of the yield response from region to region/environment to environment. The latter one is more specifically termed as the testing interaction of the genotypic effects with the environmental/places effects. The presence or absence of the so-called genotype x environment (GE) interaction, coupled with high yield, will largely dictate the average performance of the genotypes. Such genotypes are suitable for general adaptation in the range of environments considered. However, in practice, genotypes responsible for showing higher yield are less stable and vice versa. Hence, the main aim in conducting a multi-location trial research is to strike a balance between these two extremes by evolving statistical methods to reduce the component of GE interaction for identifying stable genotypes that interact less with the locations. The presence of such an interaction makes it difficult for the breeder to assess the genetic potential of a variety. To overcome this difficulty, multi-location trails are conducted to test the varieties for their stable performance over years/locations before suggesting for commercial exploitation. Before, discussing about the different statistical methods available for such a study, we shall delineate below some of the essential concepts and definitions.

i) Concepts and definitions

The performance of a particular variety is the result of its genetic constitution and the environment in which it has been grown. In practice it is quite possible that a particular variety may not exhibit the same phenotypic performance under different environments or different varieties may respond differently to a specific environment. This variation, arising from the lack of correspondence between the genetic and non-genetic effects is known as the genotype-environment

(GE) interaction. The presence of GE interaction reduces the correlation between phenotype and genotype, makes difficult to judge the genetic potential of variety and alters the relative ranking of different varieties. Because of these very reasons, before the successful release of particular variety on commercial basis, plant breeders grow different varieties at different locations over the years just to determine the magnitude of GE interactions for ensuring the stability of varieties, under varying environments. A basic pre-requisite to the planning of a breeding programme is the knowledge about the amount of genetic variability present in the population for which the programme is meant.

The Genotype is any genetic material, like inbred lines, hybrids etc. that the breeder is interested to study and the phenotype is its revealed characteristics. The phenotypic value, associated with the observed value of any characteristic like, yield/plant is the interplay of genotypic and environmental values. The predictable and unpredictable factors influencing the performance of an individual are termed as the environmental factors. The performance of a variety is determined by the cumulative effect of its genotypes and the environment in which it grows. The differential performance of a given variety over two different locations is an indication of presence of GE interaction, which needs to be confirmed through statistical procedures

Example

Consider a hypothetical data pertaining to a crop yield of two different genotypes at two different locations as below:

Variety	I	Locations	II
Genotype I	33		25
Genotype II	31		29

Here, though the genotype I recorded higher yield than II in location I, its performance is reverse in the other location. Thus the genotype-I, relative to II, performs differently in the two locations considered, suggesting about the presence of GE interaction. For discussion about different types of such interaction readers may refer to Prabhakaran and Jain (1992). We now turn our attention in discussing about the concepts and definitions of stability and adaptability.

The foregoing thoughts revealed that the main goal in any breeding program is to reduce the GE interaction component to a greater extent using sound scientific techniques. But, the much-needed reduction in the GE component cannot be achieved due to the fact that only minor part of the GE interaction can be attributed to controlled environments. Hence, apart from the average performance of the genotypes, stability characteristics of the genotype need to be considered before flagging off lines for general/specific adaptation. A genotype is considered stable if, in a given location, its yield/performance remains more or less constant over different years. On contrary, a genotype is said to be adaptable if it has little variation in performance (averaged over years) across different locations. We now discuss about the different models/techniques available for performing stability analysis in a multi-location/year data to select stable genotypes for commercial exploitation.

ii) Regression models for studying G×E interaction effects

Three different methods are commonly employed in breeding works, to perform stability analysis of genotypes tested over varying environments. The frequently used approach, *viz.,* Eberhart and Russell (1966) method is based on the following model:

$$Y_{ij} = \mu_i + b_i e_j + d_{ij}, \qquad (4)$$

where Y_{ij} is the phenotypic value of the ith genotype at the jth environment (I=1,2,...t; j=1,2...s) averaged over r replications; μ_I is the mean of the ith genotype over all the environments; bi is the regression coefficient measuring the sensitivity of the ith genotype to the environmental change and dij the deviation from regression of the ith genotype from its linear regression on the environmental index ej for the jth environment, ej being obtained as

$$e_j = (\Sigma\, Y_{ij}/t\,) - \mu, \qquad (5)$$

μ, being the general mean.

It may be noted here that the environmental index e_j , is not independent of the dependent variable , Y_{ij}. It is estimated as a function of the dependent variable. Such a situation leads to statistically invalid regression. In particular, the use of environmental values as independent measures, when in fact they are not independent of the phenotypic variable regressed on leads to statistically invalid regression [Prabhakaran and Jain (1992)]. Perkins and Jinks (1968), with little modification, too suggested a regression model, which also does not use independent measure of environmental index. Freeman and Perkins (1971), by considering a measure of environment completely independent of the phenotype, propose the following model for carrying out stability analysis.

$$Y_{ij} = \mu_i + di + \bar{\beta} z_j + \bar{\delta}_j + \beta d_i\, z_j + \delta d_{ij} \qquad (6)$$

where $z_j = (x_j\text{-}x)$, the effect of environment; $\bar{\beta}$ is the combined regression coefficient; βd_i is the difference between regression coefficient and $\bar{\beta}$; δij is the deviation of the ith genotype from its linear regression on jth environment; δj is the deviation from the combined regression line of the mean of all genotypes in the jth environment and δdij is the deviation of the ith genotype from its linear regression on z_j in the jth environment minus δj. This method overcomes the shortcoming of the earlier two approaches as it uses independent environmental index z_j obtained by replication of genotypes. This method is therefore, of much use in drawing statistically valid conclusions about the nature of genotype x environment interaction and their implications. The details of these methods are elucidated in Prabhakaran and Jain (1992), more recently from practical point of view in Venugopalan and Gowda (2003), Venugopalan and Pitchaimuthu (2009) and Venugopalan and Madhavi Reddy (2010).

d) Measures of stability

These approaches, commonly known as joint regression analysis give us the two measures of stability - the regression coefficient (b_i) and the deviation from regression mean square (S^2d_i) which are computed as below.

i. Regression coefficient (b_i): $b_i = \sum(Y_{ij} - \bar{Y}_{i.})(\bar{Y}_{.j} - \bar{Y}_{..}) / \sum(\bar{Y}_{.j} - \bar{Y}_{..})^2$

ii. Deviation from regression (S^2d_i): $S^2d_i = [\{\sum(Y_{ij} - \bar{Y}_{i.})^2 - b_i2\sum(Y_{ij} - \bar{Y}_{..})^2\} / (s-2)] - S_e^2$

Based on these measures, three groups of classification may be obtained.

Group I: Ideal genotype (suitable for wide range of environment) bi=1 and S^2d_i=0

Group II: Average stability genotype (Suitable for poor environment) bi<1 and S^2d_i=0

Group III: Above average stability (suitable for favorable environment) b_i>1 and S^2d_i=0.

Further, an easy to compute measure known as the ecovalence measure (W_i) to assess the contribution of each of the genotypes to GE interaction is given by Wricke as,

$$W_i = \Sigma_j(Y_{ij} - \bar{Y}_{i.} - \bar{Y}_{.j} + \bar{Y}_{..})^2$$

The lower the value of W_i the smaller will be the fluctuations from the predictable response in different environments. Accordingly, as an index of ranking in their order of their stability characteristic, the genotype with least ecovalence is considered to be the most stable. Thus, in any applied research wherein the researcher's aim is to assess the consistent performance of any technology/product over time or over environment, it is highly essential to apply a systematic statistical process so as to arrive at a suitable recommendation to the end users.

It may be noticed that through parametric approach contribution of each genotype to G×E interaction was assessed solely based on their performance and stability over years. However breeders are interested in assessing the contribution of each genotype to G×E interaction based on their relative performance (performance of a genotype compared to others) and stability over years. Accordingly, by considering relative performance of a genotype a suitable non-parametric technique needs to be applied to assess the stability. A number of nonparametric measures for assessing yield stability have been proposed (Thennarasu, 1995; Nassar and Huhn, 1987). These statistical measures are based on the ranks of the genotypes in each environment tested. The ranking is based on values of Y_{ij} with lowest Y_{ij} value receiving the rank 1, the next higher value 2 and so on. The nonparametric measures proposed by Nasser and Huehn (1987) $S_i^{(3)}$ and $S_i^{(6)}$, Thennarasu (1995) $NP_i^{(1)}$, $NP_i^{(2)}$ based on yield ranks of the genotypes in each environment were as below.

$$NP_i^{(1)} = \frac{1}{n}\sum_{j=1}^{n}\left|r_{ij}^{*} - Md_i^{*}\right| \tag{7}$$

$$NP_i^{(2)} = \frac{1}{n}\left[\sum_{j=1}^{n} \left|r_{ij}^{*} - Md_i^{*}\right| / Md_i\right] \tag{8}$$

$$S_i^{(3)} = \sum_{j=1}^{n} \left(r_{ij} - \bar{r}_{i.}\right)^2 / \bar{r}_{i.} \tag{9}$$

$$S_i^{(4)} = \sum_{j=1}^{n}\left|r_{ij} - \bar{r}_{i.}\right| / \bar{r}_{i.} \tag{10}$$

Recently, Ravi *et al.*, (2013) demonstrated the practical utility of these non-parametric measures in assessing crop yield stability in horticultural crops.

e) Statistical techniques in molecular breeding

In the recent past, moving away from assessing variation based on the morphological characters alone, researchers concentrated on using molecular markers to assess genetic variability and thus called for employing advance statistical tools to assess the genetic diversity. Geneticists infer the presence of genes or allelic state based on phenotypic expression. More specifically, statistical techniques can be used to isolate genes which normalize a set of quantitative traits. Further, genomic data are generated exponentially in different crop research experiments and culling out significant information hidden inside the data generated would demand the use of advance statistical tools at every experimental stage. More emphasis is being laid out by the researchers world-wide to understand the complex genetic architecture underlying the traits so as to hasten the crop research to solve imminent problems faced by farmers. With an advent of advanced computational tools for data analysis, complex techniques, which better explain the reality in the vast data generated by the researchers, are being employed. One of such areas which calls for employing advanced statistical tools is bioinformatics.

Bioinformatics, a systematic study of informatics process in biological systems, aims to increase our understanding of biological processes controlling entire crop genome. It could be viewed as a three way multi-disciplinary approach with biotechnology, statistical techniques and information technology as its three faces, with the principles of statistics applied to biological science through an interface information technology. Some of the promising areas in bioinformatics which demand greater applicability of statistical tools, among many others are, experiments based on functional genomics and genetics, diversity analysis, micro-arrays, genome –wide studies, gene silencing, genome sequencing, prediction of protein structures and studies relating to gene expression. All these tools are resourceful in Indian horticulture system for studies relating to perennial crop improvement (plant genome studies), pest management studies (transforming resistance genes to a desired crop to develop resistance varieties), studies relating to enhancing nutritional quality, shelf life of the crop and other areas which ultimately lead to increased crop productivity. However, it is suggested that while undertaking research studies about different plant systems, invariably the focus should be maintained for identifying crop specific problems that has to be solved using bioinformatics, rather applying tools of bioinformatics in different areas of crop improvement research. Though the application of statistical tools in horticultural bioinformatics is vast, we discuss and provide an over-view about some of these areas most sought after, however, utilized or yet to be exploited completely (Sarada *et al.,* 2008).

The advent of molecular markers and their extensive mapping in several species have enabled detection of genes of biometric characters known as quantitative trait loci (QTL). The process of identification of QTL's and measure its effect are considered to be the back bone of statistical genetics. We shall now present an overview of various statistical techniques used in QTL detection.

f) Statistical techniques for QTL detection

A quantitative trait is traditionally defined as a trait with a continuous distribution, in contrast to a discrete distribution. The trait values are usually obtained by measuring instead of counting. The trait is often considered to be controlled by many genes and each of the genes has a small effect on the trait; however recent findings, using the contribution of genomic mapping and the traditional quantitative genetics show that a small number of genes can produce a trait with continuous distribution.

Searching for genes controlling complex or quantitative traits plays an important role in applying genomic information to clinical diagnosis, agriculture and forestry because a large portion of the traits related to agronomic importance and human diseases are quantitative. The loci controlling quantitative traits are commonly referred to as QTLs (quantitative trait loci). The procedure of finding and locating the QTLs are called QTL mapping. The process of QTL analysis requires:

1. A suitable mapping population of phenotypically contrasting parents,
2. A linkage map of molecular markers,
3. Mapping methods and software and
4. Reliable phenotypic screening methods and generation of phenotypic data.

Generally the following three methods are adopted in such a study (Liu, 1998).

1. Single Marker Analysis (SMA)
2. Interval analysis or Interval Mapping (SIM)
3. Composite Interval Analysis (CIM).

g) Single Marker Analysis (SMA)

Here, one marker is involved at a time to find the QTL-marker association. SMA can be implemented as a simple t-test, ANOVA, linear regression.

h) Interval analysis or Simple Interval Mapping (SIM)

SIM was first proposed by Lander and Botstein (1989) and it takes full advantage of a linkage map. The method evaluates the target association between the trait values and the genotype of a hypothetical QTL (target QTL) at multiple analysis points between pair of adjacent marker loci (the target interval). It is based on the expected co-segregation models among the two flanking markers.Interval analysis can be implemented as a Linear regression (LR) approach (Knapp *et al.,* 1992).

$y_j = X_1\theta_1 + X_2\theta_2 + X_3\theta_3 + X_4\theta_4 + \varepsilon_j$

or by a Non-linear regression (NLR) approach (Knapp *et al.,* 1990).

i) Linear vs nonlinear regression models

Linear regression models assume that every unit change in the contribution of explanatory factors, result in an equal amount of change in the response factor, which is hardly possible in any biological studies and otherwise captured by non-linear regression models. NLR models which describe the growth behaviour over time are applied in many fields. Nonlinear statistical models play a very important role in almost all branches of agricultural/horticultural and biological sciences due to the existence of wide population fluctuations and complex non-linear inter-relationship among variables of interest. In the area of population biology, growth occurs in plants, animals and organisms. The type of model needed in a specific situation depends on the type of growth that occurs. In general, growth models are mechanistic in nature, rather than empirical. In the former, the parameters have meaningful biological interpretation; the latter is just like a 'black-box' where some input is given and some output is taken out. A mechanistic model usually arises as a result of making assumptions about the type of growth, writing down differential or difference equations that represent these assumptions, and then solving these equations to obtain a growth model. The utility of such models is that, on one hand, they help us to gain insight into the underlying mechanism of the system and on the other hand, they are of immense help in efficient management. Some well-known nonlinear models are (Prajneshu,2009):

(i) Malthus Model. If N(t) denotes the population size or biomass at time t and r is the intrinsic growth rate, then the rate of growth of population size is given by

$$dN/dt = rN. \qquad (11)$$

Integrating, we get

$$N(t) = No \exp(rt), \qquad (12)$$

where No denotes the population size at t=0. Thus this law entails an exponential increases for r>0. Furthermore, N(t) which cannot happen in reality.

Note. The parameter r is assumed to be positive in all models.

(ii) Monomolecular Model. This model describes the progress of a growth situation in which it is believed that the rate of growth at any time is proportional to the resources yet to be achieved, i.e.

$$dN/dt = r(K-N), \tag{13}$$

where K is the carrying size of the system. Integrating eq. (5), we get

$$N(t) = K - (K-No) \exp(-rt). \tag{14}$$

(iii) Logistic Model. This model is represented by the differential equation

$$dN/dt = rN (1-N/K). \tag{15}$$

Integrating, we get

$$N(t) = K / [1+(K/No-1) \exp(-rt)]. \tag{16}$$

The graph of N(t) versus t is elongated S-shaped and the curve is symmetrical about its point of inflexion.

(iv) Gompertz Model. This is another model having a sigmoid type of behaviour and is found to be quite useful in biological work. However, unlike the logistic model, this is not symmetric about its point of inflexion. The differential equation for this model is

$$dN/dt = rN \log e (K/N). \tag{17}$$

Integration of this equation yields

$$N(t) = [K \exp[\log e (No/K) \exp(-rt)]. \tag{18}$$

(v) Richards Model. This model is given by

$$dN / dt = rN(K^m - N^m) / (mK^m), \tag{19}$$

which, on integration, gives

$$N(t) = \mathrm{K\,N_o} / [N_o + (K^m - N_o^m)\exp(-rt)]^{1/m}. \tag{20}$$

Further, choice about suitable nonlinear methods and models differs from situation to situation. A good review and application of these NLR methods are presented in Venugopalan and Shamasundaran (2003).

j) Composite Interval Analysis (CIM)

CIM (Zeng, 1994) and MQM (multiple-QTL model or marker-QTL-marker analysis) developed by Jansen and Stam (1994) combine interval mapping for a single QTL in a given interval with multiple regression analysis on marker associated with other QTL. It considers a marker interval plus a few other well-chosen single markers in each analysis, so that (n-1) tests for interval-QTL associations are performed on a chromosome with n markers.

To summarise, for detecting QTLs, single marker analysis is the most commonly used method. Different statistical techniques like, simple t-test, ANOVA and simple regression can be used for single marker analysis data obtained from backcross and F2 progeny. However, the single marker analysis has the disadvantages of low statistical power and confounding estimates of QTL effects and locations. To overcome the disadvantages of single marker analysis, interval mapping approach can be used. But, the interval mapping also suffers from certain drawbacks, viz., the number of QTLs may not be resolved, the exact position of the QTLs may not be determined and the statistical power is relatively low. Hence, a more sophisticated approach called composite interval mapping can be used for QTL detection.

Bioinformatics and quantitative genetics facilitate the understanding of gene expression patterns. Micro array analysis concentrates on measurement of mRNA samples to identify pattern of gene expression. In quantitative genetics, mRNA samples are considered to be simply a phenotype. As in quantitative genetics, each mRNA samples may vary be correlated to other samples or subjected to environmental variability. Thus measurement of changes in gene expression, gene interaction and G×E interaction are aimed at both. Hence, we devote our attention to discuss about some of the statistical techniques used in such a micro array data analysis.

Microarrays are a new technology to investigate millions of genes simultaneously. They are miniature arrays of gene fragments attached to glass slides. The presentation of thousands of gene fragments in a single array allows investigators to define changes in gene expression within a significant fraction of the total genome. It will have a significant role in plant protection, breeding and basic science. By comparing active genes from high yielding crops with genes from low yielding crops the genes for higher yield can be identified and with the aid of statistical models researchers can select for desirable trait in high yielding crops at early stages for mass screening. Improved understanding of the gene changes occurring in diseases will lead to the development of new diagnostic tools as well as more effective pesticides.

Microarray studies often aim to identify genes that are differentially regulated across different classes of samples; e.g. finding the genes affected by a treatment, or finding marker genes that discriminate diseased from healthy subjects. Statistical analyses for differentially expressed genes the test carried out *viz.*, hypothesis tests, like t-test (parametric) or Wilcoxon's Rank Sum test (nonparametric test) to study significance between treated and untreated in an experiment or ANOVA (to identify significant changes between any pair of experiment) or general linear models (to identify significant changes between a group of experiments). For more complex data, multivariate statistical procedures are the proper tools. Principal Component Analysis (PCA) and Multi-Dimensional Scaling (MDS) provide a good way to visualize data. Clustering can be used to identify related genes or samples and portray the usage of dendrogram. There are several methods for classifying samples, each with advantages and disadvantages, including linear discriminant analysis, hierarchical clustering, k-means clustering, self organizing maps (SOM) and dynamic Bayesian networks.

k) ANN models for genomic selection

Another area which can be potentially exploited by statistician in association with crop breeders is the development statistical models for genomic selection in perennial crops. Genomic Selection (GS) is a statistical approach that utilizes all available genome-wide markers simultaneously to estimate breeding values or total genetic values. In breeding programs, statistical methods for GS will need to simultaneously estimate many marker effects from a limited number of phenotypes. A greater number of explanatory variables (markers) than observations (phenotype lines) leads to carefully handling the selection and use of the most appropriate statistical model, that is, the model that results in the highest GEBV accuracy with consideration of model complexity and computational requirements (Heslot *et al.,* 2012). GS is a statistical approach that utilizes all available genome-wide markers simultaneously to estimate breeding values or total genetic values (Kumar *et al.,* 2012).

Development of models: A classical MLR technique, by considering estimated breeding value as a function of marker's data, aims to develop a predictive modelling approach. However, a greater number of marker data than the observations recorded on phenotypic lines, nullifies the use of standard MLR modelling and calls for adopting several advanced modelling techniques such as PLS, ridge regression and nonlinear statistical modelling technique such as Artificial Neural Networks (ANN). Recently, Venugopalan and Krishnakumar (2013) employed the ANN theory for developing non-linear climate change models for thrips incidence in Capsicum by developing codes in SAS. Prediction accuracy of genomic estimated breeding value can be assessed by cross-validation approaches using repeated sampling approaches like Bootstrapping.

Though an attempt has been made in the foregoing to delineate the role of statistical tools in crop improvement, the discussion is not exhaustive, as the literature on these aspects is vast. However, the present communication can act as a spring board to amalgamate both conventional and molecular crop breeding aspects with the sole aim of achieving sustainable crop productivity.

References

Burton, G.W. 1952. Quantitative inheritance in grasses. *Proceedings of the 5th International Grassland Congress,* 1:277-283.

Burton, G.W. and Dewane, E.M.,1953. Estimating heritability in tall fescue (Festuca circuclinaccae) from replicated clonal material. *Agron., J.* 45: 478-481.

Eberhart, E.R. and Russel, W.A. 1966. Stability parameters for comparing varieties. *Crop Science*, 6:36-40.

Fisher, R.A., 1918. The correlation between relatives on the supposition of Mendelian inheritance. *Roy. Soc., Edinburgh*, 52:399-483.

Freeman, G.H. and Perkins J.L., 1971. Environmental and Genotype-Environmental components of variability; Relation between genotypes grown in different environments and measures of these environments. *Heredity*, 27:15-23.

Galton, F. 1886. Regression towards mediocrity in hereditary stature. *J. Anthro. Inst.,* 15:246-263.

Hanson, C.H., Robinson, H.F. and Comstock, R.E. 1956. Biometrical studies of yield in segregating populations of Korean lespedza. *Agronomy Journal*, 48(6):268-272.

Heffner, E.L., M.E. Sorrells, and J.L. Jannink. 2009. Genomic selection for crop improvement. *Crop Sci*., 49:1-12.

Jansen, R.C., and P. Stam. 1994. High resolution of quantitative traits into multiple loci via interval mapping. *Genetics*,136: 1447-1455.

Johnson, H.W., Robinson, H.F. and Comstock, R.E. 1955. Estimates of genetic and environmental variability of Soybeans. *Agronomy Journal*, 47: 314-318.

Knapp, S. J., Bridges, W.C. and Birkes, D. 1990. Mapping quantitative trait loci using molecular marker linkage maps. *Theor. Appl. Genet.,* 79: 583-592.

Kumar, S., Marco C.A.M. Bink, Richard K. Volz, Vincent G. M. Bus and David Chagné 2012. "Towards genomic selection in apple (*Malus* × *domestica* Borkh.) breeding programmes: Prospects, challenges and strategies", *Tree Genetics and Genomes*, 8:1-14.

Lander, E.S. and Botstein, D. 1989. Mapping Mendelian factors underlying quantitative traits using RFLP linkage maps. *Genetics,*121:185-199.

Liu, B.H., 1998. Statistical Genomics. CRC Press, Boca Raton, FL.

Lush, J. L. 1949. Heritability of quantitative characters in farm animals. Proceedings of 8th Congress of Genetics, *Heredity*, 35: 356-375.

Narain, P, Bhatia, V.K. and Malhotra, P.K. 1979. Handbook of statistical genetics. ICAR. PP 365.

Narain, P. 2010. Statistical genomics and bioinformatics. *J. Horti. Sci.,* Focus article, 5(2): 85-93.

Nassar, R. and Huehn, M. 1987. Studies on estimation of phenotypic stability; Tests of significance for non-parametric measures of phenotypic stability. *Biometrics*, 43:45-53.

Perkins, J.M. and Jinks, J.L., 1968. Environment and Genotype-Environment components of variability. III Multiple line crosses. *Heredity*, 23:339-356.

Prabhakaran, V.T. and Jain, J.P. 1992. Statistical techniques for studying Genotype-Environment interactions. SAP Pvt. Ltd. New Delhi.

Prajneshu., 2009. E-book on Models approaches to the Analysis of Agricultural Data., IASRI. Chapter 4.1. Non linear Statistical Models and their applications. pp. 1-9.

Ravi, G.S, Venugopalan, R., Padmini, K. and Gowda, D.M. 2013. Nonparametric measures for assessing yield stability in cucumber. *International Journal of Agricultural and Statistical Sciences*, 9(1):365-371.

Sarada,C, Venugopalan, R. and Alivelu, K. 2008. Application of Statistical Techniques in Agri-Bioinformatics. In "*Souvenir of 62nd annual conference of Indian Society of Agricultural Statistics"*, Tirupathi, 24-26 November, 2008. pp 28-34.

Sivasubramanian S. and Menon M. 1973. Heterosis and inbreeding depression in rice. *Madras Agric. J.,* 60: 1139-1143.

Thennarasu, K. 1995. On certain non-parametric procedures for studying genotype-environment interactions and yield stability. *Indian J. of Genet.,* 60:433-439.

Venugopalan, R. and Veere Gowda R. 2005. Stability analysis in onion: A statistical look. *J. Indian Soc. Coastal Agric. Res.,* 23(2):123-29.

Venugopalan, R. and Shamasundaran, K.S. 2003. Nonlinear regressions: A realistic modeling approach in horticultural crop research. *J. Ind. Soc. Ag. Statistics*, 56:1-6.

Venugopalan, R. and Krishnakumar, N.K. 2013. Artificial neural network models for optimising the role of climatic factors on thrips incidence in Capsicum. (In) "*New Horizons in Insect Science*", (eds) A.K. Chakravarthy *et al., Proceedings of International Conference on Insect Science*, 2013, UAS, Bangalore. pp. 72.

Venugopalan, R. and Pitchaimuthu, M., 2009. Statistical models for stability analysis in Watermelon. *J. Horti. Sci.,*4(2):153-157.

Venugopalan, R. and Madhavi Reddy, K., 2010. Stability analysis for fruit yield and attributing trials in chilli. *Veg, Sci.*, 37(2):141-145.

Wright, S. 1921. Correlation and causation. *J. Agril. Res.,* 204: 557-585.

Zeng, Z.B. 1994. Precision mapping of quantitative trait loci. *Genetics,*136: 1457-1468.

9

Rootstock Breeding

M.R. Dinesh

Rootstocks play a very important role in propagation of plants, It may modify form or stature and adopt a variety to a soil in an incompatible climatic condition and also build up the resistance to biotic and abiotic stresses and meanwhile increase the production and productivity.

Vagaries of climate and mismanagement of soil have resulted in abiotic stresses, which in turn have lowered yield and quality production of fruits. In the course of its phylogenetic development, several fruit crops have developed various physiological and morphological mechanisms enabling plants to survive under adverse conditions. In case of fruit crops, one of the most important methods is to use rootstocks to overcome such stress conditions. However, over the years in fruit crops, the rootstocks that are present do not cater to the present day needs of pest and disease tolerance. Hence, there is an urgent need to widen the genetic base of rootstocks in different fruit crops by breeding and selecting rootstocks with desirable characteristics. It is acknowledged that the timing and intensity of the response of rootstocks to soil and atmospheric conditions are synergistically beneficial in evolving new cultivars.

Number of attempts has been made in rootstock improvement programmes all over the world. One of the methods that are being followed in perennial crops like mango is the development of large progeny population through half sib families. In this case, the female parent is common and open pollinated progeny population of varieties, in which improvement is sought, is used. Rootstocks were first used in European vineyards in the late 1800's to combat devastating Phylloxera outbreak. Rootstock can be used to improve vigor, production and help to sustain the health and survival of fruit crops (Peter Cousins., 2005).

Rootstock Breeding

Rootstock breeding is slow compared to fruiting cultivars of trees. While the species used in rootstock breeding usually have a reproductive cycle that is close to that of fruiting cultivars for a given crop, rootstock breeding is still more complex. Selecting a candidate rootstock from population is more difficult than selecting candidate fruiting cultivars because the testing cycle for rootstock is longer than for fruiting cultivars.Several factors need to be takine into consideration while taking up rootstock breeding. They are ;

Physical environment of the stock system is very different from that of the fruiting portion of the tree.

1. Biotic environment in which the stock abides quite different from that of the scion, the temperature regimes, gas exchange system and moisture variables are of course quite different
2. The physiology of the root is different in many ways from that of the leafy portion of the tree
3. There are substantial variations in environmental conditions within the rhizosphere than above ground.

Keeping the above factors in mind, the source for a particular trait needs to be looked into in the crop meant for improvement. In the case of breeding for resistance, especially with regard to soil borne diseases, rootstocks play a very important role. Plant offers three types of resistance mechanisms i.e., morphological, biochemical and physiological. Morphological resistance offered depends on root structure, length, width and root hairs, biochemically by creating osmotic environment thereby escaping from stress and physiologically by synthesis of toxins, tannins, phenolic acids etc. Schmid *et al., (*2003), evaluated the resistance of various populations against the Phylloxera which revealed that the hypersensitive reactions on leaves and roots are linked. If plants showed resistance on their leaves, the roots were most of the time also resistant but the resistance is inherited at different levels, depending on the other parent and the resistance originates from more than one gene. But the susceptible parent has an important influence on proportion of resistant hybrids in the population.

Rootstock effect on plant stature

Different opinions were expressed about how rootstock will induce reduced vigour to the scion viz., restriction of water supply to the scion induced by anatomical characteristics of the rootstock. The graft union may play a central role in limiting water transport and thus induces scion dwarfing (Cohen *et al.*, 2007). Dwarfing might be caused by partial incompatibility between the scion and the rootstock which may alter the transport of minerals and hormones.

Nematodes are the microscopic worms which feed on the roots, and restrict the plant ability to take up water. Some rootstocks offer nematode resistance in which female nematodes are unable to get into the roots thereby hindering the further multiplication of nematodes in the soil. Walker and Jin, (2000), conducted a study on breeding of *Vitis rupestris X Muscadinia rotundifolia* rootstock to control *Xiphinima index* and Fan leaf degeneration. Grape fan leaf degeneration is one of the major thrusts in grape cultivation which is caused by virus snd transmitted through dagger nematode *Xiphinima index.*

Drought and salinity tolerant breeding

Pavel Pavlousek (2011) conducted a study on evaluation of drought tolerance of new grapevine rootstock hybrids. This study involved 20 new hybrids and four existing rootstock varieties. The scheme of this study involves three variant water supplies. The study showed that drought has a significant effect on shoot length, Chlorophyll content Index and visual observation. Here they have concluded that BORNER (*Vitis cinerea*) rootstock can be used as suitable genetic resource for the purpose of the breeding for drought tolerance. Accumulation of excess of Cl- and Na+ can cause specific ion toxicities. By selecting rootstocks that restrict the uptake of these ions will serve as a beneficial tool for salinity for e.g., Mandarin, Grapefruit, Rough lemon, Sour Orange, Rangpur lime etc in Citrus, *Vitis champini*, Ramsey etc in grapes and Borchard, Duke-7 and G755C in avocado.

Kadman (1985) conducted a study on the selection of avocado and mango rootstocks for calcareous soil. Three types of Mexican, Gautemalan, West Indian avocado types along with some of their hybrids were evaluated in this soil. The West Indian type cultivar Maoz showed the highest level of tolerance. In mango, 3200 seedlings of 80 different mono and polyembryonic types were evaluated, among them 13-1 showed the highest level of tolerance. The rootstock 13-1 was observed to have less salt retention capacity than re-translocation into soil; therefore it is regarded as salt tolerant rootstock in mango. Sathisha *et al.*, (2007), characterized 10 grape rootstocks for their various morphological, physiological and biochemical parameters. Significant differences were observed for most of the parameters studied. Rootstocks in the group of *Vitis berlandierii* × *Vitis rupestris*, such as 110 R, 1103 P, 99 R and B2-56, had a significantly higher content of total phenols, flavon-3- ols, flavonoids, proline and total protein. Optimum C/N ratios in the mother vines of these rootstocks helped attain better rooting percentages of hardwood cuttings. Rootstocks such as Dogridge, 110 R, B2-56 and Salt Creek recorded the highest relative water content (RWC) and water-use efficiency (WUE) at single-leaf level, which is an important mechanism in overcoming drought tolerance. The high phenolic content in rootstocks such as Dogridge, 99 R, 110R and 1103 P may help in reducing the incidence of major grape diseases in commercial table varieties if grafted onto these rootstocks. Thus, the physio-biochemical characterisation of rootstocks may help to identify particular rootstocks that could influence a desired trait in commercial table/wine grape varieties after grafting.

Biotechnological intervention in rootstock breeding

Biotechnological tools can be used in perennial fruit crops to reduce the time gap in the breeding programme. Vos and Shoeman (2000) studied *in vitro* selection and commercial release of guava wilt resistant rootstocks. In their study cell free filtrates derived from *Penicillium vermoesenii* were used to screen 30,000 guava seedlings *in vitro*. Three of the selections exhibited 100% tolerance to wilt. The roots from moderately GWD tolerant cultivar Dimple exhibited a significant higher degree of tolerance with the result, 25% of the trees lost to GWD in South Africa have been replaced by trees with tolerant rootstocks within a research period of 5 years.

Markers have played a major role in rootstock breeding. Xiang Xu *et al.*, (2010) conducted a study on marker assisted selection in citrus rootstock breeding based on major gene locus '*Tyr1*' controlling citrus nematode resistance. The nine markers chosen for evaluating the F_1 progeny of *Poncirus trifoliate* showed that single dominant gene is responsible for the resistance.

Dominique Dambier *et al.*,(2011) studied the prevalence of Citrus Tristeza Virus (CTV) and its main vector *Toxoptera citricida*, combined with abiotic constraints such as drought, salinity and alkalinity in sour orange rootstocks. New embryogenic callus lines were developed for somatic hybridization and protoplast fusions were carried out focusing on intergeneric combinations. New interesting tetraploid somatic hybrids for six combinations resulted from this study. The agronomical performance was promising from a pre-existing somatic hybrid between *Poncirus trifoliata* and *Citrus reticulata*, validated in calcareous soils in Morocco.

Rootstocks breeding in various subtropical fruits

2008	Kevin Fort and Andrew Walker (2008)	Breeding Salinity Tolerant Grape Rootstock	Ramsey, RipariaGlorie, 5C screened as salt tolerant
2003	Forner *et al.,* (2003)	Performance of Navelina Orange on 14 rootstocks in Northern Valencia	C-35 citrange performed well
2006	Garcia *et al.,* (2006)	The response of young mandrin trees grown under saline condition depends on the rootstock	Clemenules tree grafted on Carizo performed well
2007	Milan (2007)	Collection and evaluation of guava for nematode resistance	3 accessions are potentially resistant to RKN
2007	Satisha *et al.,* (2007)	Physiological and biochemical characterization of grape rootstocks	This helps to identify particular Rootstocks that could influence a desired trait in commercial table / wine grape varieties after grafting.

Newly developed Rootstocks

Citrus	F&A 418, #23, #24	Hybridization (Troyer citrange X Cleopatra mandarin)	Dwarfing Rootstocks	Ignacio-lliso *et al.,* 2004
	US 852	Hybrid Selection	CTV, C N, Phytopthora resistance	Bowman *et al.,* (2000)
Guava	Dimple, Jonelle	*In viro* Selection	GWD resistance	Vos., (2000)
	GU8	Screening/ Selection	Nematode tolerance	Milan, (2007)
Avocado	G755C13-1	Selection	Salinity tolerance	Kadman (1985)
Mango				

Rootstock breeding progress is often slow compared to breeding fruiting cultivars of trees and vine crops. The trend in rootstock breeding is to incorporate more traits into the roster of ideotypic features. Production trends are towards reduced chemical inputs, a fewer passes by hand labour crews, and higher production with more plants per unit area, more goals for breeding programmes will increase the intractability of rootstock improvement.

References

Bowman, K.D., wutsher, H.K., Kaplan, D.T. and Chaparro, J.X. 1999, A new hybrid citrus rootstock for Florida: US-852. *Proc. Fla. State Hort. Soc.*, 112: 54-55.

Cohen S., Naor A., Bennink J., Grava A. and Tyree M. 2007. Hydraulic resistance components of mature apple trees on rootstocks of different vigours. *Journal of Experimental Botany* 58: 4213–4224.

Dominique Dambier, Hamid Benyahia, Giovanni Pensabene-Bellavia, Yaldiz Aka Kacar, Yann Foerlicher, Zina belfalah, Bruno Printz, Luis Navarro and Patri Ollitrult, 2010. Somatic hybridization for citrus rootstock breeding: an effective tool to solve some important issues of the Mediterranean citrus industry. *Plant Cell Rep.*, 30: 883-900.

Forner-Giner, M. A., Alcaide, A., Primo-Millo and Forner, J.B., 2003, Performance of 'Navelina' orange on 14 rootstock in Northern Valencia (Spain). *Scientia Horticulturae* 98 :223-232.

Garcia Sanchez, F., Perez-Perez, J.G., Botia, P. and Martinez, V. 2006. The response of young mandarin trees grown under saline condition depends on the rootstock. *Europ.J. Aronomy* 24:129-139.

Ignacio Lliso, Juan B. Forner and Manuel Talón. 2004. The dwarfing mechanism of citrus rootstocks F&A 418 and #23 is related to competition between vegetative and reproductive growth. *Tree Physiology* 24: 225–232.

Kadman, A. 1985. Selection of avocado and mango rootstocks for calcareous soil. *Acta Hort.*, 158: 63-67.

Kevin Fort and M. Andrew Walker, 2008. Breeding Salinity Tolerant Grape Rootstocks, *Proceedings of the 2nd Annual National Viticulture Research Conference.*

Milan, A.R. 2007. Breeding of *Psidium Species* For Root Not Nematode Resistance In Malaysia., *Acta Hort* 735:61-69.

Pavel Pavlousek, 2011. Evaluation of drought tolerance of new grapevine rootstock hybrids. *J. Env., Biology*. 32: 543-549.

Peter Cousins., 2005. Rootstock breeding an analysis of intractability. *Hort Science* 40(7):1945-1946.

Sathisha, J., Ramteke, S.D. and Karibasappa, G.S. 2007. Physiological and biochemical characterization of grape rootstocks. *S. Afr. J. Enol. Vitic.,* 28(2): 163-168.

Schmid, J., Manty, F., and Ruhl, E.H., 2003. Utilizing the complete phylloxera resistance of *vitiscinerea* Arnold in rootstock breeding. *Acta Hort* 603: 393-410.

Vos, J.E. and Schoeman, M.H. 2000. Invitro selection and commercial release of guava wilt resistant rootstocks. *Acta Hort* 513:69-79.

Walker, M.A. and Jin, Y. 2000. Breeding*vitisrupestris* X *Muscadiniarotundifolia* rootstocks to control *Xiphinima index* and fan leaf degeneration. *Acta Hort.*, 528: 511-515.

Xiang, Xu, Deng Zhan-Ao, Chen Chun-Xian, Fred G. Gmiitter and Kim Bowman., 2010. Marker Assisted Seletion In Citrus Rootstock Breeding Based on A Major Gene Locus '*Tyr1*' Controlling Citrus Nematode Resistance. *Agricultural Sciences in China* 9(4): 557-567.

10

Mango Breeding

M.R. Dinesh

Mango, like some of the other perennial crops viz. *Citrus* has varied reproductive mechanisms i.e. both sexual and asexual. Understanding of the physiology of flowering in mango has remained largely inconclusive and tangible explanation for the irregular pattern of flowering during the recent years is not forthcoming. Mango is a highly heterozygous crop and this high heterozygosity in the cultivars which are used in hybridization and the inherent difficulties in obtaining large hybrid population have made accurate genetic analysis in mango very difficult (Iyer and Schnell, 2009). Despite the drawbacks ailing mango breeding like high heterozygosity, occurrence of only a single seed per fruit, breeding can be successful because of a number of positive attributes of the species like the wide range of available genetic variation and the ease with which a selected hybrid can be vegetatively propagated (Iyer and Schnell, 2009). However, in spite of the large variability, sufficient headway has not been made in the improvement. Naik *et al.,* (1958), opined that it is the large variability in India, too many cultivars and heavy regional preferences which have hindered the production of new commercial varieties on a large scale. The various botanical features of mango along with features of cytology and breeding are discussed below:

Polyembryony

Depending on the mode of reproduction from seeds, mango can be classified into two groups viz., monoembryonic and polyembryonic. Monoembryonic seeds contain one zygotic embryo probably of hybrid origin and only one seedling per seed is produced. Polyembryonic seeds contain more than one embryo of which one is zygotic; the other embryos originate from the nucellar tissue, surrounding the embryo sac. The seedlings from the nucellar tissues will be true- to- type and hence can be used where uniformity of seedlings are beneficial. Maheshwari and Rangaswamy (1958) suggested that the nucellar cells destined to form adventitious embryos are recognizable by their dense cytoplasm and starchy contents. They gradually push into the embryo sac cavity where they divide and differentiate into embryos. , The incidence of polyembryony is genetically determined. Leroy (1947) considered that adventive embryony reflects the effect of one or more recessive genes. Sturrock (1968) by crossing mono- and polyembryonic varieties found that polyembryony is recessive and is controlled by a single pair of genes.

Polyembryony comes in the way of using good polyembryonic varieties as female parents as isolation of zygotic seedlings from nucellar ones becomes difficult in a hybrid population. The use of polymorphic enzyme systems (isozymes) to identify zygotic seedlings (Degani *et al.,* 1990, 1992, 1993, Schnell and Knight, 1992; Truscott 1992) can be utilized and is based on the fact that nucellar seedlings should have the same isozyme alleles as the maternal parent. A variation at a locus coding for an enzyme indicates that the plant has originated by sexual reproduction. Zygotic seedlings arising from self-pollination are distinguished from nucellar seedlings by being homozygous at one or more loci at which the female parent is heterozygous. Zygotic seedlings from cross-pollination are distinguishable from those resulting from self-pollination if they express an allele not carried by the female parent. The frequency of the occurrence of zygotic seedlings varies among the polyembryonic cultivars (Degani *et al.,* 1993; Schnell and Knight, 1992). Sturrock (1968) suggested that the polyembryonic condition in mango is a recessive factor and is probably controlled by a single pair of genes. However, contrary to this, Aron *et al.,* (1998), reported that polyembryony is controlled by dominant genes. Brettell *et al.,* (2004) also observed that polyembryony is controlled by dominant genes. However, they also observed that prevalence of polyembryony varied with the location. Cordeiro *et al.,* (2006) determined in the mango *cv.* Rosinha using RAPD markers, the polyembryonic origin of plantlets.

Floral Biology

The mango inflorescence is basically terminal, although axillary and multiple panicles may also arise quite frequently from axillary buds. The hermaphrodite and male flowers occur on the same panicle. The total number of flowers in a panicle may vary from 1000 to 6000 depending on the variety (Mukherjee, 1953). Initial fruit set in mango depends on the percentage of perfect flowers and the final set need not depend on this (Iyer *et al.*, 1989). When the proportion of perfect flowers drop to 1%, then only it becomes critical for optimum set. Flowers in mango open early in the morning and anthesis is completed by noon under sunny conditions. Maximum number of flowers opens between 9 and10 a.m. The stigma receptivity is maximum during the first 6 hours, although it continues upto 72 hours after anthesis. Dinesh and Vasugi (2009) observed that yield does not depend mainly on the bisexual flower percentage *per se,* probably within a genotype there could be a threshold sex ratio below which the yield is affected adversely. In general, the main contributing factor for the density of the flowers is the number of male flowers and not the perfect flowers as bisexual flower percentage is negatively correlated with the density of flowers, though this could vary with the genotype.

Pollination

Mango is a cross-pollinated crop. Pollination is carried out by insects. Among the seven insect species observed five *viz*., three spp. of *Apis florea* and two dipterans *viz.,Eristalinus barvorum* and *Chrysomya megacephala* were observed to be the major pollinators on mango (Reddy *et al.,* 2011). Pollination by other agents like wind and gravity has been suggested to some extent. In nature more than 50% of the flowers appear to be receiving no pollen. Dijkman & Soule (1951) suggested that self-pollination can also occur quite frequently. For the variety Kensington, a night temperature below 10°C resulted in pollen grains with low viability (less than 50%); the normal meiosis taking place between a temperature of 15 & 33°C with 70 to 85% viability.

Incompatibility

The prevalence of self-incompatibility in mango was confirmed in the variety Dashehari by Singh *et al.,* (1962). However, the existence of self-sterility was suspected earlier (Dijkman & Soule, 1951). The studies carried out by Mukherjee *et al.,* (1968) and Sharma and Singh (1970) revealed the presence of self-incompatibility in Dashehari, Langra, Chausa and Bombay Green. Embryological studies revealed that in spite of normal fertilization after self-pollination, degeneration of endosperm occurs 15 days after pollination (Mukherjee, *et al.,* 1968). The self-incompatibility system operating in mango is the sporophytic type. Ram *et al.,* (1976) have reported the presence of cross-incompatibility among certain mango cultivars.

Cytology

The chromosome number in *M.indica, M. sylvatica* Roxb, *M. caloneura* Kung, *M. zeylanica* Hools and *M. odorata*. Ariff are reported to be 2n = 40

(Mukherjee, 1950). Roy and Visweswariya (1951). However, the polyembryonic variety Vellaikolumban was found to be 2n = 80 (Roy and Viswewariya, 1951), although subsequent studies showed that it is a diploid (Majumder& Sharma, 1990). Eleven types of chromosomes are there in *M. indica &M. sylvatica* and these two species differ from one another mainly in the assortment of these eleven chromosome types (Mukherjee, 1950). He also suggested that primitive type(s), which gave rise to mango varieties originated through allopolyploidy. Mukherjee (1958) reported that mango is an allopolyploid based on high chromosome number, high number of nuclear chromosomes, secondary association of bivalents, regular pairing and absence of multivalent formation and good pollen fertility.

Off-season flowering in mango

The off-season production of mango is a peculiar phenomenon in certain tracts of Tamil Nadu (Kanyakumari Dt) where, continuous flowering, fruiting and regular production of vegetative flushes are seen in most of the varieties. According to Sundararajan *et al.,* (1972) the peculiar seasonal conditions and the fairly well distributed rainfall of the area have been attributed to off - season production. The varieties Neelum, Bangalora, Calcutta or Bengal Baramasi, Kintalavenipeta, and Ali Pasand have been mentioned as varieties bearing two or more crops in succession during certain years and extend their bearing period almost till December and sometimes throughout the year. The varieties Peter and Baneshan have been reported as off- season bearers at Coimbatore while at Kodur the varieties, Neelum, Kintalvenipeta, Manoranjan and Willard behave as off season croppers. In recent times, development of off season varieties is one of the breeding objectives to tie over the problem of climate change on mango.

Breeding

Objectives in mango breeding

The objectives in mango breeding have two different sets of aims. One is from the point of view of scion breeding and another from the point of view of rootstock breeding. However, it is to be mentioned here that the breeding objectives in mango vary from region to region.

This is specially so in India. The varieties are also specific to a particular region for e.g., the variety Alphonso suffers from the internal break down 'spongy tissue'. It is popular in Southern and Western parts of India. Hence, the breeding programme undertaken in these parts has as one of the objectives as screening against this malady. With the export market increasing for mango, attractive skin colour for the hybrids is one of the foremost desired features. This is also true for Bangladesh where the fruits of majority of the varieties have green colour.

Development of dwarf mango hybrids has also been one of the objectives, as there is potential to increase productivity per unit area. As no polyembryonic rootstock has shown consistent dwarfing effect on the scion and since from monoembryonic seeds true to type rootstocks cannot be raised, transfer of this character on scion varieties has become a necessity.

In countries like Israel, rootstock breeding is aimed at evolving rootstocks resistant/tolerant to problematic soils and also to developing scion varieties with better quality, appearance and yield with longer harvest periods (Lavi *et al.*, 1993). In Australia, the main objective has been to improve the cv. 'Kensington' which bears regularly but is susceptible to anthracnose and bacterial black spot and has poor shelf life. Hence, Florida and Indian mango cultivars are used as one of the parents. The Brazilian mango breeding aims at developing varieties with improved fruit yield and quality, regular bearing and small tree size (Pinto and Byrne, 1993). So objectives in mango breeding vary with the region.

In general, a mango breeding programme can have the following main objectives.

i. developing dwarf plant types,

ii. regular bearing,

iii. better fruit quality,

iv. better shipping quality and

v. resistance to diseases and pests (Iyer, 1991).

With regard to rootstock breeding, the main desirable features are

i. polyembryony

ii. dwarf stature

iii. tolerance to adverse soil conditions (high pH, calcareous soil, etc) and

iv. good scion-stock compatibility.

However, combining all desirable traits in a single variety is difficult because of high heterozygosity and long juvenile period (Singh, 1963). Hence, breeding objectives have to be defined for specific purposes (Sharma, 1987).

Mango breeding

Early history-achievement

In India, most of the cultivars are selections made from naturally occurring, open pollinated seedlings (Singh, 1963) With the introduction of inarching technique, the process of selecting and propogating mango varieties were initiated during 1650 to 1820 of Mughal period (Mukherjee, 1953).

Breeding work was first initiated by Burns and Prayag (1921) during the year 1911 with the aim of obtaining varieties having regular bearing habit, good fruit quality, high yield and resistance to pests and diseases. Three hybrids from the cross Cowasji Patel x Pairi were isolated from 153 crosses (Sen *et al.,* 1946). Wagle (1929) made attempts to improve Alphonso by hybridization.

Systematic hybridization work was started during 1940 at Anantharajupet with special reference to South Indian varieties (Bhujanga Rao and Rangacharlu, 1958) and at Sabour with special reference to North and Eastern region cultivars (Sen *et al.,* 1946). The main objective was to breed regular bearing varieties having good quality. The work at Kodur resulted in four hybrids via. Neeleshan (Neelum x Baneshan), Neelgoa (Neelum x Yerramulgoa), Neeluddin (Neelum x Himayuddin) and Swarnajehangir (Chinna-Survarnarekha x Jehangir) which were released (Bhujanga Rao *et al.,* 1963). The breeding work started at Krishnagar, Saharanpur and Punjab did not result in the release of the any hybrid. However, the mango improvement programme carried out at Sabour resulted in two hybrids *viz*., Mahmood Bahar and Prabha Shankar, both from the combination of Bombay and Kalapady (Mallik, 1961).

Apart from India, early breeding work in mango has also been reported from Hawaii (Pope, 1929). Hybridization work was also reported in Florida (Young and Ledin 1954; Sturrock, 1969). However, the cultivars in Florida are chance seedlings and none came from controlled pollination.

Mango breeding-recent achievements

Systematic hybridization work in several centres in India resulted in as many as twenty hybrids. Several of these hybrids are slowly growing in popularity with mango growers.

In Israel, nine seedlings have been selected based on their performance in calcareous soils, with regard to rootstock breeding (Lavi *et al..*, 1993). A new cultivar 'Naomi' which had originated from the cv. 'Palmer' either by self-pollination or by cross pollination with 'Maya', Haden, 'Nimrod' or 'Irwin' was released (Tomer *et al.,* 1993). Another new variety 'Tango', which is a selection from the open pollinated seedlings of the variety 'Naomi' has been released. It has attractive skin colour with good quality. In South Africa, four new cultivars 'Heidi', Neldica 'Neldawn'.and 'Ceriese' were developed by mass selection (Marais, 1992).

Breeding work carried out at Australia, resulted in three hybrids Hybrid 4069, Hybrid 1201 and 1243. Hybrid 4069 is from the parentage Van Dyke x Kensington Pride, fruits are round in shape and weigh about 325-500g. Hybrid 1201 is from the parentage Irwin x Kensington Pride. The fruits on an average weigh about 245-530g, with ovate to round in shape. The Hybrid 1243 is also from the parentage Irwin x Kensington Pride, fruits weigh on an average about 500g.All these hybrids have very attractive skin colour.

Some of the hybrids released over the years which are promising are given below;

Mallika

This hybrid between Neelum and Dashehari was released from IARI, New Delhi. It has a strong tendency for regular bearing. The fruits on an average weigh about 350-400g with deep yellow pulp, high TSS, good odour, uniform fruits and moderate keeping quality (Singh *et al.,* 1972).

Amrapali

It is from the parentage Dashehari x Neelum. Plants are dwarf and have regular bearing habit. Fruits weigh on an average about 180-250 g and are borne in clusters. Fruits are sweet to taste and have good keeping quality.

Au-Rumani

It is a hybrid from the combination Rumani x Mulgoa released by Fruit Research Station, Kodur. It bears large fruits, good flavour, heavy yielder, pulp moderately firm (Sachan *et al.,* 1988).

KMH 1

It is a hybrid released from Fruit Research Station, Kodur. It is from the parentage Cherukurasam x Khader. Plants are semi dwarf, regular bearing, fruit-fibreless has high brix and low acidity.

Manjeera

This was released from Fruit Research Station, Sangareddy. It is from the parentage Rumani x Neelum. It produces round shaped fruits. It bears regularly and is a prolific bearer, pulp is firm and fibreless.

Ratna

It is a hybrid from the cross Alphonso x Neelum released by Fruit Research Station, Vengurla. It is regular bearing, produces medium sized fruits weighing on an average about 250 g. Pulp is orange in colour and free from spongy tissue and fibre.

Sindhu

It is a hybrid progeny derived by backcrossing Ratna x Alphonso released by Fruit Research Station, Vengurla. Fruits are borne in clusters. Fruits weigh on an average about 150-220g. Pulp is deep yellow in colour and it has good sugar acid blend. Fruits are almost seedless with very thin stone, though fruits above 200 g have well developed seeds.

Arunima

It is from the parentage Amrapali x Sensation, released by Indian Agricultural Research Institute, New Delhi. The fruits are medium sized, having attractive skin colour. The pulp is deep yellow in colour and TSS is around 20°Brix. .

Pusa Surya

It is a selection from the variety Eldon released by Indian Agricultural Research Institute, New Delhi. It bears medium sized fruits, has red peel colour similar to Sensation.

Pusa Prathibha

It is a hybrid between the cross Dashehari × Amrapali developed by Indian Agricultural Research Institute, New Delhi. It is a regular bearing variety with attractive fruit shape, bright red peel and orange pulp. It has oblong shaped, uniform sized fruits. The plants are semi-vigorous. It has good sugar: acid blend and uniform fruits.

Pusa Shresht

It is a hybrid between the cross Amrapalli × Sensation developed by Indian Agricultural Research Institute, New Delhi. Trees are semi vigorous, regular bearing with elongated fruits and attractive red peel. Pulp is orange in colour, fibreless and firm at ripening, moderate sugar: acid blend with uniform fruit size (228g). It contains good amount of Beeta -carotene and ascorbic acid.

Pusa Pitamber

It is a hybrid between the cross Amrapalli × Lal Sundari developed by Indian Agricultural Research Institute, New Delhi. The plants are semi- vigorous, regular bearing with attractive oblong fruits. Fruits turn uniform yellow on ripening. It has good sugar : acid blend and uniform size fruits.

Pusa Lalima

It is a hybrid between the cross Dashehari × Sensation developed by Indian Agricultural Research Institute, New Delhi. The plants are semi- vigorous, regular bearer, attractive oblong fruits with bright red peel on yellowish green background with orange pulp and good sugar: acid blend.

Ambika

It is a hybrid between the cross Amrapali × Janaradhan Pasand developed by Central Institute for Subtropical Horticulture, Lucknow. The fruits of this variety are medium in size with slight sinus and beak and broadly pointed apex. Peel is smooth and tough. Fruits are bright yellow in colour with dark red blush. Pulp is firm with scanty fibre. TSS of this variety is 21° Brix. It is a late maturing variety.

Arunika

It is a hybrid between the cross Amrapali × Vanraj developed by Central Institute for Subtropical Horticulture, Lucknow. Fruits of this variety are attractive with red-blush. Fruits have high TSS (24° Brix) and high carotenoids content. Pulp is firm. Variety is at regular bearer and plants are dwarf in stature.

Pant Chandra

This is a clonal selection of Dashehari from Govind Ballabh Pant University of Agriculture & Technology, Pantnagar. Plants are tall with erect growth habit. Fruit colour at maturity remains green. It is a mid-season variety. Fruit weight is up to 150 g. Fruit pulp is reddish yellow with total soluble solids of 18% and having pleasant aroma.

Pant Sinduri

This is a clonal selection of Dashehari from Govind Ballabh Pant University of Agriculture & Technology, Pantnagar. Trees are medium in height with round top canopy. Fruit colour is yellow with pink shoulder. Average fruit weight is up to 200g. Fruit pulp is yellow with pleasant aroma. Total soluble solids vary from 16-18% with average yield up to 150 kg per tree. Fruit matures during last week of May to first week of June.

PKM-1

It is a hybrid released from Horticultural Research Station, Periyakulam. It is of the parentage, Chinnaswarnarekha x Neelum. It is regular bearing, produces good quality fruits and bears in clusters.

PKM-2

It is a hybrid released from Horticultural Research Station, Periyakulam. It is from the parentage Neelum x Alphonso. It is regular bearing and produces good quality fruits in clusters.

Al Fazli

It is from the parentage Alphonso x Fazli released by Fruit Research Station, Sabour. It is superior to Fazli, does not have spongy tissue. Fruits are sweet to taste.

Jawahar

It is a hybrid between the cross Gulabkhas X Mahmood Bahar developed by Bihar Agriculture University, Sabour (Bihar). It is a mid- season variety. Fruits mature from second week of June, and have precocity and regularity in bearing. The fruits become greenish yellow on ripening. Its fruits are of medium size and average fruit weight is 215 g per fruit. Its pulp is light yellow in colour, sweet in taste and pleasant flavoured and remain firm even after ripening. The TSS, acidity and pulp percentage are 22.5, 0.14 and 79.5 per cent respectively.

Menaka

Menaka is selection from Gulabkhas seedling. It is a regular bearing variety; fruits are attractive with deep red basal portion. Pulp is deep yellow, sweet and pleasant in flavour, less fibrous and firm. Fruit shape is oblong-oblique. Average fruit weight is 300g, late maturing variety and the TSS of the fruit is 20 percent, acidity 0.14 per cent and pulp content is 75 per cent.

Subhash

It is a selection from the seedlings of Zardalu. It is a mid-season variety, The ripened fruits are bright yellow in colour as Zardalu with the shape of Langra fruit, fruits are medium in size with average weight of 220 g. The TSS and acidity of the fruits are 24 and 0.29 per cent respectively. The pulp content is 76 per cent.

Sundar Langra

It is from the cross Sardar Pasand x Langra. It is regular bearing and resembles Langra in shape and size. Fruits are medium sized and sweet to taste (Hoda and Ramkumar, 1993).

Arka Aruna

It was developed at the Indian Institute of Horticultural Research, Bangalore. It is from the parentage Banganapalli x Alphonso. It is regular bearing; pulp free from fibre and spongy tissue, pale yellow, moderately firm, good for mango bars. Fruits size is big. Plants are dwarf in stature.

Arka Puneet

It was developed at the Indian Institute of Horticultural Research, Bangalore. It is from the parentage Alphonso x Banganapalli. It has attractive skin colour, medium sized, free from spongy tissue with good keeping quality, good sugar acid blend.

Arka Anmol

It was developed at the Indian Institute of Horticultural Research, Bangalore. It is from the parentage Alphonso x Janardhan Pasand. It is regular bearing, has uniform yellow skin colour. It is free from spongy tissue, good keeping quality, and good sugar acid blend.

Arka Neelkiran

It was developed at the Indian Institute of Horticultural Research, Bangalore. It is from the parentage Alphonso x Neelum. It is regular bearing, with medium sized fruits, free from spongy tissue, good pulp colour, excellent skin, colour, semi-vigorous tree.

Methods of Breeding

The breeding methods include Selection, Hybridization, Mutation and Bio-technological approaches.

i) Selection

Most of the commercial cultivars of mango are the results of selections made from chance seedlings. Many of the present day cultivars in Florida were selected from open-pollinated seedling progenies. The variety 'Haden' was found to be a seeding of 'Mulgoba'. The cultivars 'Edward Lippens', 'Tommy

Atkins' and 'Zill' were found originated from 'Haden'. In Israel, the varieties 'Maya' and 'Nimrod' are selections from open pollinted seedlings of the same mother tree (Oppenheimer, 1967). Likewise 'Dahar' was found to be a seedling of 'Irwin' (Slor and Gazit, 1982) and 'Naomi' to be a seedling of 'Palmer' (Tomer *et al.,* 1993). Two selections MDCH-1 and STH-1 have been made at the ICAR Complex for the North Eastern region in Manipur,India. Both are dwarf and flower within one year of planting. (Singh,1996).

Clonal selection: It is one of the important tools in the improvement of perennial crops. Asexual propagation helps in the preservation of accumulated mutations. Over the years, Naik (1948), reported significant variation due to bud mutation among trees of the same clone in an orchard with respect to fruit shape, size, colour and quality. Oppenheimer (1956), after a survey of many orchards reported wide variability in performance of the same variety in the same orchard.

Singh and Chadha (1981) selected four superior clones C24, C18, C35 and C45 from the variety 'Dashehari' at Lucknow. Singh *et al.,* (1985) isolated two high yielding clones from orchards of 'Langra', One off-season clonal selection has also been reported by Deshmukh and Budrukkar, (1988). 'Paiyur-1' a clonal selection from Neelum has also been made at Tamil Nadu. Whiley *et al.,* (1993)reported the existence of strains of Kensington resistant to bacterial black spot. Rajput *et al.,* (1996) isolated a high yielding clone from Dashehari, namely Dashehari S1, reported to be a regular bearer.

ii) Hybridization

The traditional method of pollinating the flowers on a panicle for several days has now been replaced by one-time pollinating the panicle selecting only about ten flowers per panicle at the most, since it is rare that more than one fruit per panicle is carried to maturity. By doing so, fruit set can be as high as 3.85% as compared to 0.23 to 1.57% efficiency obtained by other methods (Mukherjee *et al.,*1968; Singh *et al.,* 1980).

Another improvement in the crossing has been the use of 24 x 21" perforated polythene covers of 100 gauges instead of muslin bags (Mukherjee *et al.,* 1968). Studies carried out on marker genes such as purple colour of new leaves and panicles and beaked characters of fruit by Sharma and Majumder (1985) helped in assessing the hybrid nature of F1 seedlings. Singh *et al.,* (1980), suggested that once the flower is carefully pollinated since the stigmatic surface is very small, the pollen deposited first has an advantage, there is hardly any chance of contamination from foreign pollen, the fruit set can be as high as 3.85 percent, without covering the flowers.

Male flower

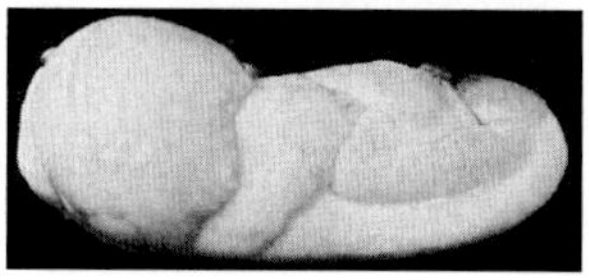

Polyembryony

Monoembryonic seedling

Pic 1

Pic 2

Pic 3

Pic 4

Pic 5

Pic 6

Flowering in mango tree

Variability for leaf colour and shape

M. andamanica

Parts of a Mango Fruit

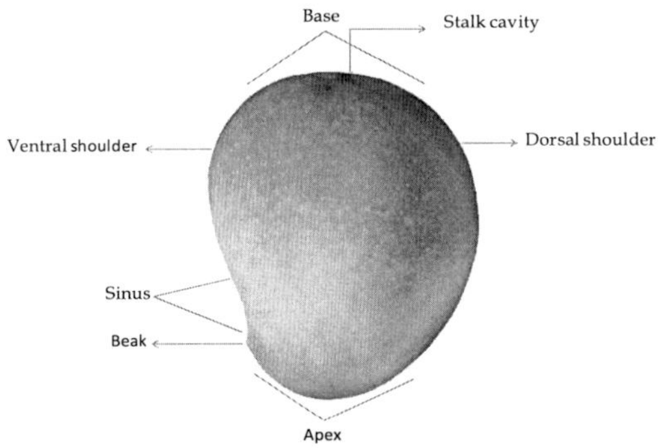

Variability for coloured varieties

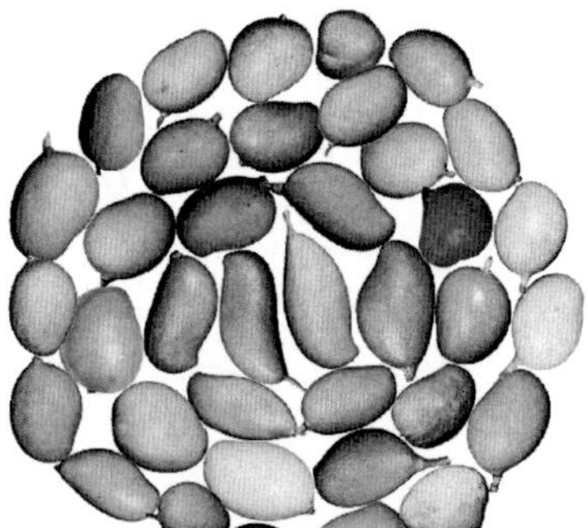

Diversity for pickling mango

Diversity

Vanaraj a coloured variety

S Jehagir F

With the discovery of self-incompatibility in some of the cultivars, a standardized caging technique for mango breeding was used (Sharma and Singh, 1970). This procedure involves the planting of self-incompatible (female) and male parents in specially prepared breeding plots, and are enclosed in an insect-proof cage into which freshly reared house flies, or any other suitable pollinator are introduced to effect cross pollination (Sharma *et al.,* 1972).

A new off-season crossing technique was suggested by Kulkarni (1986). It involves the induction of flowering in the desired parents in off-season by veneer grafting their defoliated shoots on to leafy shoots of off-season flowering cv. Royal Special and allowing open pollination between the desired parents. As no others cultivar flowers during this season, this is a safe technique.

Selection of hybrid progenies

Primary selection of hybrid progenies is mainly based on:

i. precocity

ii. fruit size and shape

iii. skin colour

iv. fruit characteristics like high pulp percentage, freedom from physiological disorders and

v. fruit quality

The selected hybrids are retained for further screening. These are then grafted and at least ten grafts per hybrid are to be evaluated for yield, regularity in bearing and for screening against diseases and pests. At least three consecutive years' data should be collected before deciding on their regional trials or release.

Interspecific hybridization

Wild species of mango can be of great value to breeding. Iyer (1991), suggested that the species other than *M. indica* could also be useful in crop improvement in two ways: i) species having edible fruits with other desirable characters, and ii) species that could act as gene donors for specific improvement like resistance to pests and diseases and other various biotic and abiotic stresses. Bompard (1993) has enumerated the potential use of *Mangifera* species in breeding. *M. laurina*, which has subglabrous and laxly flowered panicles, is well adapted to areas with perpetual wet climates and this could be of some use in incorporating resistance to anthracnose. *M. orophila*, a newly described species from Malaysia and *M. dongnaiensis* from Vietnam, which are restricted to mountain forests 1000 to 1700 metres above sea level may make growing mango in Mediterranean a distinct possibility.

Other species like *M. magnifica* is completely free from fibres. *M. rufocostata* and *M. swintonioides* have off-season bearing habit while *M. paiang* and *M.foetida* have good quality fruits. The variety 'Wani', belonging to the species *M. caesia* from 'Bali & Borneo has a distinctive taste. *M. casturi* a newly described species occurring in South Kalimanatan is a prolific bearer with small black sweet fruits. These species have good potential in breeding (Bompard, 1993; Kostermans & Bompard, 1993). Angeles (1991) reported that *M. altissima* is not affected by serious pests like hoppers, tip borers and seed borers. Yadav (1996) has also reported varied forms of *M. sylvatica*, which were polyembryonic in nature, having good fruit quality in the North Eastern Indian States.

Fairchild (1948) reported that, hybrids having pollinating quality can be produced by crossing five-stamened mango with the Indian mango having only one fertile stamen. Bhujanga Rao *et al.,* (1963), reported obtaining 32 interspecific hybrids of *M. odorata* & *M. zeylanica.* Vasugi *et al.,* (2009) reported that, the per cent fruit set was observed to be 32.69 and 25.47 for *Mangifera odorata* and *Mangifera zeylanica* respectively when used as pollen parent with the *Mangifera indica* hybrid Arka Anmol.

Very little work has been done on the evaluation of species. Hence, there is an urgent need to conserve, screen and use them in the breeding programme. Mukherjee (1963) felt that different *Mangifera species* could be inter-crossed easily, based on the success obtained from the crossing of *M. zeylanica* and *M. odorata.*

iii) Mutation breeding

Mutation is an important tool in creating variability. Although mutations and chimeras occur frequently, they are generally masked by intra somatic selections. In mango, mutants establishing as cultivars are rare. Two somatic mutants have been reported in mango. The variety 'David Haden' was reported to be a sport of 'Haden'. It is larger than 'Haden' and matures, early, (Young and Ledin, 1954). Another mutant 'Rosica' from seedless variety Peru was reported to be a mutant of 'Rosado de Ica'. It is high yielding and regular bearing (Medina, 1977). A mericlonal chimera of 'Alphonso' was observed by Roy (1950), which differed with respect to fruit shape. Roy & Viswesvariya (1951), observed mutants of 'Puthi', in which the number of palisade cell layers differed from the cultivar.

Induction of mutation was first tried using gamma irradiation by Siddique *et al.,* (1966). In the variety 'Langra', the dormant buds were exposed to ionzing radiation and grafted on to 1 year old seedlings. Some mutants obtained from the bud graft which had undergone 3 kR of radiation bore fruits which were heavier, larger and had more creamish yellow pulp than control (Siddiqui, 1985). Sharma and Majumder (1988b), irradiated bud sticks and top worked them on to 10 years old seedlings. They observed that dosages above 5 kR are lethal for mango and the LD 50 lies between 2 & 4 kR. Effective dosages of the chemical mutagens EMS and NMU were found to be 1.5% and 0.05% respectively. The spectrum of mutations induced by physical and chemical mutagens was observed to be more or less the same, indicating the high sensitivity of some loci. The mutants of Dashehari included dwarfs, changes in leaf shape and serration and TSS. In spite of these a few attempts, no worthwhile variety has been developed by mutation on mango.

iv) Biotechnological approaches

The use of various biotechnological methods to understand the crop genetics and to make an improvement is of recent origin. The use of molecular markers, *In vitro* selection and regeneration, identification of linkage for the specific traits and construction of genetic maps, gene pyramiding, allele mining *etc*., are some of the recent developments.

The discovery of molecular markers has led to the detailed genetic analysis and approaches to improvement of crop plants. It is differentiated into Protein markers, which were discovered first and DNA based markers. Protein markers had a disadvantage that the proteins are affected by the influence of environment and hence DNA markers were developed. The DNA markers are again categorized into Restriction based markers [Restriction Fragment Length Polymorphism (RFLP)] and PCR based markers [Random Amplified Polymorphic DNA (RAPD), Amplification Fragment Length Polymorphism (AFLP), Microsatellites, Inter-Simple Sequence Repeats (ISSR)].

The DNA based markers like RFLP and RAPD provide excellent tools to study the genetic diversity, eliminate duplicates in germplasm, study relationships, gene tagging, genome mapping, PVR (Plant Varietal Rights) purposes, marker assisted selection etc. The advantages of DNA markers are direct evaluation of genetic material; numerous markers can be obtained in a single population; assessment is not affected by environmental condition; powerful to pick up even minute differences and it requires little amount of plant tissue for analysis.

Molecular markers are practically unlimited in number, remain unaffected by environment and growth conditions and are simply inherited. The success of molecular markers in characterizing and analyzing genetic diversity in mango has been demonstrated by a number of studies. Schnell and Knight, (1993), Valdomiro and Paulo Sarmanho (2004) used RAPD markers to detect genetic polymorphism in the mango germplasm, suggesting that this technique may be useful tool for germplasm evaluation and management and also for directing crossings in the breeding programmes.

The RAPD technique developed by Williams *et al.,* (1990) is based on the amplification of multiple segments of genomic DNA using arbitrary primers (small DNA sequences usually 10 nucleotide long). It is one of the simplest techniques of molecular markers. According to Ferreira and Grattapalgia (1998) the development of this technique brought a real democratization in the area of molecular markers. The RAPD technique was used by Schnell *et al.,* (1995), Ravishankar *et al.,* (2000) and Hemanthkumar *et al.,* (2001) for cultivar identification and analysis of genetic relatedness in mango. Ravishankar *et al.,* (2000) grouped mango cultivars into four categories based on RAPD analysis. Hemanthkumar *et al.,* (2001) observed a moderate level of genetic diversity among south Indian cultivars by using RAPD technique. Lopez-Valenzuela *et al.,* (1997) were able to differentiate mango cultivars by embryo type and geographical origin using RAPD markers. Litz (2004) opines that the primary components of mango genetic engineering are the efficient somatic embryogenesis and plant recovery from elite material, induction of random mutations in embryogenic cultures and challenging for resistance to a specific selective agent and transformation with gene that mediates a horticultural trait. Mango cultivars are identified on the basis of morphological characters, which are not conclusive, as they are modified by environmental conditions (Tanksley *et al.,* 1989). Schnell *et al.,* (1995) studied the application of RAPD to the identification of mango cultivars. Of the 11 primers selected, no single primer gave a unique banding pattern for any of the 25 accessions examined, several different combinations of two primer banding pattern produced unique finger prints for each accession. Adato *et al.,* (1995) evaluated the usefulness of 10 different mini-and micro satellite probes for the identification of mango cultivars. Jeffrey's mini satellite probe 33.6 was the most informative, yielding well-resolved band patterns, representing polymorphic loci. Kasikush *et al.,* (2001) have used AFLP for both the identification of mango cultivars and generation of a preliminary genetic map. Valdomiro and Paulo Sarmanho (2004) used RAPD markers detecting genetic polymorphism in the mango germplasm analyzed, suggesting that this technique may be a useful tool for germplasm evaluation and management and also for directing crossings in the breeding programmes. Souzal *et al.,* (2011) evaluated genetic variability of mango (*Mangifera indica*) accessions maintained in the Active Germplasm Bank of Embrapa Meio-Norte in Teresina, Piauí, Brazil using RAPDs. Among these accessions, 35 originated from plantings in Brazil, six from the USA and one from India. Genomic DNA, extracted from leaf material using a commercial purification kit, was subjected to PCR with the primers A01, A09, G03, G10, N05, and M16. Fifty-five polymorphic loci were identified, with mean of 9.16 ± 3.31 bands per primer and 100% polymorphism. Application of unweighted pair group method using arithmetic average cluster analysis demonstrated five genotypic groups among the accessions examined. The genotypes Rosa 41, Rosa 48 and Rosa 49 were highly similar (94% similarity), whereas genotypes Sensation and Rosa 18 were the most divergent (only 7% similarity). The mango accessions were found to have considerable genetic variability, demonstrating the importance of analyzing each genotype in a collection in order to efficiently maintain the germplasm collection. Main problem in tissue culture propagation of mango is phenol exudation and explant discoloration. Curing was observed to be very beneficial for reducing the phenolic exudation. Somatic embryogenesis was induced in immature zygotic embryos and nucellus of Langra and Totapuri. Studies at Pune have revealed that plantlets could be generated from monoembryonic varieties. Leela *et al.,* (2005), were able to rescue hybrid embryos 6-8 weeks after pollination. This holds promise for increasing the number of hybrid progeny population, which is a limiting factor in mango hybridization.

In recent years, microsatellite markers are gaining importance for molecular characterization and genetic diversity studies (Fang and Roose, 1997; Moussaoui, 2005; Rafalski *et al.,* 1996 and Gupta *et al.,* 1996). Microsatellites are ubiquitous components of Eucaryotic genome (Delsney *et al.,* 1983 and Tautz and Renz, 1984). The potentials of microsatellite markers for identification of mango cultivars are suggested by Honsho *et al.,* (2005). Among 36 mango cultivars tested, 29 cultivars showed a unique pattern by six primer sets, whereas, seven cultivars could not be the identified because of genotype similarities. The observed and expected heterozygosity ranged from 0 to 0.83 and from 0.29 to 0.73 respectively. Damodaran *et al.,* (2006) in a study to find out the cause of differential flowering in mango (in the island ecosystem) reported that the specific bands obtained were responsible for genes associated with differential flowering behaviour and is due to introgression of genes during hybridization. Thus, the markers linked to off season flowering can complement the conventional breeding and expedite the mango improvement programme.

Genetic markers can help in detecting linkages which help in speeding up the breeding programme especially so in a perennial crop like mango. He-Xin Hua *et al.,* (2007) developed chloroplast intersimple sequence repeat markers for analysing the genetic relationship and diversity. Their results indicated that the cpISSR marker was a new powerful tool for the identification of mango cultivars or its relative species and their genetic relationship analysis and diversity evaluation.

'Eco-tilling' (Target Induced Local Lesions IN Genomes)- is a new technique that can identify polymorphism in a target gene by heteroduplex analysis. It is used to study the gene function and to identify point mutations in regions of interest. Microarray is one of the modern techniques to identify the genes which are responsible for a particular trait. In mango, it can be used to identify genes responsible for off season bearing. Genetic mapping in mango is not very far advanced like other crop plants; the physical mapping aims at defining the location of a particular gene (or DNA sequence) on a cloned genomic sequence.

Pre-selection indices in mango breeding

Mango, being perennial in nature, because of the long juvenile phase evaluation of progenies takes long time and this needs more area. Pre-selection indices help in overcoming this problem to a great extent. Pre-selection involves selection from seedling progenies even at juvenile stage for adult chrecteristics based on well-established correlations.

Leaf flavour was reported to be directly correlated with fruit flavour (Majumder *et al.,*1972; Whiley *et al.,* 1993). The emergence of new growth flushes simultaneously with fruiting or immediately after harvest is indicative of regular bearing (Sharma *et al.,* 1972). A higher phloem: xylem ratio was found associated with dwarf stature. If the ratio exceeds 1.0, they tend to be least vigorous; if it is from 0.6 to 1.0 they have medium vigour; these with less than 0.6 are the most vigorous (Kurian and Iyer, 1992). Higher phenolics in the apical bud have also been shown to be associated with dwarfing (Iyer, 1991). Majumder *et al.,* (1981) had earlier reported that lower stomata density is an indication of dwarf stature. However, such a correlation could not be confirmed by other workers (Iyer, 1991).

Problems in breeding

There are several problems in mango breeding, which have hindered the progress. Some of the important ones are discussed below with possible solutions.

i) Fruit drop

In spite of the large number of flowers being crossed, because of heavy fruit drop, the number of hybrid fruits generated is very low. Many recommendations are available to minimize fruit drop using growth regulators. However, these have not been of much use in breeding programme, where the number of flowers remaining in a panicle is very low. Iyer and Subramanyam (1972) have suggested the use of embryo culture to rescue hybrid embryos before they drop. However, *in vitro* techniques are yet to be standardised. The performance of exotic cultivars of mango under South Indian conditions has also shown that some of the cultivars like 'Kensington' had higher percentage of bisexual flowers. Probably these varieties if used in the breeding programme may help in developing varieties having higher fruit retention.

ii) Long juvenile phase

Mango seedlings generally take 3 to 10 years to fruit. Grafting individual hybrids to suitable rootstocks at the earliest possible stage and growing them in a location where climatic stress (particularly cold weather) induces precocious flowering. Iyer (1991) reported significant differences between seedlings and grafted plants of the same genotype for fruit size, quality and even colour. Singh (1969) suggested that young mango seedlings can be induced to flower and fruit if they are grafted into comparable shoots of a bearing tree few days before flowering. The scions are to be defoliated and girdled. Singh (1963), using this technique reported that the fruit characteristics of F1 hybrids can be determined within two years and F2 hybrids within 4 years thus eliminating at least ten years from the period required to raise and evaluate F2 populations. Ethephon has been shown to induce flowering (Chacko *et al.,* 1974). and could be tried for inducing early flowering in hybrid seedlings. Paclobutrazol was also observed to induce early flowering in the variety Dashehari (Sant Ram, 1999).

Breeding for different traits - source of genes

Apart from improving general characters like yield and quality, breeding has also been undertaken with specific objectives. Certain cultivars are known to offer genes for specific character improvement and some of the commonly used source is given below:

i) Dwarf stature

Breeding for dwarfness is important as production per unit area can be increased and management of the crop becomes easy. No rootstock so far has given a consistent dwarfing effect on the scion in mango. Some of the varieties, which could be useful in breeding for dwarfness include 'Kerala Dwarf', 'Janardhan Pasand', 'Manjeera'. 'Amrapali'.'Creeping' (Iyer and Subramanyam 1986, Yadav and Dinesh, 1999) and 'Nileshwar Dwarf' (Singh, 1990).

ii) Regular bearing

In India, alternate bearing is one of the main problems especially in most of the North Indian varieties. The main reason for alternate bearing has been the lack of initiation of vegetative growth soon after fruiting. The varieties 'Totapuri', 'Neelum', 'Banganpalli', can be used as gene donors. However, the variety 'Totapuri' imparts the prominent beak character to the progenies and has poor quality.

iii) Fruit colour

Most of the commercial varieties in the Indian sub-continent have green skin colour. Atractive skin color to the fruit is a pre-requisite if it is to fetch better price In international markets. Many of the Florida varieties can be used as donors for this trait. Some of the varieties like 'JanardhanPasand', 'Suvarnarekha', 'Fernandin' and 'Rosa' can also be used as donors.

iv) Disease and pest resistance

In mango, several diseases and pests threaten the crop at various stages of growth. Inspite of this, systematic screening of germplasm has not been carried out for various diseases and insect resistance. However, some information is available on this aspect.

The variety 'Bhadauran' is resistant to Malformation and has been used in hybridisation. However, all the F1 hybrids were susceptible when it was used as one of the parents in the breeding programme (Sharma and Majumder, 1988a). The screening of 102 cultivars by Ram Nath *et al.,* (1987) showed that the varieties 'Bhyadayam Dula', 'Samar Bahisht Rampur' and 'Mian Sahib' were found to be free from malformation. These varieties also can be tried in breeding programme.

Anthracnose (*Colletotrichum gloeosporioides* Penz.,) is one of the serious diseases affecting mango. It affects the blossom, peduncle, leaf, twig, and fruit. The Florida variety 'Tommy Atkins' was moderately tolerant. It can be used as one of the parents as it has good colour also. Yee (1958) reported the varieties Parish and Fairchild to be moderately resistant.

Powdery mildew (*Oidium mangiferae* Berthat) causes heavy loss by destroying the flowers Gupta (1976) reported the varieties 'Neelum', 'Zardalu', 'Janardhan Pasand' to be tolerant to this. These could be used in breeding programme. Bacterial canker is also a serious problem. Om Prakash and Srivastava (1987) reported that the variety 'Bombay Green' is resistant to canker.

Several insect pests are known to attack the mango crop. The most important ones are hopper (*Idiocerus*s pp), fruit fly (*Dacus dorsalis*) and Stone weevil (*Sternochetus mangiferae*). Varying degree of susceptibility by different varieties have been reported (Iyer, 1991).

Genetical studies

Some of the main problems in mango breeding has been the lack of information on the inheritance of characters, high heterozygosity in the cultivars and the few progenies realized by crossing have only compounded the problem. However, some data generated though not conclusive can be used as indicator in the breeding programme.

Genetical studies carried out in more than thousand hybrid progenies by Sharma and Majumder (1988a) showed that dwarf stature, regular bearing and precocity are controlled by recessive genes, regularity in bearing appeared to be linked with precocity and the character contributing to biennial bearing are dominant over those governing regular bearing habit. Sharma (1987) opined that flesh colour is controlled by additive genes. Iyer (1991) observed that light yellow colour is dominant over orange-yellow in the progenies of Alphonso x Neelum cross. Dinesh (2003), in an study carried out using half-sib analysis found that fruit characters like fruit weight, TSS and pulp percentage are controlled by non-additive factors and heritability is less. Lavi *et al.,* (1998) reported that parents should not be chosen on the basis of phenotype since offspring performance is quite unpredictable. With regard to skin colour, it was found that when red coloured varieties were crossed with green coloured varieties, gradation of colour in the progenies indicated that it is controlled by a number of loci (Sharma, 1987; Iyer & Subramanyam, 1987). The presence of beak on the fruit seems to be dominant as the entire progenies had beak on their fruits when 'Totapuri' was used as one of the parents (Iyer and Subramanyam, 1979). Bunch bearing was dominant over single fruiting (Sharma *et al.,* 1972).

In open pollinated progenies Lavi *et al.,* (1989) observed that there is no maternal effect on juvenile period and fertility and there is maternal effect on harvest season and fruit colour and there is slight effect of maternal parent on fruit size and taste.

Cytoplasmic inheritance was noticed for resistance to 'bacterial canker' as all the progenies were susceptible when Neelum was used as female parent, irrespective of male parent (Sharma and Majumder, 1988a). Iyer (1991) observed that internal breakdown (Spongy tissue) is mediated by recessive genes. Sharma and Majumder (1988a) found that susceptibility to mango malformation is controlled by dominant genes, as crosses with 'Bhadauran' a resistant variety did not yield any resistant hybrid.

Improvement in hybridization technique, using a few flowers in a larger number of panicles for crossing and then covering the panicles with polythene bags instead of muslin bag resulted in large number of hybrid progenies being raised. In fact, one of the main reasons for success in the later work on hybridization was the use of information on gene donors and prevention of indiscriminate crossing among varieties. The observation on these progenies has brought the following facts about the parents used. Work carried out at various centres showed that the varieties Totapuri, Neelum, Banganapalli and Lazzat Baksh are good sources of genes for regular bearing. The variety Neelum can also be used as a source for extending crop duration. However, Totapuri whenever has been used, the progenies were found to have beak in fruits which is undesirable. The varieties Creeping, Kerala Dwarf, Rumani and Mohammada Vikarabad were observed to be good source for dwarf stature; work on evaluation at various places has shown that varieties like Totapuri Red Small, Sensation, Latra, Kalapadi, Manipur and the hybrid variety, Amrapali can have good potential for dwarfing. Evaluation of germplasm for good skin colour has shown that several varieties have very good skin colour. The cultivars, Janardhan Pasand, Suvarna Rekha, Fernandin, Rosa and Sindhura can be good gene donors. The exotic varieties, Tommy Atkins, Kensington, Irwin and Sensation have also good potential as a donor source for skin colour. Work carried out extensively so far on a resistance source for some of the pests and diseases have proved futile. The varieties, which were resistant in the initial period, have not been found to be genetically resistant. Using these varieties in breeding programmes also resulted in progenies, which were susceptible. Mango, as a breeding material is difficult to handle, because of long juvenile phase, high heterozygosity and small number of hybrid progenies recovered. However, the information obtained on inheritance of characters can be summarized as follows;

1. upright habit of the tree as dominant over spreading and spreading dominant over dwarf.
2. dwarfness was found governed by recessive genes.
3. strong linkage has been found between bearing and fruit quality.
4. biennial bearing tendency was dominant over regular bearing.
5. regularity of bearing is controlled by recessive genes which have close linkage with precocity of bearing.
6. bunch bearing habit was observed to be dominant over single fruit bearing.
7. presence of beak marked with sinus was dominant.
8. heterosis and transgressive segregation for fruit size was observed in the F_1 hybrids.
9. red skin colour was dominant and gradations in skin colour of F_1 have suggested the role of duplicate genes, it was also the case with flesh colour.
10. resistance to floral malformation seems to be controlled by recessive genes.

11. cytoplasmic inheritance was noticed for bacterial canker.
12. spongy tissue was observed to be genetically controlled.

Bretell *et al.,* (2004) observed that many important fruit quality aspects such as fruit weight, fruit shape, ground skin colour, fruit width and pulp depth have high heritability, and can therefore be readily selected in a breeding programme. For non-ordered traits scored in discrete categories (blush colour, bloom, lenticel colour, embryo type and flavour), an estimate was made of data consistency from multiple scores for individual hybrids at different times and locations. Relatively high consistency value was recorded for fruit flavour, and in combinations involving Kensington Pride. The analysis of blush colour and fruit flavour in twelve families of hybrids confirmed that these characters have a strong genetic component, the high frequency of hybrids with red or burgundy blush can be recovered from crosses where one parent has an intense red blush colour. Singh *et al.,* (2004) observed wide magnitude of phenotypic coefficient of variation with high genetic advance for yield per plant and fruit weight in 31 chance seedlings of mango. Yadav and Dinesh (1999) evaluated the genotypes having dwarf stature, so that these could be further used in the breeding programme and progenies having dwarf stature could be isolated. The varieties Kerala Dwarf and Janardhan Pasand were found to be most suitable for usage as donor parents in the breeding programme. Dinesh (2003), observed that non-additive variance controls the characters studied in mango. Heritability is low and the chances of hybrid vigour manifesting for the characters in the F1 generation are bright. Selection of progenies can be made based on fruit size *i.e.,* medium sized fruits will have good TSS and selecting was more than the genotypic coefficient emphasizing the greater manifestation of characters and lesser influence of environment.

Dinesh and Vasugi (2002) based on germplasm evaluation (Table 1) observed that the following varieties could be used as probable gene donors for different characteristics in mango;

Table 1: Probable Gene Donors

Sl. No	Characters	Varieties
1.	Dwarf stature	Creeping, Kerala Dwarf, Rumani, Neelum, Kalapadi, Manjeera, MohammadaVikarabad, Arka Aruna, Amrapali
2.	Fruit size	Banganpalli, Janardhan Pasand, Kensington
3.	Skin colour	Janardhan Pasand, Sindhura, Fernandin, Suvarna Rekha, Vanraj, Rosa, Kensington, Tommy Atkins, Sensation, Gulabi, Arka Puneet, Nazuk Badan, Raspuri, Borsha
4.	Pulp content	Pairi, Arka Aruna, Mallika, Arka Anmol, Bhutto Bombay, Sora, Vanraj.
5.	TSS content	Arka Puneet, Mallika, Lazzat Baksh, Amrapali, Dashehari
6.	Keeping quality	Alphonso, Totapuri, Dashehari, Banganpalli, Kesar, Rataul, Arka Anmol, Arka Puneet.
7.	Regular bearing	Neelum, Kalapadi, Totapuri, ArkaAnmol, Kesar, Amrapali
8.	Maturity period	Early-Lazzat Baksh, Guruvam, Chandrakaran, Panakalu, Late-Vellaikolamban, Neelum, Arka Anmol, Mallika
9.	Processing	Padiri, Alphonso, Arka Puneet

Conclusion

Mango breeding is time consuming and complicated due to high heterozygosity and perennial nature of the crop. In spite of the problems, there has been some improvement in terms of the development of hybrids with good traits. The improvement brought about in the hybridization technique has also helped in realizing more number of recombinants from the crosses. The pre-selection indices, genetical studies and the biotechnological tools has helped in the selection of parents for crossing as well as for reducing the time taken for breeding. However, still there is a need to raise large number of progenies, which can be accomplished by raising large number of half sibs of desired parents so that better recombinants with desirable quality could be isolated. There is a need to study the crossability with some of the wild species so that desirable traits especially disease resistance traits if any could be imparted to the cultivated varieties.

References

Adato, A., D. Sharon, and Lavi, U. 1995. Application of DNA fingerprints for identification and genetic analyses of mango (*Mangifera indica*) genotypes. *J. Amer. Soc. Hort. Sci.* 120(2):259-264.

Angles, D.E. 1991. *Mangifera altissima.* (In) Verheij, E.W.M. and Coronel R.E. (eds) Edible Fruits and Nuts, Plant Resources of South East Asia 2. PUDOC, Wageningen, pp. 206-207.

Aron, Y.H, Czosnek, S. Gazit and Degani, H. 1998. Polyembryony in Mango (*Mangifera indica* L) is controlled by a single dominant gene. *Hort Science* 33(7):1241-1242.

Bhujanga Rao, C. and Ranga Charlu, V.S. 1958. Breeding New Mango varieties in South India. *Ind. J.Hort..*, 15: 173-183.

Bhujanga Rao, C., G.S.Swamy, M.Nagabhushanam and Rama Rao, B.V. 1963. Performance of some promising Andhra mango hybrids. *Punjab Hort. J.* 3: 124-36.

Bompard, J.M., 1993. The genus *Mangifera* rediscovered the potential contribution of wild species to mango cultivation. *Acta Horticulturae* 341: 69-77.

Brettell, R.I.S., P.R., Johnson, V.J., Kulkarni, W. Muller, and Bally, I.S.E. 2004. Inheritance of fruit characters in hybrid mangoes produced through controlled pollination,*Acta Horticulturae.* 645: 319-326.

Burns, W. and S. H.Prayag, 1921. The Book of Mango Bombay. Dept. Agric. Bull.

Chacko, E.K., R.R. Kohli and G.S. Randhawa., 1974. Investigations on the use of (2-chloroethyl) phosphonic acid (Ethephon, CEPA) for the control of biennial bearing in mango. *Scientia Hort.* 2:389-398.

Cordeiro, M.C.R., A.C.Q., Pinto., V.H.V. Ramos., F.G. Faleiro., L.M.S. Fraga., 2006. Identification of plantlet generic origin in polyembryonic mango (*Mangifera indica L)* cv. Rosinha seeds using RAPD markers. *IRevista Brasileira de Fruiticultura.* 28(3):454-57.

Damodaran, T., Medhi R.P., Gapil Dev, G., Damodaran, V., Rai, R.B. and Kavino. M. 2006. Identification of molecular markers linked with differential flowering behaviour of mangoes in Andaman and Nicobar Islands. *Curr. Sci.* 92(8) :1054-1056.

Degani,C., R. El-Bastri, and S.Gazit, 1990. Enzyme polymorphism in mango. Jour*nal of the American Society for Horticultural Science.* 115:844-847.

Degani, C., M.Cohen, R. El-Batsri, and Gazit, S.1992. PGI isozyme diversity and its genetic control of mango. *Hort. Science* 27:252-254.

Degani, C., M.Cohen, O. Reuvani, El-Batsri, R and Gazit, S. 1993. Frequency and charactristic of zygotic seedlings from polyembryonic mango cultivars determined using isozymes as genetic markers. *Acta Horticulturae,* 341:78-85.

Delseny, M., Laroche, M. and Penon. 1983. Detection of sequences with Z-DNA forming potential in higher plants. *Biochem. Biophys. Res. Commun.*, 116:113-120.

Deshmukh, P. and Budrukkar, N.D. 1988. Promising mango selections from Himayth, Aurangabad, *Acta Horticulturae* 97-99.

Dijkman, N.J. and Soule, M.J., Jr. 1951. A tentative method of mango selection, *Proceedings of the Florida State Horticultural Society* 64: 257.

Dinesh M.R. 2003. Genetical studies in mango (*Mangifera indica* L.). *J. Appl. Hort.* 5(1):27-28.

Dinesh, M.R. and Vasugi, C. 2002. Catalogue of mango germplasm. Published by IIHR

Dinesh M.R. and Vasugi, C. 2009. Bisexual flower percentage -does it have any bearing on the crop load in mango? National Seminar on Production, Post-harvest Technology and Marketing of Mango at Horticultural College and Research Institute, 23-25 September 2009, Periyakulam - 625 604, Tamilnadu.

Fairchild, D., 1948. The mango relatives of Cochin China:those with five-stamen flowers. *Proceedings of the Florida State Horticultural Society* 61:250-255.

Fang, D. Q. and M.L. Roose. 1997. Identification of closely related citrus cultivars with intersimple sequence repeat markers. *Theor. Applied Genet.*, 95: 408-417.

Ferreira, M.E. and Grattapalgia. 1998. Introducao ao uso de marcadores moleculares em analise genetic Brasilia: Embrapa, CENARGEN, p. 220.

Gupta, J.H. 1976. Reaction of mango varieties to powdery mildew (*Oidium mangiferae* Berthet) in Uttar Pradesh. *Progressive Horticulture* 8:63-64.

Gupta, P.K., Balyan, H.S. Sharma, P.C. and Ramesh, B. 1996. Microsatellite in plants: a new class of molecular markers. *Curr. Sci.*, 70: 45-54.

Hemanthkumar, N.V., Ravishankar, K.V., Shivashankar, K.S., Lalitha Anand, Narayanaswamy, P. Prasad, T.G. and Mukunda, G.K. 2001. Poster presented at the Plant and Animal Genomes XII Conference, San Diego, CA.

He-Xinhua., Guo YongZe., Li "YangRui and Ou ShiJin., 2007. Assessment of the genetic relationship and diversity of mango and its relatives by cpISSR marker. *Agricultural Sciences in China*. 6(2):137-142.

Hoda, M.N. and Ram Kumar, 1993. Improvement of mango. Proceedings of the National Seminar on Irregular Bearing in mango-Problem and Strategy (July 12-13,1991) held at Pusa, Pratan Kamal Printing Press Muzaffarpur, Bihar. India. 34-35.

Honsho, C., Hishiyama, K. Eiadthong, S. and Yonemori, K. 2005. Isolation and characterization of new microsatellite markers in mango (*Mangifera indica*). *Molecular Ecology Notes*., 5 :152-154.

Iyer, C.P.A., 1991. Recent advances in varietal improvement in mango. *Acta Horticulturae,* 291:109-132.

Iyer, C.P.A. and Schnell, R.J. 2009. In "The mango botany, production and uses, Breeding and genetics', ed Litz R.E. (CABI, Wallingford, UK), 2nd Ed, pp 67–96.

Iyer, C.P.A. and Subramanyam, M.D. 1972. Possible role of embryo culture in mango breeding. *Indian J. Hort*. 29:135-136.

Iyer, C.P.A. and Subramanyam, M.D. 1979. Improvement of mango by selection and hybridization. Annual Report of the Indian Institute of Horticultural Research, Indian Institute of Horticultural Research, Bangalore, p. 16.

Iyer, C.P.A. and Subramanyam, M.D. 1986. Creeping, a promising genotype for introduction of dwarfness in mango. *Indian J. Hort.* 43:321-223.

Iyer, C.P.A. and Subramanyam, M.D. 1987. Improvement of mango by selection and hybridization. Annual Report of the *Indian Institute of Horticultural Research*, Indian Institute of Horticultural Research, Bangalore, p. 11.

Iyer, C.P.A, M.C.Subbaiah, M.D. Subramanyam, and Rao, G.S.P. 1989. Screening of germplasm and correlation among certain characters in mango. *Acta Horticulturae* 231: 83-88.

Kasikush. K., F. Jinggui., E. Tomer., J. Hillel and Lavi, U. 2001. Cultivar identification and genetic map of mango (*Mangifera indica*). *Euphytica* 122:129-136.

Kostermans, A.J.G.H. and Bompard, J.M. 1993. The Mangoes Their Botany, Nomenclature, Horticulture and Utilization Academic Press, London.

Kulkarni,V.J. 1986. Graft induced off season flowering and fruiting in the mango(*Mangifera indica* L.) *J. Hort. Sci* 61(1):141-45.

Kurian, R.M. and Iyer, C.P.A. 1992. Stem anatomical characters in relation to tree vigour in mango. *Sciencia Hort*. 50:245-248.

Lavi, U., Tomer, E. and Gazit, S. 1989. Inheritance of agriculturally important traits in mango. *Euphytica* 44:5-10.

Lavi, U., E.Tomer, and J. Hillel, 1998. Components of genetic variance and genetic correlations of mango traits. *Sciencia Hort*. 75: 11-25.

Lavi, U., D. Sharon, E.Tomer, A. Adato, and Gazit, S. 1993. Conventional and modern breeding of mango cultivars and rootstocks. *Acta Horticulturae* 341: 146-151.

Leela Sahijram, K.T. Bollamma, Anitha Naren, Jaya R. Soneji, M.R. Dinesh and Halesh, G.K. 2005. *In vitro* hybrid embryo rescue in mango (*Mangifera indica* L) breeding, *Indian J. Hort*, 62(3):235-37.

Leroy, J.F.1947. La polyembryonic chez les Citrus son interet dans la culture *et* Lamilioration, *Revenue Internationale de Botanique Appliquee* Paris 27:483-495.

Litz, R.E., 2004. Biotechnology and mango improvement, *Acta Horticulturae,* 645: 85-92.

Lopez- Valenzuela, J.A., Martinez, O. and Parades-Lopez, O. 1997. Geographic differentiation and embryo type identification in *Mangifera indica* L, cultivars using RAPD markers. *Hort Science* 32:1105-1108.

Maheshwari, P. and Rangaswamy, N.S. 1958. Polyembryony and in vitro culture of embryos of Citrus and *Mangifera*. *Indian Journal of Horticulture* 15:275-282.

Majumder, P.K., Singh, R.N., Sharma, D.K. and Mukerjee, S.K. 1972. Preliminary studies on inheritance in *Mangifera indica* L. *Acta Horticulturae* 24:101-106.

Majumder, P.K., Sharma, D.K. and Singh, R.N. 1981. Breeding for dwarfness in mango (*Mangifera indica* L.) *National Symposium on Tropical and Subtropical Fruit Crops*, Bangalore, p.3.

Majumder, P.K. and D.K.Sharma, 1990. Mango (In) Bose T.K. and Mitra, S.K.(eds). *Fruits: Tropical and Subtropical Naya Prokash* pp. 1-62.

Malik, P.C. 1961. Breeding of mango. Fourth Hort. Res. Workers Conf. Poona.

Marais, Z.1992. Mango evaluation for breeding. *CSFRI Information Bulletin* 234: 7.

Medina, J.P. 1977. 'Rosica'- a new mango variety selected in Ica, Peru. *Fruit VarietiesJournal* 31:88-89.

Moussaoui, S. 2005. Molecular characterization of grape vine cultivars from Aragon gene bank using sequence microsatellite markers. M.Sc.Thesis, IAMZ, Spain.

Mukherjee, S.K.1950. Mango: its allopolyploid nature. *Nature* 166:196-197.

Mukherjee, S.K.1953. The mango -its botany, cultivation, uses and future improvements, especially as observed in India. *Economic Botany* 7:130.

Mukherjee, S.K. 1958. The origin of mango. *Indian J. Hort.,* 15:129-34.

Mukherjee, S.K., 1963. Cytology and breeding of mango. *Punjab Horticultural Journal* 3:107-115.

Mukherjee, S.K., Singh, R.N., Majumder, P.K. and Sharma, D.K. 1968. Present position regarding breeding of mango (*Mangifera indica* L.) in India. *Euphytica* 17:462-467.

Naik, K.C.1948. Improvement of the mango (*Mangifera indica* L.) by selecton and hybridisation. *Indian Journal of Agricultural Science* 18:35-41.

Naik, K.C.C. Bhujanga Rao, Ramani, V.S. 1958. Problems in crop improvement in mango. *Indian Journal of Horticulture* 15:159-167.

Om Prakash and Srivastava, K.C. 1987. *Mango Diseases and Their Management.* World Review. Today & Tomorrows Printers, New Delhi.

Oppenheimer, C.H. 1956. Study tour report on sub-tropical fruit growing and research in India and Ceylon. Special Bulletin No.3. State of Israel Ministry of Agriculture, Agricultural Research Station, Rehovot, Oct., 1956.

Oppenheimer, C.H. 1967. Nimrod - a new mango variety selected in Israel. *Proceedings of the Florida State Horticultural Society* 80:358-359.

Pinto A.C.Q. and Byrne, D.H. 1993. Mango hybridisation studies in tropical Savannah ('Cerrados') of Brazil Central Region. *Acta Horticulturae* 341:98-106.

Pope, W.T.1929. *Mango culture in Hawaii.* Hawaii Agricultural Experiment Station Bulletin 58.

Rafalski, J.A., J.M. Vogel, M. Morganate, W. Powell, C. Andre and Tingey, S.V. 1996. Generating and using DNA markers in plants. In: Birren, B.and E. Lai, (Eds.), Non Mammaliam Genomic Analysis. A Practical Guide. Academic Press, San Diego, pp. 75-134.

Rajput, M.S., Chadha, K.L. and Negi, S.S. 1996. Dashehari – 51, a regular bearing and high yielding clone of mango cv. Dashehari. *Programme and Abstracts, 5th International Mango Symposium*, Tel Aviv, Israel, pp. 42.

Ram Nath, Kamalwanshi, R.S. and Sachin, I.P. 1987. Studies on mango malformation. *Indian Journal of Mycology and Plant Pathology*: 17:29-33.

Ram Sant, L.D. Bist, Lakhanpal, S.C. and Jamwal, I.S. 1976. Search of suitable pollinizer for mango cultivars. *Acta Horticulturae.* 57:253-263.

Ravishankar, K.V., Anand, L. and Dinesh, M.R. 2000. Assessment of genetic relatedness among mango cultivars of India using RAPD markers. *J. Hort. Sci. Biotechnol*, 75:198-201.

Reddy, P.V.R., Dinesh, M.R., Verghese, A. and Vasugi, C. 2011. Insect pollinators : A key component to increase fruit set in mango. Abstract presented in 'Global Conference on Augmenting Production and Utilization of Mango : Biotic and Abiotic Stress', held at CISH, Lucknow from 21-24 June, pp 90-91.

Roy, B.1950. A mango chimera. *Curr. Sci* 19:93.

Roy, B. and Visweswariya, S.S. 1951. Cytogenetics of mango and banana. *Report of Maharashtra Association for Cultivation of Science*, Pune.

Sachan, S.C.P., Kotrodia, J.S., Chundawat, B.S. and Patel, M.N. 1988. New Mango hybrids from Gujarat. *Acta Horticulturae* 231:103-105.

Sant Ram., 1999. Hormonal physiology of flowering in 'Dashehari' mango *Jour. Appl. Hort.*, 1(2):84-88

Schnell, R.J. and Knight, Jr. R.J.1992. Frequency of zygotic seedlings from fivepolyembryonic mango rootstocks. *Hort. Science* 27:174-176.

Schnell, R.J. and Knight, R.J. 1993. Genetic relationship among *Mangifera* spp. based on RAPD markers. *Acta Horticulturae*. 341: 86-92.

Schnell, R.J., Ronning, C.M. and Knight, Jr. R.J. 1995. Identification of cultivrs and validation of genetic relationships in *Mangifra indica* I., using RAPD markers.

Sen, P.K., Mallik, P.C. and Ganguly, B.D. 1946. Hybridization of mango. *Indian J. Hort* 4:4.

Sharma, D.K.1987. Mango breeding. *Acta Horticulturae* 196:61-67.

Sharma, D.K. and Majumder, P.K. 1985. Studies on inheritance in mango (*Mangifera indica* L.). Paper presented at the Second international symposium on mango held at Bangalore.

Sharma, D.K. and Majumder, P.K.1988a. Further studies on inheritance in mango. *Acta Horticulturae* 231:106-111.

Sharma, D.K. and Majumder, P.K. 1988b. Induction of variability in mango through physical and chemical mutagens. *Acta Horticulturae* 231:112-116.

Sharma, D.K. and Singh, R.N. 1970. Self- incompatibility in mango (*Mangifera indica* L.) *Horticultural Research* 10:108-115.

Sharma, D.K., Majumder, P.K. and Singh, R.N. 1972. Inheritance pattern in mango (*Mangifera indica* L.) Proceedings of the Symposium on Recent Advances in Horticulture, U.P. Institute of Agricultural Sciences, Kanpur, pp. 66-68.

Siddiqui, S.H., Mujeeb, K.A. and Vasti, S.M. 1966. Evolution of new varieties of mango (*Mangifera indica* L.) through induced somatic mutation by ionizing radiations. *Proceedings of the Ist Agricultural Symposium.* Atomic Energy Commission, Dacca, pp. 34-37.

Siddiqui, S.H.1985. Induced somatic mutation in mango. *Mangifera indica* L. cv. Langra. Pakistan, *Journal of Botany* 17:75-79.

Singh, L.B.1963. A new technique for inducing early fruiting in mango hybrids based on movement of hormones. *Proceedings of the Florida Horticultural Society* 75:410-412.

Singh, L.B. 1969. Mango (In) Fererda, F.P. and Wit. F.(eds.) *Outlines of Perennial Crop Breeding in the Tropics*. H.Veenen and Zonen, Wageningen, pp. 309-327.

Singh, R.N. 1990. Mango. ICAR low-priced books series No.3, ICAR, New Delhi. Mango Hybrids; *Acta Horticulturae* 24:117-119.

Singh, H. and Chadha, K.L. 1981. Improvement of Dashehari by clonal selction *Natonal Symposium on Tropical and Subtropical Fruit Crops*, Bangalore, p. 5(Abstract).

Singh, R.N., Majumder, P.K. Sharma, D.K. and Mukherjee, S.K. 1972. Some promising mango hybrids *Acta Horticulturae* 24:117-119.

Singh, R.N., D.K.Sharma, and P.K.Majumder, 1980. An efficient technique of mango hybridization. *Scientia Horticulturae* 12:299-301.

Singh, R.N. S. Gorakh, Rao, O.P. and Mishra, J.S. 1985. Improvement of Banarsi Langra through clonal selection. *Progressive Horticulture* 17:273-277.

Singh, J., Singh, R.R. Dubey, P.S. and Singh, U.K. 2004. Studies on genetic variability and character association for yield and quality traits in mango chance seedlings. *Journal of Applied Biology.* 14(2):37-39.

Singh, S.P. 1996. New mango variety. Paper presented at the International Crop Science meet New Delhi.

Singh, R.N., Majumder, P.K. and Sharma, D.K. 1962. Self incompatability in mango var. Dashehari. *Current Science* 31:209.

Slor.E. and Gazit, S. 1982. 'Tahar', a new mango cultivar (in Hebrew). *Alon haNotea* 36.807.

Souza1, I.G.B., Valente1, S.E.S., Britto, F.B., V.A.B. de Souza and Lima, P.S.C. 2011. RAPD analysis of the genetic diversity of mango (*Mangifera indica*) germplasm in Brazil. *Genetics and Molecular Research* 10 (4): 3080-3089.

Sturrock, T.T. 1968. Genetics of mango polyembryony.*Proceedings of the Florida State Horticultural Society* 81:311-314.

Sturrock. D.1969. Final report on some mango hybrids-1969. *Proceedings of the Florida State Horticultural Society* 82:318-321.

Sundararajan, J.S., Muthuswamy, S. and Raghavan, N.1972. Off season cropping of mango. Symposium on mango and mango culture. *Acta Hort.* 24:189-191.

Tanksley, S.N., Young, A., Paterson, B. and Bonierbale, M. 1989. RFLP mapping in plant breeding; new tools for an old science. *Bio/Technol.* 7:257-264.

Tautz, D. and M. Renz. 1984. Simple sequences are ubiquitous repetitive components of eukaryotic genomes. *Nucl. Acids Res.*, 12: 4127-4138.

Tomer, E., Lavi, U., Degani, C. and Gazit, S.1993. Noami: a new mango cultivar *Hort.Science* 28:755-756.

Truscott. M., 1992. Biochemical screening of polyploid mango seedings. *CSFRI Information Bulletin* 237:17-18.

Valdomiro Aurélio Barbosa de Souza, Paulo Sarmanho da Costa Lima and Márcio Elias Ferreira., 2004. Genetic analyses of mango genotypes by RAPD markers. Poster presentation at the 7th International Mango Symposium held at Recife City, Brazil, September 22nd to 27th.

Vasugi, C., Dinesh, M.R. and Sekar, K. 2009. Studies on inter and intra specific crossability of Mangifera using cryopreserved pollen. *The ICFAI University Journal of Genetics and Evolution.*, 33-39.

Wagle., P.V. 1929. Preliminary studies on Alphonso mango. *Agric. J. India*. 24:259.

Whiley, A.W., Mayers, P.E., Saranah, J., Bartley, J.P. 1993. Breeding mangoes for Australian conditions. *Acta Horticulturae* 341:136-145.

Williams, J.G.K., Kubelik, A.R., Livak, J.J., Rafalski, J.A. and Tingey, S.V. 1990. DNA polymorphisms amplified by arbitrary primers are useful as Genetic markers. *Nucleic Acids Res.*, 18: 6531-6535.

Yadav I.S. 1996. Germplasm conservation and utilization in breeding of mango In :Tropical fruits in Asia - Diversity , Maintenance, conservation and use. R.K.Arora and V. Ramanatha Rao(eds), Published by IPGRI, Rome, Italy : P 163-169.

Yadav I. S. and Dinesh, M.R. 1999. Breeding for dwarf genotypes in mango. *J. Appl. Hort.*, 1 (1): 24-26.

Yee, W. 1958. The Mango in Hawaii. Hawaii Agricultural Experiment Series Circular 388.

Young, T.W. and Ledin, R.B. 1954. Mango breeding. *Proceedings of the Florida State Horticultural Society* 67:241-244.

11

Citrus Breeding

Awtar Singh

Introduction

Citrus is one of the most important fruit crops in the world. Citrus fruits are grown in the tropical and subtropical regions of the world. Some of the important countries where these fruits are grown include USA, Brazil, Spain, Israel, South Africa, Australia, China, India, Mexico and Morocco. Citriculture as a garden industry dates back to less than one hundred years. Indian citrus industry is the third fruit industry after mango and banana. The important groups of crops in citrus are sweet oranges, mandarins, limes, lemons, grapefruits and pummelo (Singh, 2001). Total world citrus production is about 115525.2 thousand tonnes, whereas in India, the production of citrus fruits is 8267.2 thousand tonnes (Anonymous 2012). The most important commercial *Citrus* species grown in India are the mandarin (*Citrus reticulata* Blanco), followed by sweet orange (*Citrus sinensis* Osbeck) and acid lime (*Citrus aurantifolia* Swingle). The main cultivars in the mandarin group are the Nagpur mandarin, Coorg mandarin, Khasi mandarin, Hill mandarin and the Kinnow mandarin. Maharashtra, Karnataka, Andhra Pradesh and Punjab are the major citrus producing states, while in Haryana, Uttar Pradesh, Madhya Pradesh, West Bengal, Tamil Nadu, Gujarat and Bihar, the fruit is grown to a limited extent. In Rajasthan new plantation of citrus are coming up very well and the main cultivars are Nagpur and Kinnow mandarins.

India has a varied range of climatic conditions and is recognized as one of the eight Vavilovian centers of origin of crop plants and diversity. The diverse geographical regions characterized by varying temperature and rainfall conditions have given rise to a wide range of variability in *Citrus* and related genera. Despite the vast genetic diversity, the productivity of citrus in India remained low. This emphasizes the importance of properly laid out breeding programmes for the improvement of citrus and use of its natural genetic resources. Citrus improvement activities should, therefore receive priority attention.

Origin of Different Citrus Species/Relatives

A detailed knowledge of germplasm and its centre of origin is a prerequisite for an effective collection of the germplasm. The information on the origin of different *Citrus* species is scarce and only lime is one of the important commercial citrus, found growing in wild state.

Sweet orange (*Citrus sinensis* (L.) Osbeck) is believed originated in Southern China or Indo-China border. The centre of origin of sour orange (*C. aurantium* L.) is Southeast Asia. Mandarin (*C. reticulata* Blanco) might have originated in the Philippines or Indo-China border. Lime (*C. aurantifolia* Christm.) originated in Northern India and the Malaysian archipelago, where the wild lime trees still grow. Pummelo (*C. grandis* (L.) Osbeck) and lemon (*C. limon* (L.) Burm.) are probably of South-East Asian origin, whereas the areas of Himalayas in Northern Myanmar and Southern China are suggested with respect to lemon. Citron (*C. medica* L.) probably originated in Southwest Asia and possibly in India, where it still grows in wild conditions, particularly in Nilgiris, Assam and lower Himalayas. Grapefruit (*C. paradisi* Macf.) is of recent origin either as bud mutation of pummelo or as a hybrid between pummelo and sweet orange (Zeven and de Wet, 1982).

Rough lemon (*C. jambhiri* Lush.) is found in many distinct forms in India (Assam) in semi-wild state and was believed originated in Western India. *Citrus hystrix* (DC), *C. macroptera* (Montr.) and *C. junos* (Sieb.) were supposed originated in Myanmar and *C. latipes* (Swingle) in India (Assam) (Tanaka, 1958). *C. indica* (Tan.), *C. assamensis* (Dutta and Bhatt.), *C. ichangensis* (Swingle) are also indigenous to India (Assam) (Bhattacharya and Dutta, 1956). Trifoliate orange (*Poncirus trifoliata* (L.) Raf.) is native to central and northern China, where it is widely distributed wild or cultivated throughout the country.

Genetics of Citrus and Related Genera

The *Citrus* and its related genera belong to sub-family Aurantioideae of the family Rutaceae. The main characteristics of the species within this family are:

i. presence of oil glands,

ii. ovary is raised on a floral disc,

iii. pellucid dots are present in the leaves,

iv. fruits have axile placentation (Swingle and Reece, 1967). *Citrus*, *Poncirus* and *Severinia* species have 2n=18 chromosomes (Agarwal, 1987a, 1988a, b).

However, some triploids, tetraploids and hexaploids also exist. Chromosome size is small and variable and chromosomal aberrations are quite common. The pollen fertility is variable in different *Citrus* species (Agarwal, 1984, 1987b, 1989). Self-incompatibility is found in cultivars of *C. grandis*, Clementine, Orlando, Minneola, Sukega, Robinson and Fairchild. Cross incompatibility is also found in Minneola and Orlando tangelos. Hybrids between self-incompatible cultivars are also self-incompatible and some of them may be cross incompatible. On selfing or crossing, hybrid vigour and inbreeding depression are common in citrus. Wider crosses like mandarin x grapefruit produce vigorous progenies, while narrow crosses like Ruby x Valencia orange or Eureka x Lisbon lemon result in production of many weak seedlings in the progenies. Trifoliate leaf character is dominant over monofoliate leaf and parents with trifoliate leaves are generally used as pollen parents to identify the hybrid seedlings in the progeny. Nucellar embryony is a simply inherited character and for desirable rootstocks, at least one parent should be highly nucellar. Disease resistance, nematode resistance and salinity tolerance exhibit a variable degree of inheritance in different *Citrus* and related species.

Nucellar embryony is a major hurdle in the breeding programmes, but selections with high percentages of nucellar embryony have helped in raising uniform plantations on disease free rootstock seedlings. Further, many natural hybrids have perpetuated through nucellar embryony and are contributing factors to the confusing species situation in citrus.

Poncirus is used as a rootstock for citrus scions. *Eremocitrus* appears to be freeze hardy (Barrett, 1985). *Eremocitrus* is not of commercial importance, but has been used in citrus breeding programmes to enhance drought tolerance and freeze-hardiness of the progeny. It has also been used as an interstock.

Some species of *Microcitrus,* like *M. australica* are resistant to burrowing nematode (*Radophilus citrophilus*) and *Phytophthora* fungus (Barrett, 1985). *Microcitrus* has been used as an interstock and in rootstock breeding programmes on a limited scale.

Citrus Breeding: Historical background

Citrus is an important commercial fruit tree and stable, long-term development of the citrus industry depends upon continuous supply of new superior cultivars (Liu, *et al.* 2009). From the primitive days of modern plant breeding, farmers and gardeners selected improved forms of wild plant species and established for many crops, productive cultivars giving good quality produce (Austin, 1988). The process of selection has been followed since the very old times and people used to select the best seeds/fruits for the next season crops and this process of selection for the best is still being followed. Introductions had played an important role for the cultivation of citrus in newer areas. The Jews were responsible for the distribution of citrus around the eastern Mediterranean and later cultivation of citrus extended to Italy and other parts of Europe (Webber, 1967). The sweet orange (*C. sinensis* [L.] Osbeck) was introduced in Europe by the Portuguese from China. Mandarins reached England from China and from there to Mediterranean. Columbus brought seeds of oranges, lemon and citron to Haiti in 1493. Success of Kinnow mandarin in India is a good example of the importance of introductions.

Clonal selection has been the major force in the historical development of citrus cultivars. Nearly all of the important commercial cultivars and rootstocks originated as chance seedlings, or bud sport mutations (Hodgson, 1967). Clonal selection within cultivar groups has been practised in many citrus growing regions of the world and has been particularly useful for the development of improved strains of cultivar groups like Satsuma mandarin and Navel sweet orange. The Washington Navel orange and Marsh seedless grapefruit cultivars have arisen from closely related seeded forms. Other selections in sweet orange are Shamouti, Salutiana, Marrs, Skagg's Bonanza and in grapefruit are Thompson Pink and Red blush. In Satsuma mandarin, maximum improvement has been achieved through the selection of naturally occurring somatic mutants. In India, Mudkhed and Seedless Nagpur mandarin were selected for seedlessness (Chakrawar *et al.*, 1988; Chadha and Singh, 1990). Vikram, Pramalini and Chakradhar cultivars of acid lime have been selected in Maharashtra (Bagde and Patil, 1989; Chakrawar and Rane, 1977) and Jai Devi (PKM-1) in Tamil Nadu.

Systematic breeding work was carried out at a few institutions in the world. The planned hybridization has given rise to many important rootstock cultivars namely Troyer, Carrizo, Rusk, Morton, Savage, Benton, C-32, C-35, Willits citranges and Swingle citrumelo, which were superior in many respects to the existing rootstocks (*Phytophthora* and citrus nematode resistance, fruit quality, etc.). Ford (1978) selected hybrids of Ridge Pineapple x Milam and *Poncirus* x Milam resistant to burrowing nematode. The hybridization has also resulted in many improved scion cultivars like Sunburst, Fallglo, Robinson, Osceola, Lee,

Kinnow and Pixie mandarins, Melogold and Oroblanco grapefruits and Kiyomi tangor. In India, planned hybridization was initiated at Chethalli, Rahuri and Tirupati. The breeding programme at Chethalli resulted in the selection of three promising hybrids resistant to *Phytophthora* and nematodes (Agarwal, *et al*., 1987). Sexual hybridization will continue to receive attention for improvement of rootstocks and scion cultivars in future.

Keeping in view the contribution made by naturally occurring mutations in citrus cultivar development, citrus breeders attempted induced mutagenesis for specific objectives. The first cultivar developed through induced mutagenesis was 'Star Ruby' grapefruit (Hensz, 1977). Seedless clones of Pineapple orange and Duncan and Foster grapefruit have been selected from the irradiated progeny (Hearn, 1984; 1986). In India, a Kagzi lime mutant was isolated at Parbhani, which bears fruits in clusters of 3 to 8 and produces fruits in off-season and is less susceptible to canker (Chakrawar *et al.,* 1988). Similarly LD_{50} for seeds and budwood of Nagpur mandarin was determined (Singh *et al.,* 1996).

Problems faced in citrus breeding

Despite substantial genetic diversity and interspecific fertility, the genus *Citrus* includes some of the most difficult species to breed (Gmitter, *et al*., 1992; Perea-Molphe-Balch and Ochoa-Alejo, 1998). Citrus breeding is difficult and time consuming due to following reasons:

i) Intergeneric crosses are generally difficult to make and the yield of crossed seed is very low or the seeds may not be viable.

ii) Hybrid progenies are a few, which are available for selection due to lethal, abnormal recombinants or their poor survival.

iii) The hybrids may not flower, may have very long juvenility or may be completely ovule or pollen sterile.

iv) Hybridization may bring in progeny the undesirable traits of the relatives.

v) Most *Citrus* and related species are highly heterozygous and a few important traits show single gene inheritance, therefore, F_1 hybrids tend to exhibit variability and production of higher generations and study of inheritance of characters is very difficult.

vi) The common occurrence of nucellar embryony and absence of characteristic morphological marker genes, make selections of hybrids difficult, although this situation is improving through the use of biochemical and molecular markers.

vii) The protracted juvenility period of seedlings in the field (5-15 years) makes citrus breeding a very long term, costly and land intensive proposition.

Therefore, most of the cultivars have resulted from natural mutations or hybridization and the best examples are:

- Pink and red fleshed grapefruit cultivars were selected from limb sports of common grapefruits.
- Navel oranges are probable limb sports of common oranges.
- Several early maturing Satsuma and Clementine mandarin selections have also resulted from natural mutations.
- Grapefruit and some mandarins like Temple, Murcott are most probably natural hybrids of oranges and mandarins.

Objectives for improvement of citrus cultivars

The breeding objectives are different for rootstock and scion cultivars. Underlying the breeding objectives is the need to gain genetic information on the inheritance of specific characters and combining ability of available parents. Since, citrus is a perennial crop, fixing of objectives is a critical factor. These objectives vary from crop to crop and according to the immediate and long term needs. Recent research has been focused on producing seedless mandarins for new markets and the ability to set parthenocarpically is an important characteristic for future citrus production (Sykes and Lewis, 1996). The major objectives for the improvement or breeding of new cultivars of citrus are:

i. Build and enhance a germplasm base responsive to current and future needs of the citrus industry for new cultivars.

ii. Understand the inheritance of genetic basis for key characteristics.

The common objectives in citrus improvement programme are according to the desirable attributes of scion and rootstock cultivars, which are as under:

Desirable attributes of the scion cultivars

i. Vigour and longevity of the tree with sufficient amount and regularity of the crop.

ii. Reduced tree size without the reduction of yield per unit canopy volume to facilitate easy handling of plants.

iii. Desirable fruit size, shape, exterior appearance, tight skinned with pleasant flavour.

iv. Deep orange or red orange rind colour for oranges and mandarins.

v. Fruit with smooth rinds and without stem-end necks or blossom end nipples.

vi. Ease of peeling to facilitate eating out-of-hand.

vii. Higher ratio of soluble solids to acid for oranges and mandarins.

viii. Season of ripening (early/late), storage life and adaptability to specific environments.

ix. Seedless or less seeded fruits.

x. Resistance / tolerance to major diseases.

xi. Spreading type of plants with internal bearing instead of bearing on the periphery of the canopy.

xii. High foliage density and fruits relatively resistant to sunburn.

xiii. Feasibility of mechanical harvesting.

xiv. Compatibility with important rootstocks.

Desirable attributes of rootstock cultivars

i. Compatibility with the important scion cultivars.

ii. Reduction of tree size without affecting the yield or scion health.

iii. Resistance to root rot diseases caused by *Phytophthora*, which seem to take a heavy toll of seedlings in the nursery and grown up trees in the field.

iv. Resistance to citrus and burrowing nematodes.

v. Resistance to diseases caused by viruses and virus like pathogens.

vi. Tolerance to excess salts, drought, cold and water logging.

vii. High yield and quality of the fruits.

viii. Higher number of seeds/fruit and high rate of nucellar embryony.

ix. Rapid growth and lack of branching, so that it can be handled easily in the nursery.

Methodology of citrus breeding

Artificial control of reproduction and inheritance is called breeding. Usually, it involves the selection of parents and selection of individuals among the progeny. Simple selection for desirable forms being followed is also a form of breeding, but controlled pollination is usually a basic requirement to a breeding programme. Mutation breeding by the use of irradiation and chemical mutagens is also an important approach, but collection of germplasm from all over the world and its evaluation in the target environment is an integral and important part of any breeding programme. If variability for desirable characters is present in the existing population, selection is the best approach, but when sufficient variability is not existing, then one has to go for hybridization, because this form of breeding leads to the recombination of genes and we may get desirable recombinants in the progeny. Sometimes, this approach is not applicable due to less variability, poor heritability of the desired character, self and cross incompatibility of the parents to be used in hybridization. Under such circumstances, we have to go for mutation breeding, because, in this approach, we may get entirely different genes in the progeny. Again, in this approach, most of the mutations are undesirable and only a few are favourable. Now a days, non-conventional breeding approaches are also being used, which include the use of tissue culture / biotechnology. These approaches include somaclonal variation and selection, anther culture, protoplast fusion, embryo culture, triploid production, genetic transformation and molecular markers.

The plant breeding methods used for the improvement of citrus are similar to those used for other horticultural and field crops. Since, citrus is propagated asexually, any improvement once achieved will last for a longer period, because, the clones are theoretically immortal. Before deciding the method of improvement to be followed, it is necessary to understand the problems in citrus breeding and objectives to be considered for improvement.

Any approach for citrus breeding should have the basic assumption that the main goal of such programme is to obtain new and better performing varieties. Good traits should be kept or incorporated and the bad traits have to be excluded or minimized in the new genotype. Scion and rootstock cannot be considered separately and graft compatibility, fruit quality and productivity are essential traits to be evaluated (Machado, *et al*., 2011).

Exploitation of the existing genetic variability

There is substantial genetic variability among the true *Citrus* and related genera. *Severinia* sp. has been reported resistant / tolerant to *Phytophthora*, citrus nematode, salinity and excess boron (Hutchison and Grimm, 1973; Hutchison and O' Bannon, 1972; Swingle and Reece, 1967). It is graft compatible to *Citrus*. *Citropsis gilletiana* Swingle and M. Kell is immune to *Phytophthora*, tolerant to burrowing nematode and is also compatible with *Citrus* (Ford and Feder, 1960; Swingle and Reece, 1967). Because, most of these species are distantly related to C*itrus*, they may exhibit resistance or even immunity to the pathogens that attack citrus.

Clymenia, Eremocitrus, Fortunella, Poncirus and *Microcitrus* are closely related genera to true citrus fruits. All are interfertile with *Citrus* except *Clymenia* and hence provided a base for citrus improvement like the divergent relatives. They possess different characters of importance like, the tolerance to salinity, drought, burrowing and citrus nematode and short juvenility (Barrett, 1981, 1985). Among them, *Poncirus* is especially important due to its resistance against *Phytophthora*, citrus nematode and citrus tristeza virus. This genus transmits these characteristics in the hybrid progeny when involved in breeding programmes.

Wide variation among and within different *Citrus* species exists for growth habit, vigour, vegetative morphology, juvenility, fruit size, tolerance / resistance to insect-pests and pathogens and physiological stresses. Fruit size varies from very small (< 3 cm diameter) in limes and some mandarins to very big (upto 30 cm diameter) in pummelos and citrons. Fruit shape, flesh and juice colour, flavour and aroma, TSS and acidity are variable among and within different *Citrus* species.

The genetic bases of some species like sweet orange and grapefruit are quite narrow, though the genetic diversity within the genus is broad. They are not true *Citrus* species, but unique interspecific hybrids (Scora, 1975; Barrett and Rhodes, 1976) and different cultivars within these species are the results of somatic mutations.

It is believed that most species under the genus *Citrus* are native to tropical and subtropical regions of the Southeast Asia, particularly, India, China and the region between these two countries. The North-Eastern region of India is considered as one of the natural homes of citrus - atleast for a few species. Certain non-edible *Citrus* species like *C. indica, C. ichangensis, C. macroptera, C. latipes* are found growing in wild and semi-wild state in the North - Eastern region. Bhattacharya and Dutta (1956) considered them as indigenous to that area. Also, different strains of citron (*C. medica*), sour pummelo (*C. megaloxycarpa*), rough lemon (*C. jambhiri*) and sour orange (*C. aurantium*) have been found growing widely in semi-wild conditions in different North-Eastern states of India (Verma and Ghosh, 1979). It is possible that since very early days there might have been spread of different *Citrus* species in the contiguous regions of their natural home and there are possibilities of intercrosses, particularly between China and India, so far as carrying of citrus germplasm from one country to the other is concerned.

In South India, the indigenous types include Gajanimma (*C. pennivesiculata* Tan.), Kichli (*C. maderaspatana* Hort. Ex. Tan.) and some wild mandarin types, viz. Kodakithuli, Billikichili, Nakoor lemon, Mole Puli (sour orange type). In the foothills of Western Himalayas, the Hill lemon or Galgal (*C. limon* var.

decumana) and Attani (*C. rugulosa* Hort. Ex. Tan.) are common. Citron, Citron-lemon, Karna Khatta, rough lemon, Rangpur lime, acid lime, hybrid pummelo and different types of mandarins are found all over the country.

Out of the indigenous germplasm of citrus, useful selections were made for desirable horticultural traits. Strains of rough lemon and Rangpur lime screened against *Phytophthora,* exhibited varying degrees of resistance. In rough lemon, strains (Katajamir, Sikkim No. 4 and Sohmyndong) of North-Eastern India were found to be highly susceptible, while one strain collected from South India (Coorg) showed high degree of survival (94%) in inoculated samples.

In acid lime, one seedless type and a variety known as 'Nepali Oblong' were resistant to canker. Another acid lime selection (Selection 49) made in a university in Maharashtra (Rahuri) showed 30 per cent disease incidence against an average infestation of 85 per cent in other varieties. Screening against citrus greening disease (Kapur *et al.,* 1977) covering 15 *Citrus* species and 234 cultivars showed that Gajanimma (*C. pennivesiculata* Tan.) and Karna Khatta (*C. karna*) were free from the disease. Incidence of greening was also low when scion cultivars were grafted on Rangpur lime rootstock. Rangpur lime also exhibited greater salt tolerance, where accumulation of higher sodium, even upto one percent level did not exhibit injury to the plants. Differential uptake and accumulation of salts mainly influenced the salt tolerance. The tolerant rootstock has the capacity to exclude chloride uptake from the system.

Rootstock varieties have been evaluated for their tolerance to drought. At Tirupati (South India), Rangpur lime was highly tolerant to drought, which maintained water balance even at low soil moisture. In other states also, Rangpur lime was more drought hardy when compared to standard rootstocks like rough lemon.

Against the citrus nematode, rough lemon strains namely rough lemon Poona (M.P.) and lemon 8782 were moderately tolerant (Mani, 1989). All the local *Citrus* species / varieties, viz. *C. assamensis* (Adajamir), *C. aurantifolia* (Kagzi lime), *C. jambhiri* (Katajamir, Nemutenga), *C. karna* (Karna Khatta), *C. limonia* (Rangpur lime), *C. macroptera* (Satkara), *C. megaloxycarpa* (Bortenga), *C. nobilis* (Janerutenga), *C. pennivesiculata* (Gajanimma), *C. reshni* (Billikichilli), *C. reticulata* (Coorg, Nagpur), *C. sinensis* (Mosambi, Sathgudi), tested against citrus nematode, *Tylenchulus semipenetrans* Cobb. were found susceptible.

Introduction and collection of citrus germplasm

The real value of germplasm can be realized only if it is collected and evaluated properly. The germplasm, which may not be important today, can be beneficial in the future. Particularly with the danger of extinction, the germplasm may be collected and conserved as soon as possible. If any germplasm is lost at this stage, it may not be possible to recover it in future. So, one has to properly collect and conserve the gene pool available today without its further loss. From collection and conservation point of view, IPGRI (Anonymous, 1986b) has recognized some categories of citrus germplasm and these are:

i. Currently cultivated scion and rootstock varieties.

ii. Germplasm which either has been used in breeding new cultivars or possess characteristics which are of interest.

iii. Germplasm, which may be required for future breeding, work.

The first two categories are being taken care presently, which is evident from large-scale collections and introductions of germplasm. All the cultivated varieties are being maintained in various centres in the world. Some species like trifoliate orange and kumquats have been used successfully for improving the resistance of the varieties against biotic and abiotic stresses like cold resistance, disease resistance and nematode resistance. There are many hybrids of trifoliate orange with citrus which are being used as rootstocks throughout the world and prominent among them are Carrizo, Troyer, Willits, Rusk, C-35, C-32, Benton citranges, X-639 and Rangpur lime X Troyer citrange. Some trifoliate orange strains like Rich 16-6 and Flying Dragon are also used commercially as rootstocks. Many hybrids have also been produced through protoplast fusion from *Citrus* and related genera (Grosser and Gmitter, 1990 a,b; Grosser *et al.* 1996), which are under different stages of evaluation.

The third category mentioned above poses a real challenge for the breeders as they are not cultivated presently and they possess very important traits which will be required in the future breeding programmes. The germplasm under this category includes the following species:

Citrus indica (Tan.): This species originated in Eastern Himalayas and still grows wild in Meghalaya in Garo hills. It is known as Indian wild orange and may be useful for breeding purposes.

Citrus tachibana (Mak.) Tan.: Swingle and Reece (1967) considered it to be a satellite species of *C. reticulata*, because it is very much similar to it. It is found in wild state from Southern Taiwan to the South West Japan. Its natural crossing with *C. reticulata* might have given rise to Shekwasha mandarin, which is a potential rootstock.

Citrus ichangensis (Swingle): It is the most cold resistant of all the evergreen species in the Aurantioideae subfamily and is of interest for breeding. It is growing in wild in central and south China.

Yuzu: It is probably a natural hybrid between *C. ichangensis* and *C. reticulata* and is cultivated in China and Japan for its fruits as well as a rootstock.

Ichang lemon: It is probably a cross between *C. ichangensis* x *C. grandis* and is grown as an ornamental plant in China. Its fruits are used as deodorizer. It is a hardy plant and requires further studies to be used as a rootstock.

C. latipes (Swingle) Tan. It is found in wild state in North-Eastern India and Northern Myanmar. It is extremely tolerant to cold and is useful in breeding hardy citrus rootstocks.

C. celebica var. southwickii (Wester): It is found in wild state in Philippines and Indonesia. It might be a hybrid between *C. celebica* and *C. grandis* (Swingle and Reece, 1967). Its fruits are of fair size, smooth and thin-skinned and hence it is very promising for breeding purposes.

C. macroptera (Montr.): It is found in wild state in Thailand, Indo-China, Philippines, Papua, New Guinea, New Caledena and Polynesia. The tree is vigorous. It is immune to wither tip and resistant to canker. It is supposed to be a good choice as rootstock or for breeding new citrus rootstocks (Swingle and Reece, 1967).

Clausena lansium (Lour.): It is widely cultivated in tropical and subtropical regions, but it is native to Southern China and Indo-China.

Severinia buxifolia(Poir.): It is also called as Chinese box orange. It is distributed in Southern China, North Vietnam, Taiwan, Hongkong, Cambodia and Laos. It is tolerant to salinity and is useful as a rootstock or as a parent in rootstock breeding programmes.

Limnocitrus littoralis(Miq.) Swingle: It is also called as Swamp orange. It is found growing wild in Java, Bali, Vietnam, Annam and Cochin China. It has high tolerance to salinity and may be useful as citrus rootstock or in citrus rootstock breeding programmes.

Eremocitrus glauca (Lindl) : It is known as Australian desert lime and grows in wild state in Australia. It can withstand severe drought and hot dry winds. It also has tolerance to salinity and boron. It can be used as a rootstock for sour orange (*C. aurantium*) or other *Citrus* species.

***Microcitrus* spp.** (Swingle): In this group six species are there, five of which grow wild in Queens land and Northern New South Wales in Australia and one in south-eastern Papua, New Guinea. It is very vigorous. *Microcitrus australis* x *M. australasica* might be useful as a citrus rootstock in soils, which are low in organic carbon.

Swinglea glutinosa (Blanco) Merr.: It is called as tabog or boyag and is found in Philippines and Luzon islands. It grows well at high temperatures and has a potential as a citrus rootstock. It is susceptible to citrus canker.

Feronia limonia (L.) Swingle: Woodapple is the only species in *Feronia* genus. It is found in India, Sri Lanka, Myanmar, Indo-China, Malaya Peninsula and Indonesia. As a rootstock for citrus, it stimulates early flowering, which can be used for evaluation and hybridizing newly introduced/ evolved varieties.

Feroniella oblata (Scheff.): It is grown in Columbia, Laos, South Vietnam and Eastern Thailand. It also stimulates early flowering when used as a citrus rootstock.

Feroniella lucida (Scheff.) Swingle: It is native to central Java. It may be a possible rootstock for dwarfing and for stimulating early flowering in the scion.

Citropsis(Swingle): This genus has 11 species which are native to tropical and subtropical Africa. *C. gilletiana* is the most vigorous and is also resistant to *Phytophthora* and hence can be used as a rootstock or in breeding programmes.

Hesperthusa crenulata (Roxb.) : It is grown in West Pakistan, India, Myanmar, Thailand, South Western China, South Vietnam, Laos and Cambodia. It is an ornamental owing to its feathery foliage. It can be used as a rootstock for lemon.

***Atalantia* spp.** (Correa): These species are found in South Western Asia, India, Sri Lanka Myanmar, Cambodia, Laos, North and South Vietnam, South China and Sumatra. The trees are small. The fruits of *A. monophylla* are used for medicinal purposes. Its graft unions with citrus are slow but good. It can be tried as a rootstock for citrus.

***Pleiospermium* spp.** (Swingle): These species are distributed in Southern India, Sri Lanka, Sumatra, Java and Borneo. Lemons grafted on *Pleiospermium* do very well and hence it may be worthwhile trying as a rootstock for citrus.

Citrus assamensis: A distinctive species having crushed leaf aroma similar to ginger or eucalyptus smell. It has a medium-sized tree with thick glossy leaves. Fruits are medium sized, almost spherical and smooth surfaced. The fruits are of limited home use with acidic juice. The species was described as a new indigenous species from Assam by Bhattacharya and Dutta (1956), but it is same as *C. pennivesiculata* (Tanaka), the Gajanimma of South India.

Characterization and evaluation of citrus germplasm

More detailed characterization and evaluation of data are needed to adequately access the actual amount of genetic diversity present. This data should include both descriptive and molecular data and molecular level genetic analysis of the citrus germplasm.

The current direction in characterization and evaluation is to collect data on morphological characteristics like shoot growth, fruit quality, leaf shape, vegetative life cycle, etc. Molecular characterization is in initial stages (Fang and Roose, 1997; Fang *et al.*, 1997; 1998; Federici *et al.*, 1998; Gmitter *et al.*, 1999). Some of the phylogenetic relationships appear to be same as morphological characters being evaluated. There have been markers reported for cold-acclimation-responsive loci (Cai *et al.*, 1994), nematode resistance (Ling *et al.*, 1994), citrus tristeza virus resistance (Gmitter *et al.*, 1999; Fang *et al.*, 1998; Fang and Roose, 1999), and fruit quality (Fang *et al.*, 1997). The molecular characterization will increase the efficiency of evaluation of accessions and their utilization in breeding programmes.

The characterization and classification mainly rely on the morphology and anatomy in combination with isozyme analysis. The evaluation of field performance is important to recognize selection of rootstock variety based on economic traits and resistance to insects, pests and environmental stresses by long term field tests. Biochemical and biophysical tests have been conducted to screen the suitable or promising cultivars and rootstocks. Many varieties have been released after evaluation such as Mosambi, Pineapple, Valencia, Newhall, Skagg's Bonanza and Navelina, which are the leading citrus varieties in the world today. Evaluation studies have shown that the large leaf gold bean (*Fortunella hindsii*) has high *Phytophthora parasitica* resistance, Goutou Chen (*C. aurantium* x spp.) exhibits high tristeza resistance, Anjiang hongju and Zhuhong ju possess high salt tolerance and Ziyang xian chen, high alkaline tolerance.

Citrus germplasm conservation

Citrus germplasm is mainly conserved in field genebanks. Due to high level of heterozygosity, adequate biodiversity is seldom maintained in the field grown plants, which may not represent the entire species wide variability available in citrus. There could be a substantial loss of variability *ex situ* due to several biotic and abiotic stresses. Cryogenic technology offers great promise for conserving citrus in the form of pollen (Ganeshan and Rajshekharan, 1999).

Cryopreservation techniques are considered one of the most promising alternatives to the traditional field conservation of live plants. Cryopreservation of embryogenic cultures of citrus is one of the most valuable tools not only for the conservation of genetic resources of citrus, which has great economic importance, but also to conserve plant materials in a form ready for further genetic manipulations.

Management of citrus diversity in the field genebanks has proved to be difficult, labour intensive and has a great threat of many serious diseases (Chadha *et al.*, 1970; Singh, 1989). Therefore, there is an urgent need to conserve genetic diversity of citrus using alternative *ex situ* approaches in the genebank and cryobank for long term conservation.

Presence of high degree of polyembryony in many citrus species provides the opportunity of conserving germplasm in the form of seeds without any risk of genetic instability (Singh and Chadha, 1993). Seeds of many *Citrus* species display recalcitrant or intermediate storage behaviour and, therefore, cannot be stored using conventional storage methods (King and Roberts, 1979). The role of seed coat has extensively been studied and found to be inhibitory for germination in many citrus species (Mumford and Grout, 1979; Radhamani *et al.*, 1991). Seeds of *C. limon*, *C. aurantifolia* and *C. aurantium* have been reported to be showing orthodox behaviour with or without testa (King *et al.*, 1981). Many species of citrus were characterized as intermediate, recalcitrant and orthodox on the basis of various reports (Hong and Ellis, 1995).

Cryopreservation techniques were extensively used for long term conservation of various citrus species and allied genus, *Poncirus trifoliata* using range of explants such as zygotic embryos/embryogenic axes (Norman and Hamidah, 1992; Radhamani and Chandel, 1992), shoot apices, cell suspension cultures (Engelman *et al.*, 1994, Sakai *et al.* 1990; 1991), somatic embryos (Marin and Duran-Villa, 1988) with varied degree of success. Out of more than 150 species of *Citrus* reported (Tanaka, 1961), approximately 25 species and allied genera have been attempted for long, medium and short term conservation in various laboratories/genebanks around the world.

Genetic resources should be conserved *in situ*, but here the maintenance of genetic resources is very difficult. Hence, *ex situ* conservation is necessary for genetic resources, because, these are more accessible to researchers and necessary for characterization and evaluation. Maintenance of disease free germplasm is also possible under *ex situ* conservation. The main types of germplasm conservation followed in the case of *Citrus* are as follows:

a) ***In situ* conservation**: Attempts to preserve the native habitat of *C. indica*, a progenitor of mandarin were initiated in 1978 by NBPGR. Its genepool has potential to create new species. To establish the gene sanctuary -

National Park, the prevalence of *Citrus* species including its wild relatives in time in different parts of the country was studied. The natural diversity of *C. indica* was observed in the forest of Garo hills in Meghalaya showing plant characteristics varying from bush to the climber with high frequency of distribution in dense forest and showing resistance to diseases and pests. Thus, a gene sanctuary for *C. indica* was established in Tugra Range in Garo hills. This is the first genepool sanctuary for the conservation of endangered *Citrus* species (Singh, 1981).

b) ***Ex situ* conservation**: Genepool of citrus is conserved in field genebanks. Conservation in the form of seeds or *in vitro* conservation of root or shoot apical meristem have potential (Chaturvedi *et al.* 1974 a, b, Bhansali and Arya, 1978). Generally, the following approaches are being followed for conservation of citrus germplasm under *ex situ* conservation.

   i) **Field genebank**: The germplasm is collected from its area of origin and is planted in research stations or botanical gardens. In these field genebanks, besides low priority on germplasm conservation, problems of virus diseases and dieback syndrome have been major causes for the loss of many accessions. Some of the old and declining plants have been re-established by seed planting, but it has not been possible in seedless and monoembryonic cultivars.

   ii) **Seed conservation**: Many *Citrus* species have high degree of polyembryony, in which seed conservation can be used without the risk of genetic instability. But citrus seeds are recalcitrant and lose viability within 2-4 weeks of storage due to desiccation and increased polyphenols (Bajpai *et al.*, 1963, Bhekasut *et al.*, 1976). However, evidences are available to show that although intact seeds of lemon are recalcitrant, but these become orthodox if their testas are removed (Mumford and Grout, 1979). There is also a need to modify the methodology of germination of stored seeds, which may have to be allowed to germinate at 30^{0}C for 5-7 weeks (Roberts *et al.*, 1984). Thus there appears to be a possibility of conserving many *Citrus* species as seed in long or medium term storage. Citrus embryos have been successfully cultured (Roberts *et al.*, 1984), which has opened the possibility of its utilization for germplasm conservation.

   iii) ***In vitro* conservation**: Tissue culture in citrus has been successful, efforts to conserve the germplasm *in vitro* need more research with respect to genetic stability, duration of storage, etc. (Chaturvedi et al., 1974a, b). Proliferation in callus was observed even after several years (Bhansali and Arya, 1978). *In vitro* root cultures continued to retain the ability to form shoot buds, which were maintained at 25°C for 8-10 months, which later were regenerated to plants (Anonymous, 1989-90b).

Utilization of germplasm for breeding purposes

The germplasm collected has been utilized as breeding material as well as for identification of promising germplasm or as elite selection. As a result of the national and international projects on citrus breeding, several hybrid cultivars were released. Wide variation among and within different *Citrus* species exists for growth habit, vigour, vegetative morphology, juvenility, fruit size, tolerance to biotic and abiotic stresses. Out of the germplasm collections, useful selections have been made for desirable horticultural traits. Strains of rough lemon and Rangpur lime, screened for *Phytophthora*, exhibited varying degree of resistance. In lemons, Nepali Oblong was resistant to canker. Gajanimma and Karna Khatta were resistant to greening disease. Rangpur lime and Cleopatra mandarin exhibited greater salt tolerance. Rangpur lime was also highly tolerant to drought. Rough lemon (M. P. and 8782) were resistant to citrus nematode. Genetic resources have also been used for the improvement of citrus rootstock varieties and many hybrids like Carrizo, Troyer, C-35, C-32, Benton citranges and Swingle citrumelo were developed, which have varying degree of resistance to *Phytophthora*, citrus nematode and other pathogens.

Variation in fruit characteristics in Rangpur lime

Role of biotechnology in germplasm management

Recent developments in biotechnology, including cell culture and molecular biology have increased the focus on citrus germplasm. Citrus is one of the most established culture systems in crop species, because embryogenic callus line could be induced, maintained and utilized. Callus lines have been induced in Satsuma and other mandarin species, sweet orange, sour orange, grapefruit and limes. The calli also maintain their embryogenic abilities after repeated subcultures for a long time (Kobayashi *et al*., 1983: Hidaka and Omura, 1989). Also the preservation by slow growth subculture system and cryopreservation have been established (Kobayashi *et al*., 1990). Thus embryogenic callus lines could be a possible alternative for germplasm preservation *in vitro*. However, the preservation and utilization of callus system depend on the ability to produce the embryogenic callus. But for becoming an alternative form of preservation, fine definitions of field characteristics should be registered with formatted descriptors.

Shoot tip culture is developed in Japan (Omura and Hidaka, 1992). Combined with virus elimination from cultured shoots, slow growth for preservation has been investigated (Iwanami *et al.*, 1993).

Isozyme analysis and DNA markers have been developed and applied to the identification and classification of citrus germplasm. These markers include RFLP, RAPD and other DNA fingerprinting (Matsuyama *et al.*, 1992). DNA and isozyme markers give valuable information on the phylogenetic relationships. Furthermore, DNA markers made possible minor variations generated by selfing or hybridization to be identified. However, in most cases, they could not discriminate nucellar variations and bud mutants. Hence, standardization of techniques is required for an effective DNA typing of the *Citrus* species.

Germplasm preservation, obtaining information and preserving citrus DNA are very important, because soon the genetic engineering by gene introduction to citrus cells will be performed as routine procedures (Kobayashi *et al.*, 1990; Hidaka *et al.*, 1990; Hidaka and Omura, 1993). This information will also contribute on genotyping of citrus germplasm, for new taxonomic analysis based on the function of genome. Because of rapid increase in the information concerning the genus, setting up of DNA banks, more genebank projects will be necessary.

Citrus species are able to hybridize with each other, without sexual isolation, and produce numerous hybrids. Thus, it is very difficult to supplement new accessions in germplasm lists for international communication. The standardization of fine description lists and molecular typing should contribute for a revision of citrus taxonomy. For this, international collaborative works should be required to communicate the specimens, description and molecular information. The germplasm should be collected as large quantity as possible, because once any germplasm is lost, it may not be available in future. Efforts should be made to collect the germplasm from their natural homes, because, there citrus has evolved naturally alongwith biotic and abiotic stresses and in those areas, we may get sources of resistance to both biotic and abiotic stresses. Moreover, citrus is sexually compatible with many of its related genera, which are resistant or even immune to many biotic and abiotic stresses. Best example is the trifoliate orange, which is resistant to *Phytophthora*, citrus nematode, citrus tristeza virus and is also cold hardy.

Selection

This is the most efficient way to get the new varieties in the species with high rate of somatic mutations such as sweet orange, grapefruit and Satsuma mandarin, but this approach is entirely causal, uncontrolled and has unknown mechanism (Machado, *et al.*, 2011). Commercially grown *Citrus* species are propagated only vegetatively (buds for scion cultivars and nucellar seedlings for rootstocks), thus citrus has massal selection of mutants as the fastest way of breeding. Traditional programmes of genetic improvement using hybridization and recurrent selection always met genetic difficulties (high heterozygosity, incompatibility, long selection cycle, etc.) and botanical barriers (mainly adventitious nucellar apomixes and juvenility) characteristic of citrus group (Machado, *et al.*, 2011). In this method of breeding, superior cultivars are selected from the natural variability present in the orchards under cultivation. This involves the survey of different areas, collection of germplasm and their evaluation in the target environment and then selection of the desirable ones. Most of the present day cultivated cultivars are not the result of planned hybridization programmes, but they have come through the selection of natural bud sport mutants or the chance seedlings (Shamel, 1943).

Selection for highly heritable characteristics such as plant height, days to flowering, colour, disease resistance, etc. is easy and effective, even on the basis of individual plants, but for characters like yield, evaluation in replicated trials is must, which are to be conducted in the target environment. For sustaining the production, survey of the orchards, which are neglected, is important and the plants producing under these conditions are selected, which are further evaluated under the minimum use of chemical fertilizers and pesticides. In old orchards, occurrence of such plants is common, and there is need to isolate such plants. If a desirable plant is found, it will be multiplied and used as a new variety, which requires minimum use of off- farm inputs.

The method of clonal selection resulted in the selection of Yuvraj Blood Red from Blood Red (Singh, 1964). Mudkhed Seedless mandarin was selected as a bud mutant of Nagpur mandarin (Chakrawar and Rane, 1977). Pramalini and Vikram, acid lime varieties, bearing fruits in clusters and give 30-35 percent more yield than the existing Kagzi lime variety were selected at Marathwada Agricultural University, Parbhani (Anonymous, 1985-86). Chakradhar lime, a seedless and thornless selection of Kagzi lime was selected by Bagde and Patil (Bagde and Patil, 1989). PKM-1 was developed at Periyakulum (Anonymous, 1987). A seedless selection of acid lime was also made in Himachal Pradesh (Badiyala *et al.*, 1990). Selections namely Dabwali Baramasi and PAU Selection were made in lemon and they produced the heaviest crop (Anonymous, 1987). Similarly, clonal selection resulted in the selection of early maturing and commercially seedless clones of the Nagpur mandarin (Singh and Singh, 2001). A few clones were also collected in Nagpur mandarin.

Low seeded clone of Nagpur mandarin

Early maturing clone of Nagpur mandarin

A view of variability in Nagpur

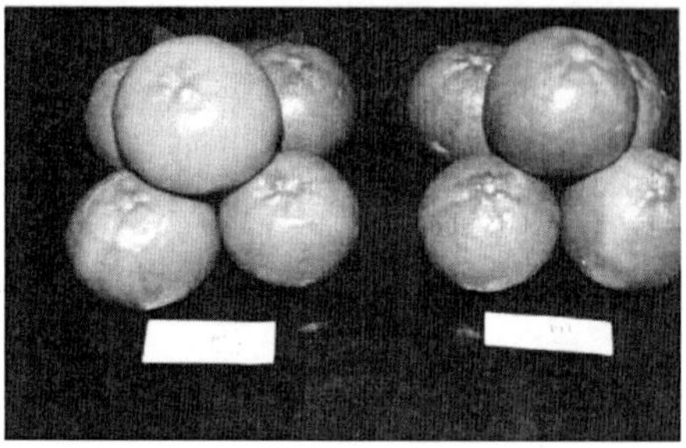

Early and late maturity in Nagpur mandarin

This is the earliest and easiest method for identification of improved citrus cultivars. Most of the present day cultivars, both scions and rootstocks, have originated as bud sports or chance seedlings (Hodgson, 1967). Clonal selection has been very effective within the cultivar groups, particularly in Satsuma mandarin and Navel orange. Most of the grapefruit coloured cultivars can be traced back to a single variety and all these have originated through chance seedlings or bud mutations. Spontaneous mutations occur frequently in citrus. They are often observed as limb sports or sectors on fruits, also detected in nucellar seedlings or other budded progeny. In sexual seedlings, mutations can seldom be distinguished from gene recombination. These mutations are largely unfavourable, showing poor yield, atypical leaf characters or abnormal fruits, but a few had characters, which could be useful. Washington Navel orange, Marsh grapefruit, Shamouti, Salustiana, Marrs, Skagg's Bonanza orange, Thompson Pink grapefruit etc., are spontaneous mutations and are the results of clonal selection. Satsuma is another group where the spontaneous mutations and clonal selection had played an important role. Here, changes in rind colour and time of ripening are of particular interest. According to Shamel (1943), clonal selection has been used to isolate superior bud sources of the cultivars and to eliminate inferior propagated material. This method is very much effective in improving the existing cultivars in relation to yield, early/late maturity and fruit quality. The pre-requisite for clonal selection is the fixing of the objectives in a justified manner and thorough survey of the citrus belt in a particular area.

Hybridization

Conventional hybridization or cross-breeding is simply a way of combining the best features of two parents by crossing one as a female and the other as a male and then selecting for the best hybrid within the resulting progeny. Hybridization recombines the traits of both the parents with most progeny having characteristics mid-way between the parents. However, it depends on how the characteristics are inherited and how they are expressed in the progeny. Citrus hybrids are all unique and different from their parents and the aim is to generate a variety that captures and combines the best characteristics of the parents. Almost all the programmes for improving citrus cultivars were based on the processes of selection and hybridization followed by selection (Hearn, 1985). In scion breeding, parents with high combining ability and heritability of desirable characters are selected and the parents having high zygotic embryony are commonly used as seed parents. Seeds from the hybrid fruits are extracted and raised in the nursery and finally the zygotic seedlings are planted in the field for detailed evaluation of the plant and fruit characteristics. Simultaneously, during this period, the plants are also evaluated for resistance/ tolerance to biotic and abiotic stresses. The desired plants are selected and budded on to the commercially used rootstocks for their detailed studies in multilocational trials. After this evaluation, the scion variety is released for commercial cultivation.

In citrus breeding programmes, sexual hybridization normally faces several reproductive impediments, such as apomixes and polyembryony which result in a cumbersome identifications of hybrid seedlings (Frost and Soost, 1968). These characteristics are considered undesirable since the correct identification of hybrids is hindered by these conditions (Golein, *et al.* 2011).

Almost similar procedure is followed for rootstock breeding (Barrett, 1985, Hutchinson, 1985), but, here among the selected parents, at least one parent must be highly nucellar, so that the hybrids can be propagated through seeds. Here, the seedlings are screened for majority of the soil borne diseases like foot rot, nematodes and other factors like salinity tolerance. The desired plants selected are also tested for scion performance budded onto them. If any plant, which is highly nucellar, have resistance to biotic and abiotic stresses and produces quality fruits and longer life of the scion, is released for commercial plantations.

Many scion varieties have been developed through hybridization and selection and the best examples are Kinnow, Honey, Sunburst, Fallglo, Robinson and many tangors and tangelos. Similarly, the planned hybridization programmes for rootstocks have resulted in many improved rootstocks, like Carrizo, Troyer, Rusk, C-32, C-35, Benton, Willits citranges, Swingle citrumelo, X-639, Shekwasha mandarin × rough lemon, Rangpur lime × Troyer citrange (dwarf hybrid) and Trifeola. Now days, these rootstocks are being used throughout the world for citrus plantations, because these rootstocks impart resistance to both biotic and abiotic stresses in addition to production of quality fruits and higher yields.

Hybridization involves the controlled crossing of two genetically different parents to obtain a desirable variety. Reshuffling the genes in the existing germplasm creates new varieties. For any planned programme of hybridization, a collection of germplasm is must. Different varieties with desirable characters are utilized for hybridization for achieving specific objectives like producing early-maturing varieties with high yield and quality of fruits, developing rootstocks having resistance to biotic and abiotic stresses and wider adaptability.

Trifoliate orange and Kumquats, which are related to citrus are extensively used in crossing with citrus and a few broad groups like citranges (sweet orange x trifoliate orange), citrumelos, (grapefruit x trifoliate orange), citradias (sour orange x trifoliate orange) and limequats (kumquat x West Indian lime), have been developed. Other hybrids like Tangelos (mandarin x grapefruit), lemonimes (lemon x lime) and tangor (mandarin x sweet orange) have also been developed. Kinnow mandarin is a well-known hybrid between King orange and Willow-leaf mandarin commercially successful in North-India, Karnataka and Tamil Nadu.

Hybridization really suffered from several problems in scion and rootstock improvement programmes. This is mainly due to longer juvenility period, which delays the evaluation of hybrid seedlings. Further, most of the cultivated varieties are highly polyembryonic and due to this, the hybrid seedlings produced are weak and a few in number and further, it is very difficult to identify them from the nucellar ones. Such types of problems become very prominent in scion breeding programmes owing to the absence of specific morphological markers. This situation is now improving by the use of isozyme and molecular markers. In rootstock breeding, in most of the cases, trifoliate orange was employed as male parent, because, not only it carries a dominant marker gene, the trifoliate leaf character, but also has the factors like *Phytophthora*, citrus nematode and citrus tristeza virus resistance. Hence, the seedlings with trifoliate leaf are all hybrids in the progeny. On the other hand, nucellar embryony is helpful in the asexual reproduction by seeds. Many cultivars of *C. grandis* and Clementine mandarin are monoembryonic and produce only the zygotic seedlings. Moreover, the nucellar seedlings are vigorous, uniform and free from most of the viruses carried by the mother plant.

In hybridization programme, parents with complementary characters are selected. The unopened but mature flower buds are emasculated on the seed parent in the early hours and pollination is done in the next morning after the anthers dehisce. The selected buds on both male and female parents, the emasculated buds and the flowers after pollination are bagged. The bags from the pollinated flowers are removed after a week and the fruits are allowed to mature. Where synchronization of male and female flowering is not there, the pollen storage is required at low humidity and temperature. Procedures for pollen storage are now standardized (Iyer *et al.*, 1983; Soost and Cameron, 1954). Adjusting the irrigation schedules (Iyer *et al.*, 1983) can also synchronize flowering in *P. trifoliata* and *Citrus* species.

From the matured fruits, seeds are extracted and sown immediately in the nursery. The hybrid seedlings are separated from nucellar ones and planted in the field. These seedlings are grown till fruiting and their seeds are evaluated for resistance to various diseases, insect-pests, nematodes and for suitability as scion or rootstock. For diseases and insect- pest resistance, the hybrid seedlings can also be evaluated at an early age by propagating them asexually by cuttings. This will save a lot of time. Now a days, owing to the availability of isozyme and molecular markers, the hybrids from parents without some morphological markers can also be identified with a greater accuracy.

The desirable rootstock hybrids should have high percentage of nucellar embryony and resistance to different diseases and nematodes. After this evaluation, the desirable hybrids are tested for performance of different scion varieties budded on them. Similarly for scion hybrids, the hybrid seedlings are propagated on the commercial rootstocks for their detailed evaluation related to tree size, vigour, yield, weight of fruit, juice and seeds/ fruit, TSS and acidity. The selected hybrids are tested in multilocation trials on different rootstocks.

In India, canker resistant hybrids in acid lime, like N-52, Hybrid-2 and Hybrid-4 have been produced by crossing it with different species like lemon, citron, kumquat and Rangpur lime. The hybrids made by Agarwal (1987b) and Agarwal *et al.* (1987) at Bangalore, resulted in the selection of three nematode resistant hybrids viz. CRH-3, CRH-5 and CRH-41 (Anonymous, 1987-88). One precocious bearing hybrid was evolved which can be used in the development of precocious bearing scion varieties (Agarwal, 1987b).

From the hybridization programme for the improvement of citrus rootstocks, many hybrid plants were planted in the field and simultaneously screened for *Phytophthora* and nematode resistance. These plants have started fruiting in the field and at various stages of evaluation (Singh, 2005). They exhibited the segregation in the F_1 progeny (Dass, *et al.*, 1998).

Fruits and leaves of a Rough lemon x Troyer citrange hybrid (centre) and the parents

The plant of Rough lemon x Troyer citrange

A block of citrus rootstock hybrids exhibiting segregation in the F_1 progeny

Mutation Breeding

Mutation breeding relies on the fact that some traits can be modified by altering the genetic makeup of an individual. Thus, it is possible to alter just one characteristic without changing others. Mutation breeding can be used to improve an established variety by one key characteristic such as seedlessness, which has been a popular target in citrus mutagenesis research. Citrus and its related genera are extremely prone to mutations. Mutations result in the form of mutated buds or sports, part of the trees or the entire tree. Most of these are undesirable, but many of them are extremely useful. Most of the present day scion and rootstock cultivars are the results of these natural mutations selected from the commercial plantations.

Spontaneous mutations have played an important role in the development of new cultivars. Many spontaneous mutants characterized by better fruit color, longer periods to fruit maturity and seedlessness have been found, particularly in apple and citrus (Donmez, *et al.*, 2013). Keeping in view the usefulness of natural mutations, artificial mutations were also induced in citrus with the help of chemical and physical mutagens. Seeds or budwood was irradiated with gamma rays and the seeds were found more tolerant to irradiation than the budwood (Hearn, 1986; Singh *et al.*, 1996). Development of a deep red and seedless cultivar, 'Star Ruby', is an example that how mutations are useful for development of new varieties (Hensz, 1981). Majority of the mutations are undesirable and the resultant desirable mutants may not be stable. Sometimes, they produce chimeric effects and it is very difficult to isolate them.

Bud variation is an abnormal variation in stems, leaves and fruits that can be carried from one generation to another through budding. Such variations arise in the vegetative tissues of the plant and are termed as bud sports on bud mutations. Bud variations originate first as single changed cells, which may develop into fruits, limbs or entire tree depending upon the place of origin of such cells. Colour, size, shape, texture, rate of growth or other physical characteristics can identify these. Many of the fruit variations in citrus have been the result of individual fruit variations in otherwise normal trees. There is ample scope to locate the desired bud variations and perpetuate them vegetatively; however; majority of bud variations are not useful.

Mutations have been recognized as one of the main driving forces of evolution (Micke *et al.,* 1987). Many of the improved citrus cultivars originated as natural budsports, indicating the potential value of induced mutations. Earlier fruiting of the mutants in comparison to zygotic seedlings is certainly a great advantage in providing a shorter breeding cycle. Chimera formation is common following the mutagenic treatment (Micke *et al.*, 1987). Irradiation could prove a valuable breeding technique in increasing mutation frequency in citrus, thus enhancing the possibility of producing new cultivars. Genetic variability masked by dominant genes at heterozygous loci could be revealed, while the disadvantages of hybridization e.g. the disruption of favourable gene combinations, nucellar embryony, incompatibility, as well as attainment of selection limit, could be circumvented.

Citrus and related genera are extremely prone to mutations, particularly the bud mutations. The induction of mutations and selection of desirable mutants are easy approaches to the improvement of citrus cultivars having one or a few undesirable traits. Seedless clones of normally seeded cultivars of Pineapple orange, Duncan and Foster grapefruit (Hearn, 1984; 1986) and Monreal Clementine (Russo *et al.,* 1981) have been selected from irradiated seeds or budwood derived plants. Changes in fruit quality and maturity period have also been induced via induced mutagenesis. As the demand for fruit quality varies, the induced mutagenesis will play an important role in future.

Mutagenesis is not targeted to a specific locus and therefore requires laborious screening of large populations to obtain individual plants exhibiting desired phenotypes. Progress in the Citrus genome project coupled with recent advances in molecular biology techniques should aid the development of faster, targeted and more efficient tools to expedite citrus breeding (Vardi, *et al.* 2008). DNA markers for traits associated with seedlessness, such as parthenocarpy, male and female sterility and self-incompatibility could be used as tools for directed breeding. These markers could be used in map-based or candidate gene approaches for the cloning of genes that encode important traits, thus expanding the list of genes that can be transformed to induce seedlessness.

Polyploidy breeding

This technique will be very useful in the production of larger and seedless fruits. Crossing tetraploid with diploid species can produce the triploids, which are generally seedless because of improper meiosis. The development of Melogold and Oroblanco grapefruit hybrids are good examples of the triploid varieties.

Biotechnology

The development of biotechnological tools has made it possible to overcome some the problems being faced in conventional breeding methodologies. Somatic hybridization (Boscariol, *et al.*, 2003; Grosser and Gmitter, 1990a, b; Mendes, *et al.*, 2001) and genetic transformation (Boscariol, *et al.*, 2003; Pena and Navarro, 1999; Costa *et al.*, 2002) have been applied in many countries for the improvement/development of citrus cultivars. Recent advances made in the field of biotechnology will be helpful in partially solving or eliminating at least some of the limitations involved in the conventional breeding procedures. In this the use of tissue culture for selection of somaclones will definitely play an important role in future breeding programmes, because, somaclonal variations arise spontaneously, or changing the composition of tissue culture medium can induce them. These variations can be used in the similar ways as to that of conventional breeding methods. Somaclones in citrus have been selected for salinity tolerance and they are under different stages of evaluation (Kochba *et al*., 1982). *In vitro* mutagenesis will definitely play an important role in eliminating particular defects in the scion or rootstock cultivars. Anther culture can be used to produce the haploid androgenic plants, which on doubling of chromosomes will produce the homozygous diploid plants. These homozygous plants will be very useful as parents in breeding programmes. Nucellus culture and *in vitro* shoot tip grafting have been utilized to produce virus free clones. Somatic hybridization/ protoplast fusion and embryo rescue techniques are used for the recovery of distant hybrids in the parents, which have pollen or ovule sterility or self and cross incompatibility. This will be useful especially in transferring the genes for biotic and abiotic stresses from related species.

Isozyme analysis is useful for distinguishing hybrids from the nucellar seedlings. Molecular markers like, RFLP and RAPD's are helpful in the separation of zygotic seedlings and specific genes can be isolated which govern resistance to biotic and abiotic stresses or for the effects on fruit quality and fruit maturity characters (Roose, 1988). Simple Sequence Repeats (SSR's) are very helpful in DNA fingerprinting and the elimination of redundancy in germplasm collections. The genes identified by using molecular markers can be directly transformed in the desired variety using *Agrobacterium* mediated transformation (Moore *et al*., 1989) or direct DNA delivery methods like biolistic gene gun (Kobayashi and Uchimiya, 1989). Bt-gene used in cotton for insect pest control, the introduction of such genes in citrus will definitely sustain the production of citrus.

Plant biotechnology is a rapidly growing area in plant biology and is a potential tool towards further increasing the horticultural production. Biotechnology will have its main impacts in horticulture by providing new strains of plants, supply of planting material, more efficient and selective pesticides and improved fertilizers. The genetic manipulation and modification of plant using new genetic engineering principles are priority areas of research (Pandey, 1993).

Recent developments in biotechnological methods of genetic manipulation helped in solving some of the limitations of the conventional breeding programmes. The various techniques used for improvement of citrus are as under:

Regeneration of whole plants and micropropagation

The amenability of citrus to regenerate via somatic embryogenesis and organogenesis is well known. The nucellar embryos and the ovules are the valuable sources for plant regeneration and plants have been regenerated from callus, induced from various tissues, such as endosperm (Gmitter *et al.,* 1990), anthers (Chaturvedi and Sharma, 1985) and stem sections (Chaturvedi and Mitra, 1974). Bowman *et al.,* (1991) have reported that regeneration of plants from juice vesicles, undeveloped ovules or seed chimeras should allow genetic transmission of previously chimerical traits. Root explants have been successfully used for whole plant regeneration in certain *Citrus* species. The establishment of embryogenic calli from the nucellar tissue of cultured ovules (Kochba *et al.*, 1972; Vardi *et al.,* 1982; 1986b; Kobayashi *et al.,* 1984) and the utilization of these calli as a source of protoplasts, which have the capacity to divide and regenerate plants (Vardi *et al.,* 1975; 1982; 1986b; Kobayashi *et al.,* 1985) paved the way to novel approaches in citrus breeding. Citrus nucellar callus has the characteristics that it does not require the auxin for sustained cell division and it has chromosomal stability (Vardi *et al.,* 1982). It also maintains embryogenic capacity for a longer time. The advantage of micropropagation are rapid clonal multiplication of elite plants, controlled conditions and year round production, freedom from pests and diseases and amenability to various manipulations, but it has certain limitations like requirement of specialized facilities, skilled personnel and is also quite costly. The reproducible techniques for the regeneration of functional plants from calli and isolated protoplasts enabled searching for somaclonal variation and comparison with nucellar variation as expressed by nucellar seedlings. In addition, this regeneration system can be exploited as an efficient genetic screening and selection system in order to obtain resistance to diseases and stresses. Sources of biotic and abiotic stress resistance exist in citrus relatives, but sexual hybridization between most of these genera with citrus proved difficult (Barrett, 1977) in addition to other problems like nucellar embryony, heterozygosity, sterility and juvenility. Cell manipulation techniques like protoplast fusion for the production of somatic hybrids (Ohgawara, *et al.,* 1985), cybrids (Vardi *et al.,* 1987) and transgenic plants may have an important and direct impact on future citrus improvement.

Virus free clonal lines

Many viruses, virus like pathogens and mycoplasmas cause serious diseases in citrus. These organisms can be eliminated from the propagative planting material by using nucellar seedlings, which are initially virus free and true to type. But owing to juvenility in the nucellar seedlings, this method becomes a long-term approach. Virus free plants, true to type to the mother plant can also be obtained via nucellar culture in purely zygotic embryo producing cultivars, because plants can be regenerated from the nucellar tissue culture (Kochba *et al.*, 1972; Vardi *et al.*, 1982; 1986b; Kobayashi *et al.,* 1984). A method of shoot tip grafting *in vitro* on a vigorous rootstock seedling is now being used for elimination of these pathogens from citrus (Navarro *et al.*, 1975; Nauer *et al.*, 1983), because viruses do not penetrate to the extreme tip of shoot and thus, virus free plants can be obtained.

Somaclonal variation and somatic hybridization

In tissue culture, plant regeneration is routed through callus, single cells or protoplasts, it may be accomplished by minor or extensive changes in the plant phenotypes termed as somaclonal variation. Somaclonal variation is now widely used for crop improvement and particularly for recovering plants resistant / tolerant to biotic and abiotic stresses.

Somaclonal variation has been observed in citrus plants regenerated from nucellar callus (Navarro *et al.,* 1985), but variation appears to be rare (Kobayashi, 1987; Vardi and Galun, 1988). Salt tolerant cell lines have been isolated and regenerated into plantlets (Kochba *et al.,* 1982). Till date, no cultivar arising through somaclonal variation has been reported. Cell culture techniques and especially *in vitro* selection can provide alternative methods for screening and selecting plants which exhibit resistance and / tolerance to biotic and abiotic stresses. Using this technique, a large number of cells can be screened rapidly in a relatively small area under controlled environment and nutrient conditions. In addition, the use of tissue culture provides a useful tool for the study at the cellular level, of mechanisms of tolerance in the isogenic cell lines. A better understanding of the cellular processes involved in stress and disease tolerance could provide information how to improve the selection criteria and shorten the selection cycle, both *in vitro* and at the whole plant level. The *in vitro* selection can be used for screening of disease resistance by using the culture filtrate of the pathogen in the selection medium, however, it is not sure that whether the results obtained *in vitro* will be similar to the results obtained *in vivo*. It indicated that for disease resistance, standardization of the procedure is very important.

The citrus systems allowed to test parallelism of results *in vivo* and *in vitro* in a wide range of citrus genotypes at different levels of organization. Salt tolerant lines have been obtained by Kochba *et al.,* (1982) in case of *C. sinensis* and *C. aurantium*. Starrantino *et al.,* (1990) reported that the performance of *in vitro* derived nucellar budlings of Navelate sweet orange was superior to other budlings for fruit quality and yield. It was possible to select stable salt tolerant callus lines of *Citrus sinensis* (Ben-Hayyim and Kochba, 1982), but regeneration of whole, salt tolerant plants has been very difficult (Ben-Hayyim and Goffer, 1989). Somatic embryos, retaining resistance to mal secco (*Phoma tracheiphila*) toxin were produced from *Citrus limon* embryogenic culture lines (Nadel and Spiegel-Roy, 1987). However, culture filtrates of *Phoma tracheiphila* and *Phytophthora citrophthora* proved ineffective as selection agents *in vitro* for disease resistance (Vardi *et al.*, 1986a; Gentile *et al.*, 1990).

Protoplast manipulation

Somatic hybridization via protoplast fusion is a powerful alternative for circumventing some of the reproductive barriers such as male or female sterility and the polyembryony (Liu, *et al.*, 2009). Protoplasts are cells without cell wall and are prepared by the digestion of cell wall with the help of enzymes. These protoplasts are capable of regeneration. Embryogenic callus in citrus is the only appropriate source of protoplasts, which can be regenerated into normal plants. Protoplasts to tree system have been developed for many *Citrus* and related species (Vardi, 1981; Vardi *et al.,* 1975; 1986b; Kobayashi *et al.*, 1984; 1985).

Protoplast fusion is an important technique to combine different types of genotypes, which are not sexually compatible. It can result in two types of plants viz. somatic hybrids in which the nuclei as well as the organelles of the two fusion partners are contained in the initial fused cell as well as in the regenerated plants (tetraploids) and cybrids, the donor - recipient protoplast fusion method (Galun and Aviv, 1986). In cybrids, the nucleus of one partner is deactivated by mutagens and the fused product has the nucleus of one parent and the cytoplasm of both the partners. Using protoplast fusion, somatic hybrids of citrus rootstocks having resistance to diseases have been produced.

Majority of the *Citrus* species of commercial importance have been regenerated from protoplasts (Kobayashi and Ohgawara,1988; Kobayashi *et al.,* 1988; Grosser *et al.,* 1989; 1990; Ohgawara *et al.,* 1989; Tusa *et al.,* 1990; Grosser *et al.*, 1992; Louzada *et al.,* 1992).

Somatic hybridization has been reported in citrus with other *Citrus* species and related genera. The techniques generally rely on polyethylene glycol (PEG) induced fusion of protoplasts (Ohgawara *et al.,* 1985; Grosser and Gmitter, 1990a, b). The regenerated plants after somatic hybridization are cybrids. Hybridity of these regenerated plants has been confirmed by chromosome counts, isozyme tests, morphology and DNA analysis. Many intergeneric hybrids involving *Citrus,Poncirus, Fortunella, Severinia* and *Citropsis* have been produced by somatic hybridization.

Anther culture

Androgenic plants, derived from haploid microspores of *in vitro* cultured anthers have a great potential for citrus improvement, because this will produce hemizygous plants, which will lead to homozygous citrus trees with specific advantageous traits. Such plants will be very useful in species like trifoliate orange that possesses resistance to *Phytophthora,* nematodes, tristeza and cold. Such homozygous plants will be of special value in the production of hybrids of trifoliate orange for the particular character. Androgenic plants have been produced in *Citrus* and *Poncirus* (Hidaka *et al.,* 1979; 1982; Hidaka 1984a; 1984b).

Genetic transformation

Citrus trees are frequently threatened by biotic and/or abiotic stresses that cause tremendous loss to crop loading and limit geographic distribution. Therefore, production of germplasm with enhanced stress tolerance will be of critical importance for citrus breeding and genetic improvement in the long run. Several strategies exist for the genetic improvement of citrus including conventional breeding and genetic transformation (Dutt and Grosser, 2010; Pena, *et al.* 2007). The genetic nature of a plant species can be altered and genetic diversity can be increased by transferring DNA from one plant cell to the cells of another. It can be accomplished by *Agrobacterium* mediated genetic transformation or by direct DNA delivering methods like biolistic gun.

Genetic transformation is a non-conventional method for genetic manipulation of plants. Once a transformation system is in place for a given cultivar or species, any gene can be inserted in the genome. The list of characters that could potentially be manipulated by genetic transformation of citrus is identical with the list of breeding objectives for this crop. Moore *et al.,* (1989) have developed an *Agrobacterium* mediated transformation system for *Citrus* and *Citrus* x *Poncirus* hybrids similar to the protocols that have been successful with other plant species. Carrizo citrange was extensively used for transformation studies, because it is easily regenerable from seedling stem segments (Moore, 1986) and produces high frequency of nucellar seedlings (Moore and Castle, 1988) and seeds are readily available.

Kobayashi and Uchimiya, (1989) transformed citrus protoplasts by direct DNA transfer methods. Hidaka *et al.*, (1990) transformed embryogenic callus by co-cultivation of sweet orange cell suspensions with two *Agrobacterium* strains, harbouring binary and co-integrating vector plasmids. All these experiments indicated that it is possible to transform citrus using different transformation methods.

Transformation efficiencies are generally low and protocols are dependent on species or even cultivar dependent (Donmez, *et al.* 2013). Development of effective genetic transformants requires specific studies on *in vitro* regeneration conditions for each genotype.

Biochemical and molecular markers

Knowledge of genetic variation and genetic relationship among genotypes is an important consideration for classification, utilization of germplasm resources and breeding. Without determining the diversity reliably, it would not be possible to identify molecular markers or quantitative trait associations. Moreover the viability and purity of rootstocks can be analyzed through the utilization of fingerprints based on molecular markers (Baig, *et al.*, 2009). The phylogeny and taxonomy of citrus are complex, confusing and controversial due to the polyembryonic nature and the long generation time needed to carry out selection and recombination (Nicolisi, *et al.*, 2000), therefore, analysis of the genetic diversity of citrus fruit is crucial.

Identification of hybrids resulting from the crosses between polyembryonic parents is extremely difficult when neither parent has a convenient dominant character. Scientists have developed different procedures, including morphologic (Hearn, 1977), chromatographic (Tatum, *et al.*, 1974) and isoenzymatic (Torres, *et al.,* 1982; Anderson, *et al.*, 1991) methods to recognize different hybrids.

Isozyme analysis is useful and reliable for distinguishing hybrids from nucellar seedlings. This method is quite accurate for identifying hybrids, including most interspecific hybrids, but is unable to distinguish between closely related cultivars. RFLP method is used to develop maps of the citrus genome that may be useful in locating genes that have a specific function. Marker genes are identified on specific chromosomes, some of which are tightly linked to other genes responsible for desirable traits like disease resistance. By mapping these markers, the genes can be located on the chromosomes that will then further be used for identification of the desired hybrids. After the location of a specific gene is ascertained, it can be easily transformed in other cells via bacterial plasmids.

For discrimination among citrus cultivars, gas chromatography analysis of leaf oils, qualitative analysis of flavanones, high performance liquid chromatography, analysis of flavanone glycosides and isozyme analysis have been reported. However, such analysis cannot be used for plants in which the markers are common or similar between the parents. PCR- based markers have been used to discriminate citrus cultivars and species (Deng *et al.,* 1995; Omura *et al.*,1993). Sugawara *et al.*, (1995) used RADP markers to identify citrus chimeras.

Due to high frequency of bud mutations in citrus, sports may be propagated vegetatively and maintained by budding, which can lead to small, mutation-based differences among varieties within cultivated species. Little genetic variation was detected within important cultivated species, *C. sinensis* and *C. paradisi* when examined by microsatellite based markers (Kijas *et al.*, 1995;

Luro *et al*., 1995; Fang and Roose, 1997). Organellar RFLPs have been used to study phylogenetic relationships of citrus (Green *et al.*, 1986) and citrus and its relatives (Yamamoto *et al.,* 1993). RAPDs have been used for gene mapping (Cai *et al*., 1994) and cultivar identification (Omura *et al*., 1993; Deng *et al*., 1995) in citrus. Using RFLP and RAPD analysis, Federici *et al*., (1998) reported that several accessions have been probably assigned to the wrong species. Using molecular markers, Fang and Roose (1999), studied the inheritance of citrus tristeza virus resistance in the Chandler pummelo. Using RAPD markers, Machado *et al*., (1996) reported that Mediterranean mandarins are true hybrids of common mandarins. Motahashi *et al*., (1992) used RFLP to investigate genetic relationship between hybrids and parents of *Citrus* species. Though RAPD markers proved to be useful for germplasm characterization and diversity analysis in citrus, use of other molecular techniques such as AFLP, ISSR and SSRs should be considered for finer molecular analysis of genotypes and to solve discrepancies left by unresolved by RAPDs (Baig, *et al*., 2009).

Conclusion

Advances in citrus breeding will be slow and will require long range planning and efforts, due to the quantitative nature of character inheritance, the long generation time and the presence of nucellar embryony. Information on several of these problems is now available. Significant advancements have already been made in the improvement of citrus using selection and hybridization methods. New information is now available for desirable genetic manipulation in citrus. Knowledge on the nature and range of inheritance of desirable characters such as *Phytophthora*, nematode and cold resistance is building up. The number of sexual hybrids available as parents has been greatly increased. There is free exchange of germplasm within and among countries, which is broadening the genetic base for citrus improvement. Much progress has not been made in reducing the long juvenile period.

The use of isozyme analysis had greatly improved the ability to identify zygotic seedlings from nucellar ones. DNA RFLP's as genetic markers should greatly enhance the identification of zygotic seedlings and early identification of hybrid seedlings with desirable characters. *In vitro* culture has contributed in recovery of embryos that otherwise fail. Somatic hybridization is expanding the range of germplasm availability to the breeders and is likely to provide new sources of disease resistance for rootstock breeding and tetraploids for the production of triploids in scion breeding. Genetic transformation will provide the breeders with the ability to correct defects of other characters. Mutation breeding is very promising in developing seedless cultivars without affecting other characters. For Indian conditions, rootstock breeding for resistance to *Phytophthora,* nematodes, stress tolerance and dwarfness will be the priority research areas.

The experiments based on cytological differences in chromosomes of hybrids, triploids, self and cross incompatibility will be the targets for future studies. Slowness to fruit of the seedlings and separation of zygotic and nucellar seedlings are the most important research areas, which require immediate attention. For early fruiting, the genes present in other varieties or related species can be transformed into the desired variety with the help of biotechnological techniques. Separation of hybrid seedlings will definitely improve with the help of molecular markers, although, they can be separated using chromatography, spectrophotometry or electrophoresis. Cross incompatibility can be overcome by protoplast fusion or embryo rescue techniques. For creating *de novo* variability, the techniques of tissue culture, particularly somaclonal variation and site- directed *in vitro* mutagenesis will be very helpful. Identification of genes for bitterness, fruit colour, flavour, earliness, biotic and abiotic stresses should be easy with the help of molecular markers and marker- assisted selection will be effective in selecting the desirable parents and / hybrids. Ploidy manipulation will be another approach to have seedless cultivars, because, triploids are known to produce seedless fruits.

As the traditional breeding methodologies are hampered by some unique barriers, modern biotechnology, particularly cell engineering, are showing their increasingly important role in citrus cultivar improvement. Novel germplasm originated from such state-of-the-art techniques will be of paramount importance for citrus cultivar improvement (Liu, *et al.* 2009). The newer methods developed in recent years will not replace the conventional plant breeding methods of hybridization and selection, but they will complement them. The breeding process will comparatively be faster and the distant parents can be crossed.

References

Agarwal, P.K. 1984. Meiosis in an interspecific mandarin hybrid Kinnow (*Citrus nobilis* Lour. x *C. deliciosa* Tan.). *Curr. Sci.* 53: 547-548.

Agarwal, P.K. 1987a. Karyotype of *Citrus tamurana* Tan. *Chrom. Info. Serv.* 42: 3-5.

Agarwal, P.K. 1987b. Cytogenetical investigations in Rutaceae II. Meiotic studies in three intergeneric hybrids of *Citrus*. *Cytologia* 52: 757-760

Agarwal, P.K. 1987c. Improvement of citrus rootstocks by breeding - Evaluation of eleven rootstock hybrids. *Indian J. Hort.* 44:165-168.

Agarwal, P. K. 1988a. Cytomorphy of *Citrus indica* Tan. *Second All India Conf. Cytol. Genet., Warangal* p 2.

Agarwal, P.K. 1988b. Pachytene morphology of *Severinia buxifolia* Tan. *Chrom. Info. Serv.* 44:24-26.

Agarwal, P.K. 1988c. Short juvenility in *Citrus- Poncirus* hybrid. *Indian J. Hort*. 45: 249-251.

Agarwal, P.K. 1989. Cytogenetical investigations in Rutaceae V. Cytomorphology of three intergeneric hybrids of *Citrus*. *Cytologia* 54: 705-708.

Agarwal, P. K., Reddy, P.P. and Sawant, S. D. 1987. Varietal evaluation and improvement of citrus rootstocks. *Workshop All India Coordinated Research Project on Tropical Fruits. Research Report on Citrus, Banana, Pineapple and Papaya. 27-30 July*, 1987.p. 15-22. College of Agriculture, Punjabrao Krishi Vidyapeeth Nagpur/ Indian Instt. Hort. Research, Bangalore.

Anderson, C.M., Castle, W.S. and Moore, G.A. 1991. Isozymic identification of zygotic seedlings in Swingle citrumelo, Citrumelo x *Poncirus trifoliata* nursery and field populations. *J. Amr. Soc. Hort. Sci*. 116: 322-326.

Anonymous. 1985-86. *Annual Report* 1985-86, Department of Agricultural Research and Education, Ministry of Agriculture, Govt. of India, New Delhi, p. 45.

Anonymous, 1986 b. Genetic resources of tropical and subtropical fruits and nuts (excluding Musasa). IBPGR Rome. 105-115 pp.

Anonymous. 1987. *Workshop of the All India Coordinated Research Project on Tropical fruits. Research Report on Citrus, Banana, Pineapple and Papaya. 27-30 July* 1987, p. 4.

Anonymous, 1989-90b. Annual Report. NBPGR, New Delhi. 108 pp.

Anonymous, 2012. Citrus fruit: fresh and processed. Annual Statistics *CCP:/CT/ST/2012*, Food and Agriculture Organization of United Nations Organization.

Austin, R.B. 1988. New opportunities in breeding. *Hort Sci*.23 (1): 41-45.

Badiyala, S. D., Lakhanpal, S. C. and Bhargava, J. N. 1990. Seedless Kagzi lime (*Citrus aurantifolia* Swingle)- A new promising selection. *Abstracts, International Seminar New Frontiers Horticulture, Bangalore*, p. 69.

Bagde, T. R. and Patil, V. S. 1989. Chakradhar lime - A new thornless and seedless selection in lime (*Citrus aurantifolia* Swingle). *Annals Plant Physiol*. 3: 95-97.

Baig, M.N.R., Grewal, S. and Dhillon, S. 2009. Molecular characterization and genetic diversity analysis of citrus cultivars by RAPD markers. *Turk. J. Agric*. 33: 375-384.

Bajpai, P. N., Trivedi, R. K. and Prasad, A. 1963. Storage of citrus seeds. *Sci. Cult*. 29: 49.

Barrett, H.C.1977. Intergeneric hybridization of *Citrus* and other genera in citrus variety improvement. *Proc. Int. Soc. Citriculture*2: 585-589.

Barrett, H.C. 1981. Breeding cold-hardy citrus scion varieties. *Proc. Int. Soc. Citriculture*1:61-68.

Barrett, H. C. 1985. Hybridization of *Citrus* and related genera. *Fruit Varieties J*. 39: 11-16.

Barrett, H. C. and Rhodes, A. M. 1976. A numerical taxonomic study of affinity relations in cultivated *Citrus* and its close relatives. *Syst. Bot*. 1: 105-136.

Ben-Hayyim, G. and Goffer, Y. 1989. Plantlet regeneration from a NaCl selected salt tolerant callus culture of Shamouti orange (*Citrus sinensis* [L.] Osb.). *Plant Cell Reports* 7:680-683.

Ben-Hayyim, G. and Kochba, J. 1982. Growth characteristics and stability of tolerance of citrus callus cells subjected to NaCl stress. *Plant Science Letters*27: 87-94.

Bhansali, R. R. and Arya, H. C. 1978. Tissue culture propagation of citrus tree. In: *Proc. Int. Soc. citriculture*, Griffith, Australia. 135-140.

Bhattacharya, S. C. and Dutta, S. 1956. Classification of citrus fruits of Assam. Government India Press, Delhi, *Sci. Monog*. No. 20. pp110.

Bhekasut, C., Singh, R. and Sharma, B. B. 1976. Germination studies in citrus seeds. *Indian J. Hort*. 33: 37-40.

Boscariol, R.L., Almeida, W.A.B., Derbyshire, M.T.V.C., Mourao Filho, A.A. and Mendes, B.J.M. 2003. The use of the PMI/mannose selection system to recover transgenic sweet orange plants (*Citrus sinensis* L. Osbeck). *Plant Cell Reports*22 (2): 122-128.

Bowman, K. D., Gmitter, F. G. Jr., Moore, G. A. and Rouseff, R. L. 1991. Citrus fruit sector chimeras as a genetic resource for cultivar improvement. *J. Am. Soc. Hort. Sci*. 116 (5):888-893.

Cai, Q., Guy, C. L. and Moore, G. A. 1994. Extension of the linkage map in *Citrus* using random amplified polymorphic DNA (RAPD) markers and RFLP mapping of cold acclimation-responsive loci. *Theor. Appl. Genet.*89: 604-614.

Chadha, K. L. and Singh, H. P. 1990. Citriculture scenario of India. (In) Citriculture in North Western India. *Proc. Citrus Show-cum Seminar, Prospects, Problems Kinnow cultivation* (K. S. Gill; J. S. Kanwar and R. Singh, eds.). pp 21-64.

Chadha, K. L., Randhawa, N. S., Bindra, O. S., Chohan, J. S. and Knorr, L.C. 1970. *Citrus decline in India-Causes and control*. PAU, Ludhiana, Tech. Bull. P. 97.

Chakrawar, V. R. and Rane, D.A. 1977. Bud mutations in Nagpur mandarin orange. *Res. Bull. MAU.* 1(2): 22-23.

Chakrawar, V. R., Patil, V. K. and Shinde, G. S. 1988. Horticultural development in Maharashtra region. *Scientific Seminar*, Aurangabad. pp 13.

Chaturvedi, H.C. and Mitra, G.C. 1974. Clonal propagation of *Citrus* from somatic callus cultures. *Hort Sci.* 9: 118-120.

Chaturvedi, H.C. and Sharma, A.K. 1985. Production of androgenic plants of *Citrus aurantifolia*. *J. Plant Physiol.* 119: 473-477.

Chaturvedi, H.C., Chowdhury, A.R. and Mitra, G.C. 1974a. Morphogenesis in stem callus tissue of *Citrus grandis* in long term culture-a biochemical analysis. *Curr. Sci.* 43 (5): 139-142.

Chaturvedi, H.C., Chowdhury, A.R. and Mitra, G.C. 1974b. Shoot bud differentiation in stem callus tissue *Citrus grandis* and correlated changes in its free-amino acid content. *Curr. Sci.* 43 (17): 536-537.

Costa, M.G.C., Otoni, W.C. and Moore, G.A. 2002. An evaluation of factors affecting the efficiency of *Agrobacterium*-mediated transformation of *Citrus paradisi* (Macf.) and production of transgenic plants containing carotenoid biosynthesis genes. *Plant Cell Reports* 21(4):365-373.

Dass, H.C., Awtar Singh and Vijayakumari, N. 1998. Rootstock breeding-variations in leaf morphology in citrus rootstock hybrid progeny. *Indian Journal of Horticulture* 55(1):16-19.

Deng, Z.N., Gentile, A., Nicolosi, F., Domina, F., Vardi, A. and Tribulato, F. 1995. Identification of *in vivo* and *in vitro* lemon mutants by RAPD markers. *J. Hort. Sci.*70:117-125.

Donmez, D., Simsek, O., Izgu, T., Kacar, Y.A. and Mendi, Y.Y. 2013. Genetic transformation in citrus. *The Scientific World Journal*, Volume 2013, Article ID 491207, pp. 8.

Dutt, M. and Grosser, J.W. 2010. An embryogenic suspension cell culture system for *Agrobacterium*-mediated transformation of citrus. *Plant Cell Reports.* 29(11): 1251-1260.

Engleman, F., Dambier, D. and Ollitraut, P. 1994. Cryopreservation of cell suspension and embryonic calluses of *Citrus* using a simplified freezing process. *Cryo-Letters.* 15: 53-58.

Fang, D.Q. and Roose, M. L. 1997. Identification of closely related *Citrus* cultivars with inter-simple- sequence repeat markers. *Theor. Appl. Genet.* 95:408-417.

Fang, D.Q. and Roose, M.L. 1999. A novel gene conferring citrus tristeza virus resistance in *Citrus maxima* (Burm.) Merrilly. *Hort Sci.* 34 (2): 334-335.

Fang, D.Q., Federici, C.T. and Roose, M.L. 1997. Development of molecular markers linked to a gene controlling fruit acidity in citrus. *Genome*. 40: 841-849.

Fang, D.Q., Federici, C.T. and Roose, M.L. 1998. A high-resolution linkage map of the Citrus Tristeza Virus resistance gene region in *Poncirus trifoliata* (L.) Raf. *Genetics*. 150: 883-890.

Fang, D.Q., Krueger, R.R. and Roose, M.L. 1998. Phylogenetic relationships among selected citrus germplasm accessions revealed by inter-simple sequence repeat markers. *J. Am. Soc. Hort. Sci.* 123: 612-617.

Fang, D.Q., Roose, M.L., Krueger, R.R. and Federici, C.T. 1997. Fingerprinting trifoliate orange germplasm accessions with isozymes, RFLPs and inter-simple sequence repeat markers. *Theor. Appl. Genet.* 95: 211-219.

Federici, C. T., Fang, D. Q., Scora, R.W. and Roose, M. L. 1998. Phylogenetic relationships within the genus *Citrus* (Rutaceae) and related genera as revealed by RFLP and RAPD analysis. *Theor. Appl. Genet.* 96: 812-822.

Ford, H. W. 1978. Burrowing nematode resistant hybrids of *Poncirus trifoliata*, Milam and Ridge pineapple. *Proc. Fla. State Hort. Soc.*91: 20-21.

Ford, H. W. and Feder, W. A. 1960. *Citropsis gilletiana*, a citrus relative resistant to the burrowing nematode in laboratory tests. *Proc. Fla. State Hort. Soc.* 73: 60-64.

Frost, H.B. and Soost, R.K. 1968. Seed production: development of gametes and embryos. (In) H. W. Reuther, H. J. Webber and L. D. Batchelor (eds.). *The Citrus Industry*, University of California, USA. 2: 290-324.

Galun, E. and Aviv, D. 1986. Organelle transfer. (In) A. Weissbach, and H. Weissbach (eds.). *Methods in Enzymology*. Academic Press, Orlando 188: 595-611.

Ganeshan, S. and Rajashekharan, P. E. 1999. Conservation of nuclear genetic diversity (NGD) in citrus genepool: implications in citrus breeding and genetic conservation. In: High-Tech Citrus Management, Proc, Int. Symp. Citriculture, 23-27 Nov, Nagpur, India. 85-90.

Gentile, A., Vardi, A., Tribulato, E. and Continella, G. 1990. Responses of citrus calli and protoplasts to *Phoma tracheiphila* culture filtrate and toxin. *XXIII International Hort. Congress, Florence* P. 134 (Abstract).

Gmitter, F.G., Jr., Grosser, J.W. and Moore, G.A. 1992. *Citrus,* (In) *Biotechnology of Pernennial Fruit Crops*, F.A. Hammerschlag and R.E. Litz (eds.), CAB International, Walligford, UK.

Gmitter, F.G. Jr., Ling, X.B. and Deng, X.X. 1990. Induction of triploid *Citrus* plants from endosperm calli *in vitro*. *Theor. Appl. Genet.* 80: 785-790.

Gmitter, F.G. Jr., Krueger, R. R. and Roose, M. L. 1999. Citrus germplasm characterization by phenotype and molecular markers. pp 46-62. (In) Broadbent *et al.* (1999).

Golein, B., Fifaei, R. and Ghasemi, M. 2011. Identification of zygotic and nucellar seedlings in citrus interspecific crosses by inter simple sequence repeats (ISSR) markers. *African J. Biotech.* 10 (82): 18965-18970.

Green, R. M., Vardi, A. and Galun, E. (1986). The plastome of *Citrus*: Physical map, variation among *Citrus* cultivars and species, and comparison with related genera. *Theor. Appl. Genet.* 72: 170-177.

Grosser, J. W. and Gmitter, F. G. Jr. 1990a. Protoplast fusion to citrus improvement. *Plant Breed. Rev.* 8: 339-374.

Grosser, J. W. and Gmitter, F. G. Jr. 1990b. Somatic hybridization of Citrus with wild relatives for germplasm enhancement and cultivar development. *Hort Sci.*25: 147-151.

Grosser, J.W., Gmitter, F. G. Jr., Seto, F., Deng, X. X. and Chandler, J. L. 1992. Six new somatic citrus hybrids and their potential for cultivar improvement. *J. Am. Soc. Hort. Sci.*117: 169-173.

Grosser, J. W., Gmitter, F. G. Jr., Tusa, N. and Chandler, J. L. 1990. Somatic hybrid plants from sexually incompatible woody species: *Citrus reticulata* and *Citropsis gilletiana*. *Plant Cell Reports* 8: 656-659.

Grosser, J.W., Moore, G. A. and Gmitter, F. G. Jr. 1989. Interspecific somatic hybrid plants from the fusion of Key lime (*Citrus aurantifolia*) with Valencia sweet orange (*Citrus sinensis*) protoplasts. *Sci. Hort.* 39: 23-29.

Grosser, J. W., Mourao-Fo, F. A. A., Gmitter, F. G. Jr., Louzada, E. S., Jiang, J., Baergen, K. Quiros, A., Cabasson, C., Schell, J. L. and Chandler, J. L. 1996. Autotetraploid hybrids between *Citrus* and seven related genera produced by somatic hybridization. *Theor. Appl. Genet.* 92: 577-582.

Hearn, C.J. 1977. Recognition of zygotic seedlings in certain orange crosses by vegetative characters. *Proc. Int. Soc. Citriculture*2: 611-614.

Hearn, C. J. 1984. Development of seedless orange and grapefruit cultivars through seed irradiation. *J. Am. Soc. Hort. Sci.*109: 270-273.

Hearn, C. J. 1985. Citrus scion improvement programme. *Fruit Varieties J.* 39: 34-37.

Hearn, C. J. 1986. Development of seedless grapefruit cultivars through budwood irradiation. *J. Am. Soc. Hort. Sci.*111: 304-306.

Hensz, R. A. 1977. Mutation breeding and the development of the 'Star Ruby' grapefruit. *Proc. Int. Soc. Citriculture* 2: 582-585.

Hensz, R. A. 1981. 'Star Ruby', a new deep-red-fleshed grapefruit variety with distinct tree characteristics. *J. Rio Grande Vall. Hort. Soc.* 25: 54-58.

Hidaka, T. 1984a. Induction of plantlets from anthers of Trovita orange (*Citrus sinensis* Osbeck). *Japan. J. Breed.*34: 416-420.

Hidaka, T. 1984b. Effects of sucrose concentration, pH of media and culture temperature on anther culture of *Citrus*. *Japan. J. Breed.* 34: 416-422.

Hidaka, T. and Omura, M. 1989. Control of embryogenesis in *Citrus* cell culture. Regeneration from protoplast and attempts to callus bank. *Bull. Fruit Tree Res. Stn.* B-16: 1-17.

Hidaka, T. and Omura, M. 1993. Transformation of *Citrus* protoplasts by electroporation. *J. Japan. Soc. Hort. Sci.* 62: 371-376.

Hidaka T., Omura, M., Ugaki, M. Tomiyama, M., Kato, A., Oshima, M. and Motoyoshi, F. 1990. *Agrobacterium*-mediated transformation and regeneration of *Citrus* species from suspension cells. *Japan. J. Breed.* 40: 199-207.

Hidaka, T., Yamada, Y. and Schichijo, T. 1979. *In vitro* differentiation of haploid plants by anther culture in *Poncirus trifoliata* (L.) Raf. *Japan. J. Breed.*29: 248-254.

Hidaka, T., Yamada, Y. and Schichijo, T. 1982. Plantlet formation by anther culture of *Citrus aurantium* L. *Japan. J. Breed.* 32: 247-252.

Hodgson, R. W. 1967. Horticultural varieties of citrus. In H. W. Reuther, H. J. Webber and L. D. Batchelor (eds.). *The Citrus Industry*. vol. 1. Univ. California Press, Berkeley pp. 431-591.

Hong, T. D. and Ellis, R. H. 1995. Interspecific variation in seed storage behaviour within two genera- Coffee and Citrus. *Seed Sci. Technol.*26: 165-191.

Hutchison, D. J. 1985. Rootstock development, screening and selection for disease tolerance and horticultural characteristics. *Fruit Varieties J.* 39: 21-25.

Hutchison, D. J. and Grimm, G. R. 1973. Citrus clones resistant to *Phytophthora parasitica* : 1973 screen results. *Proc. Fla. State Hort. Soc.* 86: 88-91.

Hutchison, D. J. and O'Bannon, J, H. 1972. Evaluating the reaction of citrus selections to *Tylenchulus semipenetrans*. *Plant Dis. Rep.*56: 747-751.

Iwanami, T., Hidaka, T. and Omura, M. 1993. Shoot tip culture of *Citrus* III. Elimination of Citrus Tatter leaf Virus from cultured shoots. *Bull. Fruit Tree Res. Stn.* 24: 61-67.

Iyer, C.P.A., Singh, H.P., Jalikop, S.H., Naidu, R. and Srivastava, K. C. 1983. Breeding for rootstock improvement in citrus. *Proc. Int. Citrus Symp., Bangalore*, pp.179-88.

Kapur, S. P., Anand, S. A., Cheema, S. S. and Kapoor, S. K. 1977. Prevalence of greening marker substance in different species of *Citrus*. *Indian J. Hort.* 34: 205-208.

Kijas, J.M.H., Fowler, J.C.S., and Thomas, M. R. 1995. An evaluation of sequence tagged microsatellite site markers for genetic analysis within *Citrus* and related species. *Genome* 38: 349-355.

King, M.W. and Roberts, E. H. 1979. The storage of recalcitrant seeds- achievements and possible approaches. International Board of Plant Genetic Resources, Rome.

King, M.W., Soetisna, V. and Roberts, E.H. 1981. The dry storage of citrus seeds. *Ann. Bot.* 48: 865-872.

Kobayashi, S. 1987. Uniformity of plants regenerated from orange (*Citrus sinensis* Osbeck) protoplasts. *Theor. Appl. Genet.* 74:10-14.

Kobayashi, S. *et al.* 1983. Plant regeneration from Trovita orange protoplasts. *Japan. J. Breed.* 33: 119-122.

Kobayashi, S. and Ohgawara, T. 1988. Production of somatic hybrid plants through protoplast fusion in *Citrus*. *Japan Agric. Res. Quarterly* 22 (3):181-188.

Kobayashi, S. and Uchimiya, H. 1989. Expression and integration of a foreign gene in orange (*Citrus sinensis* Osbeck) protoplasts by direct DNA transfer. *Japan. J. Genet.* 64: 91-97.

Kobayashi, S., Ikeda, I. and Nakatani, M. 1984. Induction of nucellar callus from orange (*Citrus sinensis* Osbeck) ovules and uniformity of regenerated plants. *Bull. Fruit Tree Res. Station, Series E* (Okitsu) 5: 43-54.

Kobayashi, S., Ikeda, I. and Uchiyama, H. 1985. Conditions for high frequency embryogenesis from orange (*Citrus sinensis* Osbeck). *Plant Cell, Tissue, Organ Culture* 4: 249-259.

Kobayashi, S., Ohgawara, T., Ohgawara, F., Oiyama, I. and Ishii, S. 1988. A somatic hybrid plant obtained by protoplast fusion between navel orange (*Citrus sinensis*) and satsuma mandarin (*C. unshiu*). *Plant Cell Tissue Organ Culture* 14: 63-69.

Kobayashi, S., Sakai, A. and Ohyama, I. 1990. Cryopreservation in liquid nitrogen of cultured navel orange (*Citrus sinensis* Osb.) nucellar cells and subsequent plant regeneration. *Plant Cell, Tissue and Organ Culture*. 23: 15-20.

Kochba, J., Ben-Hayyim, G, Spiegel-Roy, P., Neumann, H. and Saad, S. 1982. Selection of stable salt tolerant callus lines and embryos in *C. sinensis* and *C. aurantium.Z. pflanzenphysiol.* 106: 237-245.

Kochba, J., Spiegel-Roy, P. and Safran, H. 1972. Adventive plants from ovules and nucelli in *Citrus*. *Planta* 106: 237-245.

Ling, P., Gmitter, F. G. Jr., Duncan, L.W. and Xiao, S.Y. 1994. Inheritance of citrus nematode resistance and its linkage with RAPD markers in citrus. *Hort Sci.*29: 483.

Liu, J.H., Fu, X.Z., Huang, X.S., Wang, J. and Fan, Q.J. 2009. Citrus germplasm enhancement via cell engineering and genetic engineering- progress report. Geijskes *et al.* (eds.) *Proc. VIth IS on In Vitro Cult. and Hort. Breed. Acta Hort.* 829: 133-138.

Louzada, E.S., Grosser, J.W., Gmitter, F. G. Jr., Deng, X. X., Tusa, N., Nielsen, B. and Chandler, J. L. 1992. Eight new somatic rootstocks with potential for improved disease resistance. *Hort Sci.* 27: 1033-1036.

Luro, F., Laigret, F., Bove, J.M. and Ollitrault, P. 1995. DNA amplified fingerprinting, a useful tool for determination of genetic origin and diversity in *Citrus.Hort Sci.* 30:1063-1067.

Machado, M. A., Coletta Filho, H. D., Targon, M. L. P. N. and Pompeu, J.Jr. 1996. Genetic relationship of Mediterranean mandarins (*Citrus deliciosa* Tenore) using RAPD markers. *Euphytica* 92: 321-326.

Machado, M.A., Cristofani-Yaly, M. and Baseianel, M. 2011. Breeding, genetic and genomic of citrus for disease resistance. *Rev. Bras. Fruitic., Jaboticabai*, SP. Volume Especial, E. 158-172.

Mani, A. 1989. Reaction of Rangpur lime and rough lemon strains to citrus nematode, *Tylenchulus semipenetrans* Cobb. *Intern. Nematol. Network Newsl.* 6(3):22-23.

Marin, M.L. and Duran-vila, N. 1988. Survival of somatic embryos and recovery of plants of sweet orange (*Citrus sinensis* (L.) Osbeck) after immersion in liquid nitrogen. *Plant Cell, Tissue and Organ Culture*. 14:51-57.

Matsuyama, T., Motohashi, R., Akihama, T. and Omura, M. 1992. DNA fingerprinting in *Citrus* cultivars. *Japan. J. Breed.* 42: 155-159.

Mendes, B.M.J. Mourao-Filho, F.D.A.A., Farias, P.C.D.M. and Benedito, V.A. 2001. Citrus somatic hybridization potential for improved blight and CTV resistance. *In Vitro Cellular and Development Biology- Plant*37 (4): 490-495.

Micke, A., Donini, B. and Maluszynski, M. 1987. Induced mutations for crop improvement- a review. *Trop. Agric.* (Trinidad) 64: 259-278.

Moore, G. A. 1986. *In vitro* propagation of *Citrus* rootstocks. *Hort Sci.* 21:300-301.

Moore, G.A. and Castle, W.S. 1988. Morphological and isozymic analysis of open pollinated *Citrus* rootstock populations. *J. Hered.* 79:59-63.

Moore, G.A., DeWald, M.G. and Cline, K. 1989. *Agrobacterium*- mediated transformation of *Citrus*. *J. Cellular Biochem*. Supplement 13D, p. 255 (Abstract).

Motohashi, R., Matsuyama, T. and Akihama, T. 1992. DNA fingerprinting in *Citrus* cultivars. *Proc. Int. Soc. Citriculture* pp 121-124.

Mumford, P. M. and Grout, B. W. W. 1979. Desiccation and low temperature (-196^0 C) tolerance of *Citrus limon* seed. *Seed Sci. Technol*. 7: 407-410.

Nadel, B. and Spiegel-Roy, P. 1987. Selection of *Citrus limon* cell culture variants resistant to the mal secco toxin. *Plant Sci.* 53: 177-182.

Nicolisi, E., Deng, Z.N., Gentile, A. and La Malfa, S. 2000. Citrus phylogeny and genetic origin of important species as investigated by molecular markers. *Theor. Appl. Genet.* 100:1155-1166.

Nauer, E.M., Roistacher, C.N., Carson, T. L. and Murashige, T. 1983. *In vitro* shoot tip grafting to eliminate citrus viruses and virus-like pathogens produces uniform bud lines. *Hort Sci.* 18: 308-309.

Navarro, L., Ortiz, J.M. and Juarez, J. 1985. Aberrant citrus plants obtained by somatic embryogenesis of nucelli cultured *in vitro*. *Hort Sci.* 20: 214-215.

Navarro, L., Roistacher, C. N. and Murashige, T. 1975. Improvement of shoot tip grafting *in vitro* for virus-free citrus. *J. Am. Soc. Hort. Sci.* 100: 471-479.

Norman, M. N. and Hamidah, S. 1992. Cryopreservation of *Garcinia mangostana* L. and *Citrus halimii* embryos. *Reproductive Biology and Plant Breeding Book of Poster Abstract. XIIIth EUCARPIA Congress*, Angers, France pp. 423-424.

Ohgawara, T., Kobayashi, S., Ohgawara, E., Uchiyama, H. and Ishii, S. 1985. Somatic hybrid plants obtained by protoplast fusion between *Citrus sinensis* and *Poncirus trifoliata. Theor. Appl. Genet.* 71: 1-4.

Ohgawara, T., Kobayashi, S., Ishii, S., Yoshinaga, K. and Oyiama, I. 1989. Somatic hybridization in *Citrus*: Navel orange (*C. sinensis* Osb.) and grapefruit (*C. paradisi* Macf.). *Theor. Appl. Genet.*78: 609-612.

Omura, M. and Hidaka, T. 1992. Shoot tip culture of Citrus. I. Culture conditions. *Bull. Fruit Tree Res. Stn*. 22: 23-36.

Omura, M., Hidaka, T., Nesumi, H., Yoshida, H. and Nakamura, I. 1993. PCR markers for *Citrus* identification and mapping. In : T. Hayashi, M. Omura, N. S. Scott (eds.) *Techniques on gene diagnosis and breeding in fruit trees*. Fruit Trees Research Station, Okitsu, Japan pp. 66-73.

Pandey, R. M. 1993. Biotechnology in Horticulture. In : *Souvenir, Golden Jubilee Symp., Horticultural Research-Changing Scenario, Bangalore, 24-28 May, 1993* pp. 21-28.

Pena, L., Cervera, M., Ghorbel, R. et al. 2007. Genetic transformation. (In) *Citrus Genetics, Breeding and Biotechnology*, I.A. Khan (ed.), pp. 329-344.

Pena, L. and Navarro, L.1999. Transgenic citrus. In: Transgenic Trees, YPS, Bajaj (ed.), Springer, Berlin, pp. 39-54.

Perea-Molphe-Balch and Ochoa-Alejo, N. 1998. Regeneration of transgenic plants of Mexican lime from *Agrobacterium rhizogenes*- transformed tissues. *Plant Cell Reports*17 (8):591-596.

Radhamani, J. and Chandel, K. P. S. 1992. Cryopreservation of embryogenic axes of trifoliate orange (*Poncirus trifoliata* (L.) Raf.). *Plant Cell Reports*. 11: 371-374.

Radhamani, J., Malik, S.K. and Chandel, K. P.S. 1991. Seed coat characteristics in relation to the physiology of seed germination in *Citrus* and its allied genus. *Seed Sci. Technol*. 19: 611-621.

Roberts, E.H., King, M.W. and Ellis, R.H. 1984. Recalcitrant seeds: their recognition and storage. In: *Crop Genetic Resources: Conservation and Evaluation* (H. H. W. Holden and Williams, Eds.). George Allen and Unwin, London. 38-62 pp.

Roose, M. L. 1988. Isozymes and DNA restriction fragment length polymorphism in *Citrus* breeding and systematics. *Proc. Int. Soc. Citriculture*1: 57-67.

Russo, R., Donini, B. and Starrantino, A. 1981. Mutagenesis applied for citrus improvement. *Proc. Int. Soc. Citriculture*1: 91-94.

Sakai, A., Kobayasi, S. and Oiyama, I. 1990. Cryopreservation of nucellar cells of navel orange (*Citrus sinensis* Osb. Var. brasilliensis. Tanaka) by vitrification. *Plant Cell Reports*. 9 (1): 30-33.

Sakai, A., Kobayashi, S. and Oiyama, I. 1991. Survival by vitrification of nucellar cells of navel orange (*Citrus sinensis* var. *brasilinensis* Tanaka) cooled to -196°C. *J. Plant Physiol*. 137: 464-470.

Scora, R. W. 1975. On the history and origin of citrus. *Bull. Torrey Bot. Club*103: 369-375.

Shamel, A.D. 1943. Bud variation and bud selection. In : H. J. Webber and L.D. Batchelor (eds.). *The Citrus Industry*, vol. 1. Univ. of California Press, Berkeley: 915-952.

Singh, Awtar 2001. Citrus Improvement. In: *Citrus*, Singh, S. and Naqvi, S.A.M.H. (eds.), International Book Distributing Co., Lucknow, India, pp. 83-98.

Singh, Awtar. 2005. Role of exotic germplasm in citrus cultivar improvement. *Indian Journal of Plant Genetic Resources*18 (1): 77-78

Singh, Awtar, Dass, H. C. and Vijayakumari, N. 1996. Radiosensitivity for seed and budwood of Nagpur mandarin (*Citrus reticulata*). *Indian J. Agric. Sci*. 66 (10):597-598.

Singh, Awtar and Singh, Shyam 2001. Clonal selection in Nagpur mandarin. *Annals Biol.*17(1): 65-69.

Singh, Balwant. 1964. Citrus breeding - A review. *Punjab Hort. J*. 4: 188-220.

Singh, B. 1981. Establishment of first gene sanctuary in India for *Citrus* in Garo hills. New Delhi Publishing Company. 182 pp.

Singh, H. P. 1989. Present status of citrus decline in India. In: Symp. Plant Disease Problems and their Management in South India, 7-9 Dec. 1989, Bangalore. P. 30.

Singh, H. P. and Chadha, K. L. 1993. Genetic resources of citrus. In: *Advances in Horticulture* (K. L. Chadha and O.P. Pareek, Eds.). 1 (Fruit Crops): 95-121.

Soost, R. K. and Cameron, J. W. 1954. Production of hybrids by use of stored trifoliate orange pollen. *Proc. Amer. Soc. Hort.*63: 234-238.

Starrantino, A., Russo, F. and Martelli, S. 1990. Initial observations on a nucellar population of Navelate sweet orange obtained *in vitro*. *XXIII International Hort. Cong., Florence* (Abstract 3055).

Sugawara K., Oowada, A., Moriguchi, T. and Omura, M. 1995. Identification of *Citrus* chimeras by RAPD markers. *Hort Sci.*30(6): 1276-1278.

Sykes, S. R. and Lewis, W. J. 1996. Comparing Imperial mandarin and Silverhill satsuma mandarin as seed parents in a breeding programme aimed at developing new seedless citrus cultivars for Australia. *Austr. J. Expt. Agric*. 36: 731-738.

Swingle, W. T. and Reece, P. C. 1967. The botany of *Citrus* and its wild relatives. (In) W. Reuther, H. J. Webber and L. D. Batchelor (eds.). *The Citrus Industry*, vol.1 Univ. California Press, Berkeley: 190-430.

Tanaka, T. 1958. The origin and dispersal of citrus fruits having their centre of origin in India. *Indian J. Hort*. 15: 101-115.

Tanaka, T. 1961. Semi-centennial commemoration papers on citrus studies. Cytologia, University of Osaka, Prefecture.

Tatum, J.H., Berry, R.E. and Hearn, C.J. 1974. Characterization of Citrus cultivars and separation of nucellar and zygotic seedlings by thin layer charomatography. *Proc. Fl. St. Hort. Soc.*87: 75-81.

Torres, A.M., Soost, R.K. and Mau-Lastovicka, T. 1982. Citrus isozymes. *J. Heredity*73: 335-339.

Tusa, N., Grosser, J. W. and Gmitter, F. G. Jr. 1990. Plant regeneration of 'Valencia' sweet orange (*Citrus sinensis*), diploid and tetraploid 'Femminello' lemon (*C. limon*), and the interspecific somatic hybrid following protoplast fusion. *J. Am. Soc. Hort. Sci.*115 (6): 1043-1046.

Vardi, A. 1981. Protoplast- derived plants from different *Citrus* species and cultivars. *Proc. Int. Soc. Citriculture* 1:149-152.

Vardi, A., Breiman, A. and Galun, E. 1987. Citrus cybrids: Production by donor- recipient protoplast fusion and verification by mitochondrial-DNA restriction profiles. *Theor. Appl. Genet.* 75: 51-58.

Vardi, A., Epstein, E. and Breiman, A. 1986a. Is the *Phytophthora citrophthora* culture filtrate a reliable tool for the *in vitro* selection of resistant *Citrus* variants? *Theor. Appl. Genet.* 72: 569-574.

Vardi, A., and Galun, E. 1988. Recent advances in protoplast culture of horticultural crops: *Citrus*. *Scientia Hort*. 37: 217-230.

Vardi, A., Hutchison, D. J. and Galun, E. 1986b. A protoplast to tree system in *Microcitrus* based on protoplasts derived from a sustained embryogenic callus. *Plant Cell Reports* 5:412-414.

Vardi, A., Levin, I. and Carmi, N. 2008. Induction of seedlessness in citrus: from classical techniques to emerging biotechnological approaches. *J. Amer. Soc. Hort. Sci.* 133 (1): 117-126.

Vardi, A., Spiegel-Roy, P. and Galun, E. 1975. Citrus cell culture: Isolation of protoplasts, plating density, effect of mutagens and regeneration of embryos. *Plant Sci. Lett*. 4: 231-236.

Vardi, A., Spiegel-Roy, P. and Galun, E. 1982. Plant regeneration from citrus protoplasts: Variability in methodological requirements among cultivars and species. *Theor. Appl. Genet*. 62: 171-176.

Verma, A. N. and Ghosh, S.P. 1979. Citrus germplasm collection from NE Hills and their evaluation. *Indian Hort*. Oct.-Dec. 2-4.

Webber, H. J. 1967. History and development of the citrus industry. (In) H. W. Reuther, H. J. Webber and L.D. Batchelor (eds.). *The Citrus Industry.* Vol. 1. Univ. California Press, Berkeley: 1-39.

Yamamoto, M., Kobayashi, S., Nakamura, Y. and Yamada, Y. 1993. Phylogenic relationships of *Citrus* revealed by diversity of cytoplasmic genomes. (In) T. Hayashi, M. Omura, and N. S. Scott (eds) *Techniques on gene diagnosis and breeding in fruit trees*. Fruit Trees Research Station, Okitsu, Japan pp 39-46.

Zeven, A.C. and de Wet, J.M.T. 1982. Dictionary of cultivated plants and their regions of diversity. Wageningen. 259 p.

12

Banana Breeding

A. Rekha

Banana is one of the most important major fruit crops grown in India and is as old as Indian civilization. It seems that it is one of the earliest fruit crops grown by mankind at the dawn of civilization. In India, bananas are so predominant and popular among people that it is liked both by poor and rich alike. It is the cheapest among all other fruits in the country. Considering the year round availability of fruits, unlike the seasonal availability of other tree fruits, it has become an inevitable necessity in any house hold in India, for all cultural occasions.

The bananas were grown in Southern Asia even before the prehistoric periods and the world's largest diversity in banana population is found in this area. Hence, it is generally agreed that all the edible bananas and plantains are indigenous to the warm, moist regions of tropical Asia comprising the regions of India, Myanmar, Thailand and Indo China. Historical evidences show that the Arabs have introduced the banana from India to Palestine and Egypt perhaps in seventh century A.D. (Hayes, 1953). It soon became popular in those areas and later spread to East coast of Africa at a very early date and subsequently spread throughout the African continent. Bananas were reported to have introduced to Central America in 1516 A.D., where it spread rapidly and attained commercial significance. The spread of bananas to West Indian Islands was through Christian missionaries and wherever it reached it assumed economic importance due to its greater adaptability and commercial value.

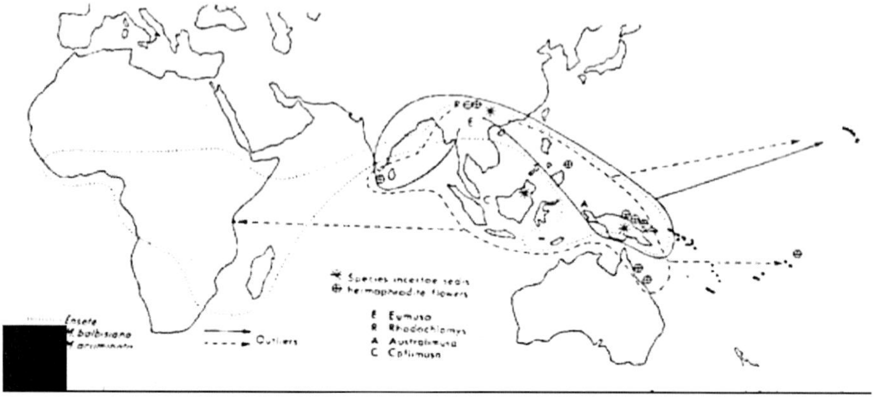

Source: Simmonds 1962

Bananas are put into a variety of uses in India, especially in South India. Almost every part of the plant is used someway or other hence popular as 'Kalpatharu'. In all auspicious occasions in the life of a South Indian, banana has got important place either as a table fruit or as an offering to God in temples. In India, though many (about 70-80) banana varieties exist, only nine to ten cultivars have attained the status of commercial importance. The fruits are used for dessert as well as culinary purposes. The varieties like Poovan, Rasthali, Grand Naine, Robusta, Dwarf Cavendish, Ney Poovan (Elakki Bale), Virupakshi (Hill Banana) are grown for table purpose. The Plantains and Monthan group yield fruits which are high in starch and hence they are used as cooking types. In India, Nendran (Plantain) is grown for table purposes and in processing, for making 'Chips'. The banana male buds are harvested soon after the female phase is over and are sold as vegetable. Similarly, the tender stem, which bears the peduncle, extracted by removing the leaf sheaths of harvested pseudostem axis is also sold as vegetable. The lamina or the banana leaves are widely used as a substitute for plates and become part of regular income for the growers. The dried leaf sheaths give an excellent fiber and are used in various purposes including paper/cloth making.

Botanical Classification

Bananas (*Musa* spp) are placed in the order Zingiberales (Scitamineae) and family Musaceae, with two genera (*Musa* and *Ensete*). The Musa genus is subdivided into four sections Australimusa, Callimusa, Rhodoclamys and Eumusa, based on the number of chromosomes. The genome with 11 chromosomes is characteristic of Eumusa and Rhodoclamys, while the 10 chromosome genome characterizes the Callimusa and Australimusa.

The edible bananas in the section Eumusa had their origin in only two wild species *Musa acuminata*, contributor of 'A' genome and *Musa balbisiana* contributing 'B' genome. Simmonds and Shepherd (1955) used a scoring method to indicate relative contributions of the two wild species by using 15 characters and the cultivars are classified as AA, AAA, AB, AAB, ABB, AAAA, AAAB, AABB and ABBB genomic groups, based on 1-5 numerical scale. The score of 15-21 was placed as AA/AAA, 26-41 as AAB group, 48 as AB group and 59-65 in ABB group. The classification is somewhat complicated because of polyploidy. Edible bananas have 22, 33 or 44 chromosomes, the basic chromosome number in this section (Eumusa) × n=11, hence the cultivars are diploid, triploid and tetraploid respectively.

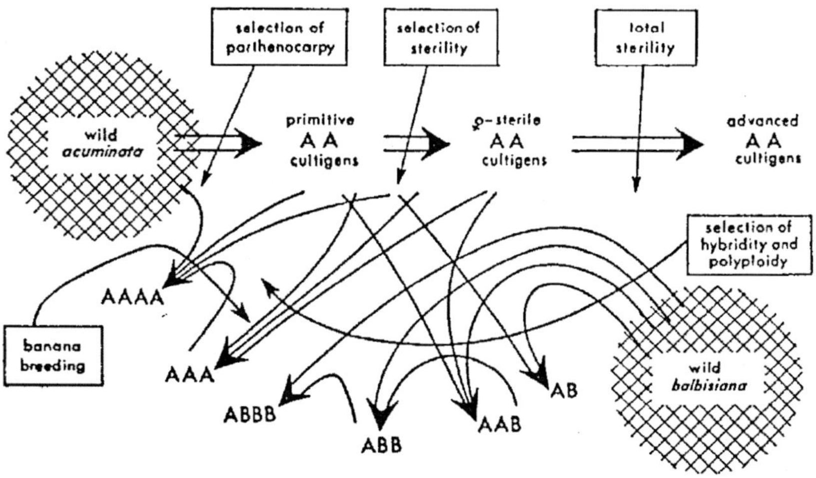

Source: Simmonds 1962

Varieties

Cavendish type cultivar Grand Naine is most popular commercial and export cultivar. However there are other important cultivars of banana which are grown in different parts of the country based on regional preferences. Some of the important ones are Dwarf Cavendish, Robusta and Red banana of AAA Cavendish group. Ney Poovan or Elakki Bale of AB group, Poovan, Rasthali, Hill Banana of AAB group and Karpuravalli of ABB group. The cultivar Nendran of AAB plantain group has special importance because of its dual purpose use (Dessert and cooking) and in preparation of processed products like chips and banana powder. Variety Monthan of ABB group is a cooking type of banana. There are several synonyms to these above mentioned cultivars of banana.

Banana and plantain breeding

Breeding through conventional methods is difficult as the cultivars are genetically male and female sterile, parthenocarpic, polyploid and vegetatively propagated. Fertility among the cultivars varies depending on various environmental and physiological factors.Hence normal methods of conventional breeding are difficult to follow in developing hybrid banana cultivars.

Generation of banana hybrids and selection has been carried out since the 1920s (Rowe, 1984; Ortiz and Vuylsteke, 1995) but till early in the 21st century, none of these synthetic cultivars and radiation mutants could be widely adapted for cultivation (Jain and Swennen, 2004). Rowe (1984) summarized sexual methods for the breeding of bananas at that time. Vuylsteke *et al.,* (1997) and Ortiz (1997) reviewed the wide array of breeding schemes combining conventional and innovative approaches, and producing potential cultivars from primary tetraploids, secondary triploids and other populations.

Seed set in triploid banana accessions is very low, some triploid cultivars have residual fertility, and hand-pollination with diploid parents gives some seed typically averaging less than one to 20 or more seeds per fruit bunch (Ortiz and vuylsteke, 1995), and was able to produce useful hybrids following *in vitro* germination of seed from triploid x diploid crosses. Oritz (1997) reported the production of 2n pollen in both male-fertile diploid and polyploid accessions of Musa, but 2n pollen production may be a character that evolved late in the domestication of cultivated Musa. The occurrence of 2n pollen may be useful for transfer of desired attributes to the polyploidy level through sexual polyploidization, may be more efficient.

Breeding objectives

1. Yield and quality especially finger length, finger pedicel strength and flavor.
2. Pseudostem height.
3. Resistance to diseases especially Fusarium wilt and Black sigatoka disease.
4. Tolerance to nematodes, weevils.
5. Tolerance to BSV, BBMV.
6. Tolerance to abiotic stress (Water & Salinity)

Until the late 1970s there were only two major banana breeding programmes, in Honduras and Jamaica, both focused on improving the export banana. The Jamaican programme does not exist at present, whereas the United Fruit

Company transferd its banana breeding programme to the Honduran government in 1984. From 2004, FHIA is supported by Bioversity to assist breeding through ProMusa, the Global Programme for Musa Improvement.

The ongoing banana breeding programmes are at Fundacion Hondurena de Investigacion Agricola FHIA in Honduras (Rowe and Rosales, 1993) based on selection of improved diploids that can be crossed to eventually generate triploids, a number of which have been released as varieties. Other programmes at the International Institute of Tropical Agriculture (IITA) in Nigeria and Uganda, in Brazil by EMBRAPA (Empresa Brasiliera de Pesquisa Agropecuaria) , Centre de Recherches Régionales sur Bananier et Plantain (CARBAP) in Cameroon, and Centre de Coopération Internationale en Recherche Agronomique pour le Développement (CIRAD-FLHOR) in France and Guadeloupe in the West Indies also have active crossing and selection projects involving crossing AA diploids with AAAB to generate triploid hybrids with improved disease resistance and agronomic characters.

The first breeding strategy developed by FHIA, and now adopted by other programs, relies on the production of improved diploids. The diploids possess useful resistance characteristics introduced from wild sources in an improved genetic background, and are then used as male parents in crosses with the desired triploid as female parent. The improved diploids developed by FHIA are now being used for the improvement of many different types of bananas, including dessert varieties, plantains, cooking bananas, and East African highland bananas. The main constraints of this breeding strategy are: (a) the low fertility rate of the desired female parents, and (b) the possibility of residual fertility in the tetraploid progeny could lead to seed-bearing fruit under certain circumstances. To overcome this latter constraint, efforts are now being focused on the production of secondary triploids.

A different breeding strategy for the production of improved triploids is being developed by CIRAD-FLHOR and CARBAP. This is based on colchicine treatment for doubling the chromosome number of desired diploids, which are then used in tetraploid × diploid crosses for the production of triploids.

Internationally available hybrids

FHIA: at Honduras, has created hybrids of different banana types, and tetraploid hybrids of Sweet (FHIA-02, FHIA-23, FHIA-17, SH-3640), Sweet acid (FHIA-01, FHIA-18), Cooking (FHIA-03, FHIA-25), and Plantain type (FHIA-21) have been released.

IITA: at Nigeria & Uganda, is mainly dedicated to generation of cooking bananas, plantains and east African highland bananas. Cooking tetraploid hybrids

(BITA-2 and BITA-3) and secondary triploid (PITA-16) resistant to BLSD are available. More hybrids developed are not released because of episomal form of BSV.

CARBAP: at Cameroon is developing triploid plantain type bananas with resistance to BLSD and nematodes. Four hybrids have been developed and a few dwarf or semi-dwarf diploid and tetraploid hybrids resistant to BSLD are under preliminary evaluation.

CIRAD-FLHOR: in Guadeloupe is mainly dedicated to generation of triploid sweet bananas using molecular tools to assist breeding. A number of hybrids have been generated.

TBRI: at Taiwan, has developed somaclonal variants from export Giant Cavendish cultivar with enhanced tolerance to Fusarium wilt disease.

EMBRPA: at Brazil also developed a number of hybrids, which are at various stages of evaluation.

Following strategies adapted for new generation hybrids for the new millennium:

i) AA improved (2x=2n) x *Musa* cv. AAA, AAB, ABB (2n=3n)

↓

AAAA, AAAB,AABB hybrids (2x=3n+1n=4n) x AA improved (2x=2n)

↓

AAA, AAB,ABB hybrids (2n=3n)

Constraint – low fertility rate in desired female parent and the possibility of seed bearing fruits in the progeny. Hence the present focus is on production of triploids.

ii) Aa improved + Colchicine treatment

↓

AaAa (2x=2n+2n=4n) x AA improved (2x=2n)

↓

AaA hybrid (2x=2n+1n=3n)

The second strategy helps in widening the genetic base and availability of improved varieties for breeding.

Elakki bale AB type

FHIA-1 of AAAB

FHIA-2 of AAAA

FHIA-3 of AABB

Karpooravalli - ABB

M. acuminata

M. balbisisana

Mysore AAB

Nendran AAB

Robusta AAA

Seed Germination musa hybrid seedlings

Seeds

In India, banana breeding was attempted as early as 1949 at Aduthurai Tamil Nadu. As a result, TNAU released the AAB (Pome) hybrid CO1, which resembled Hill Banana (Virupakshi) and retained its flavour when cultivated in plains and has developed AAB hybrid resistant to *Fusarium* wilt disease in 2009. Kerala Agricultural University (KAU) released two hybrids BRS 1 (Agnishwar × Pisang Lilin) and BRS 2 (Vannan × Pisang Lilin) for cultivation. NRCB, Trichy released a clonal selection of cultivar 'Karpooravally' called "Udayam" which is a very high yielder. Because of the difficulties involved in conventional breeding, none of the hybrids have successfully replaced the present day cultivars.

Mutation breeding

Breeding through mutation is an important approach for *Musa*, and as with many mutation programmes are particularly valuable to compensate for defined weaknesses in existing cultivars like producing a disease resistant cultivar of Rasthali retaining the fruit quality. Roux (2004) has summarized the mutation breeding history of the two commercially released banana lines derived from gamma-ray-induced mutations. Both have useful agronomic traits: 'Novaria', released in 1995, flowers about 10 weeks earlier that the original parental clone ('Grande Naine' in the 'Cavendish' group), while the Thai variety 'Klue Hom Thong' KUI has large bunch size and a cylindrical shape with larger banana fingers. A number of other traits derived from mutation treatments or from somaclonal variants. Including dwarfism, *Fusarium* wilt and Black Sigatoka resistance, are currently being evaluated.

Transgenics

A number of alternative ways to generate transgenic bananas have been developed. Transformation with alien genes and selectable markers are usually low-efficiency and dependent on the genotype and state of material. Both particle bombardment (Sagi *et al.,* 1995) and Agrobacterium mediated (May *et al.,* 1995; Khanna *et al.,* 2004) strategies are applied. As well as marker genes, anti-fungal, anti-nematode (Atkinson *et al.,* 2004), virus-resistance genes and control of fruit ripening are some of the targets for transformation. The use of banana transformed with fruit promoters and appropriate protein genes have also been considered to produce edible vaccines (Mason *et al.,* 2002). Presently efforts are being made to produce vitamin A Iron rich bananas throgh GM technology.

Cytological studies

Karyological and cytological studies are useful in establishing the relationship between species and within the genus, to detect the unstable areas and to monitor the chromosomal irregularities. Little is known about the structure of *Musa* chromosomes; most of the information has been derived from the studies on chromosome pairing during meiosis (Shepherd 1999). Local wild seeded species and subspecies showed regular chromosome pairing with 11 bivalents, indicating the absence of detectable structural heterozygosity. On the other hand, diploid parthenocarpic clones showed aberrant chromosome pairing with univalents, trivalents, and multivalents, indicating heterozygosity for one or more translocations or inversions (Faure *et al.,* 1993).Studies on chromosome number among different banana cultivars were reported by Valsalakumari and Sivaraman Nair (1990) among 98 cultivars. Somatic chromosomes of 53 *Musa* land races and hybrids were reported by Osuji *et al*., (1996) and 16 *Musa* species and land races by Dolezel *et al*., (1998). Chromosome counts and ploidy levels were determined for 60 *Musa* accessions, including plantain somaclonal variants by Osuji et.al., (1996) and chromosome number and karyoanalysis among 40 accessions including *Musa* species were carried out by Rekha in 2001.

Knowledge of ploidy and exact genome compositions of the parental material is essential for *Musa* breeding. A study was conducted to determine the ploidy levels and genome composition of the *Musa* germplasm collection, constituting over 300 accessions, at the International Institute of Tropical Agriculture in Nigeria and Uganda. Flow cytometric analysis of nuclear DNA content was used to estimate ploidy levels, while genome composition was ascertained with RAPD markers that are specific for the A and B genomes of *Musa*. It was determined that at least 8% of the plants in the germplasm collection were miss-classified in terms of ploidy and/or genome composition. The cultivars 'Pisang awak', 'Foulah 4' and 'Nzizi', previously classified as triploids, were tetraploids by flow cytometry and conventional root tip chromosome counts. Similarly, cultivars that were previously classified as diploids including 'Too', and 'Toowoolee' were triploids by analysis. (Pillay *et al.,* 2006).

Molecular markers

Application of molecular genetic analysis may dramatically improve breeding efficiency. The studies involved comparison of pedigree and DNA marker methods of estimating genetic relationships across and within generations among diploid, triploid and tetraploid accessions of plantain and banana. There was no association between marker-based similarity and pedigree relationships. While DNA markers may provide a more accurate description of genetic relatedness, the study suggested that pedigree-based analysis may prove useful for the selection of prospective parental combinations in Musa breeding (Tenkouano *et al*., 1999).

Identification of useful genotypes

Understanding the genetic diversity in relation to the character required for improvement is the pre-requisite of any breeding program. Gomes, et.al., in (2006) tried to identify genotypes that adapt to the saline soil in the northeast region of Brazil with RAPD primers, to assess the genetic variability among ten banana genotypes (*Musa* spp): 'Pacovan', 'Nanicão', 'Caipira', 'FHIA-18', 'Calcuttá', 'SN/2', 'Borneo', 'M-53', 'Microcarpa' and 'Lidi', correlating their tolerance to saline stress. Twenty-five primers were tested. The varieties of higher tolerance to saline stress, such as 'Pacovan' and 'SN/2', proved to be genetically distant when compared to the most salt sensitive, such as 'Calcuttá' and 'Lidi' and hence could be used in the improvement program.

Javed (2004) analyzed suckers collected from different populations of *Musa acuminata ssp. Malaccensis*, found highly resistant to race 4 of *Fusarium oxysporum f. sp. Cubense* (FOC) in comparison to the cultivars. Seedlings from these wild banana plants segregated for resistance to the pathogen. The infected seedlings were characterized based on external and internal symptoms and the variable response to FOC was mainly due to the genetic factors. Using the technique of random amplified polymorphic DNA (RAPD), 96 major amplification products from 15 primers were identified. Only 10 out of 96 markers were monomorphic and shared among the seed progenies, whereas the remaining 86 were highly polymorphic. Three primers showed banding patterns which were specific to resistant or susceptible seedlings. These results could conclude and confirm the great potential of the wild *Musa acuminata ssp. Malaccensis* as a source for banana improvement and also for the synthesis of segregating populations for linkage mapping, gene cloning and DNA markers related to FOC resistance. In another study, RAPD markers linked to susceptibility and resistance of *Musa* to *Fusarium oxysporum* were reported. The RAPD patterns were obtained, with primer OPK-03 detected a major fragment of 485 bp was observed in the susceptible genotypes, but absent in resistant/tolerant genotypes (Zambrano *et al.*, 2005).

Somaclonal variation

Application of micropropagation for large-scale production of elite clones is an effective alternative to conventional multiplication of banana. *In vitro* propagation has many advantages, such as higher rates of multiplying clean (pest and disease-free) planting material and the small amount of space required to multiply large number of plants. A major problem associated with *in vitro* culture is the occurrence of genetic variation resulting from micropropagation, i.e., somaclonal variation (Larkin and Scowcroft, 1981). Evans (1984) reported that the somaclonal variation results from both pre-existing genetic variation within the explant and variation induced during tissue culture. Stover (1987) reported somaclonal variation in 'Grand Naine' and 'Saba' banana in nursery and field. Vuyleteke *et al.,* (1988) also confirmed phenotypic variation among *in vitro* propagated plantain, *Musa* cv. AAB. They also observed that the degree of somaclonal variation was highly genotype specific for banana. Israeli *et al.,* (1991) reported variations of banana Cavendish cv. AAA in 'Williams' and 'Grand Naine'. Many variants like dwarf, extra dwarf, giant, deformed lamina and different coloured pseudostems were observed. Several factors like age of the culture, different hormonal concentrations and ploidiness were attributed for cause of somaclonal variation (Damasco *et al.,* 1998; Leela *et al.,* 2003). However, there are some beneficial effects of somaclonal variations. In Taiwan, Cavendish banana cultivar variant, resistant to *fusarium* wilt was selected from micropopagated plants (Hwang and Ko 2004)

Genomics

Banana whole genome sequence has been published and is a stepping stone for improvement of banana which is a challenge. It helps in identifying genes responsible for important agronomic characters, such as fruit quality, disease and pest resistance as they are affecting banana industry. The genomic studies are of immense use to understand the evolution of plants and in particular monocotyledons. Since banana is known to be an important source of starch, it is essential to evolve varieties/cultivars resistant to various biotic and abiotic stresses. The various markers, transcriptomes and related proteins would help in gene identification and better understanding of the plant - host interactions.

Future line of work

Breeding in banana has been a difficult task due its present day commercial cultivars being triploid, parthenocarpic and sterile conditions. Transferring traits is difficult and hence genetic understanding is very poor. It is very important to conserve the huge variable germplasm of banana all over the world involving the wild and cultivated types. There is a need to generate trait specific populations among the seeded types using the vast gene pool available. Present day biotechnological tools of molecular markers would help in identifying the traits and mapping of QTL's. It would also help in generating an ideal cultivar with resistance to biotic or abiotic stress and with good yield.

Characters used in taxonomic scoring of banana cultivars

Sl. No.	Characters	*M.acuminata*	*M. balbisiana*
1.	Pseudostem colour	More or less heavily marked with brown or black blotches	Blotches slight or absent
2.	Petiolar canal	Margin erect or spreading, with Scarious wings below, not clasping pseudostem	Margin closed, Not winged below clasping pseudostem
3.	Peduncle	Usually hairy	Glabrous
4.	Pedicels	Short	Long
5.	Ovules	Two regular rows in each loculus	Four irregular rows in each loculus
6.	Bract shoulder	Usually high (ratio<0.28)	Usually low (ratio>0.30)
7.	Bract curling	Bracts reflex and roll back after opening	Bracts lift but not roll
8.	Bract shape	Lanceolate or narrowly ovate, tapering sharply from the shoulder	Broadly ovate, not tapering sharply
9.	Bract apex	Acute	Obtuse
10.	Bract colour	Red, dull purple or yellow outside; Pink, dull purple or yellow inside	Distinctive brownish purple outside; brightcrimson inside
11.	Colour fading	Inside bract colour fades to yellow towards the base	Inside bract colour continuous to base
12.	Bract scars	Prominent	Scarcely prominent
13.	Free tepal of male flower	Variably corrugated below tip	Rarely corrugated
14.	Male flower colour	Creamy white	Variably flushed with pink
15.	Stigma colour	Orange or rich yellow	Cream, pale yellow or Pale pink

References

Atkinson, H.J., Grimwood. S., Johnston, K. and Green, J. 2004. Prototype demonstration of transgenic resistance to the nematode *Radopholus similis* conferred on banana by a cystatin. Transgenic Research, 13:135–142, Kluwer Academic Publishers. Netherlands.

Bose. T.K and S.K. Mitra. (ed.) 1990. Fruits: Tropical and Sub-Tropical. Naya Prokash, Calcutta.

Damasco, O.P., Aokins, S.W., Godwin, I.D., Smith, M.K. and Drew, R.A. 1998. Use of a SCAR based marker for early detection of dwarf off-types in micropropagated Cavendish bananas. *Acta Horticulture* 461:157-164.

Evans, D.A., Sharp, W.R. and Medina. Filho, H.P. 1984. Somaclonal and gametoclonal variation. *American Journal of Botany* 71:759-774.

Fauré, S., Bakry, F. and Gonzalez De Leon, D. 1993. "Cytogenetic studies of diploid bananas" in: Breeding Banana and Plantain for Resistance to Diseases and Pests, (Ganry, J., Ed.), CIRAD-FLHOR, Montpellier, 77-92.

Gomes, E.W.F., WilladinoI, L., Martins, L.S.S., Oliveira e Silva, S. and Camara, T.R. 2006: Genetic variability of banana (*Musa* spp) genotypes subjected to saline stress. *Revista Brasileira de Engenharia Agrícola e Ambiental* (BRA), 9:171-177.

Gowen S. 1995 (ed.). Bananas and Plantains. Chapman and Hall, London.

Hwang, H.C. and Ko, W.C. 2004. Cavendish Banana Cultivars Resistant to Fusarium Wilt Acquired through Somaclonal Variation in Taiwan. *Plant Disease* 88:580-88.

Israeli, Y., Reuveni, O. and Lahav, E. 1991. Qualitative aspects of somaclonal variations in banana propagated by in vitro techniques. *Scientia Horticulture*, 48:71-88.

Javed, M.A., Chai, M. and Othman, R.Y. 2004: Study of resistance of *Musa acuminata* to *Fusarium oxysporum* using RAPD markers. *Biologia Plantarum* 48:93-99.

Khanna. H,D. Becker, Kleidon. J. and Dale. J..2004. Centrifugation AssistedAgrobacterium tumefaciens-mediated Transformation (CAAT) of embryogenic cell suspensions of banana (*Musa* spp. Cavendish AAA and Lady finger AAB). *Molecular Breeding*.14:239–252.

Larkin, P.J. and Scowcroft, W.R., 1981. Somaclonal variation - a novel source of variability from cell cultures for plant improvement. *Theoretical. Applied Genetics*. 60:197-214.

Leela, S, Jaya, R., Sonejl and K.T. Bollamma. 2003. Invited review. Analyzing somaclonal variation in micropropagated bananas (*Musa* spp.) (2003). *In vitro cell. Dev. Biol. Planta.*

Mason. H.S., H. Warzecha, T. Mor and C.J. Arntzen., 2002. Edible plant vaccines: applications for prophylactic and therapeuticmolecular medicine.TRENDS in Molecular Medicine Vol.8 No.7.

May. G.D., R Afza, H.S. Mason, A. Wiecko, Novak F.J. and. Arntzen C. J., 1995. Generation of Transgenic Banana (*Musa acuminata*) Plants via Agrobacterium-Mediated Transformation. *Nature Biotechnology*13:486- 492 (1995).

Osuji, J.O., B.E.Okoli, and R.Oritz., 1996. An improved procedure for mitotic studies of the eumusa section of the genus *Musa* L (Musaceae). *Infomusa* 5(1):12-14.

Peter. K.V. (Ed.) 2007. Horticultural Science Series. Fruit crops.

Pillay M., Ogundiwin, E., Tenkouano, A. and Dolezel, J. 2006. Ploidy and Genome Composition of Musa Germplasm at the International Institute of Tropical Agriculture (IITA). *African Journal of Biotechnology*. 5:1224-1232.

Rekha. A. 2001. Studies on Karyotype, Nuclear DNA content and RAPD markers in some *Musa* species and banana cultivars. PhD. Thesis submitted to Karnatak University, Dharwad, Karnataka, India.

Roux 2004. Mutation induction in *Musa* – review. Proceedings from a meeting, Leuven (BEL), 2001/09/24-28. Science Publishers, Enfield (USA),

Ságі. L., Panis, B., Schoofs, H., De Smet, K., SwennenBruno, R. and Cammue, P.A. 1995. Genetic Transformation of Banana and Plantain (*Musa* spp.) via Particle Bombardment *Nature Biotechnology*13:481-485.

Shepherd, K., 1999. Cytogenetics of the genus *Musa*. INIBAP, Montpellier.

Simmonds, N.W. 1962. Bananas. Longman Publications, London.

Singh, H.P. and Chadha, K.L. 1993. Genetic resources of Banana in: Advances in Horticulture Vol.1 Fruit Crops.

Stover. R.H and Simmonds, N.W. 1987. Bananas.Longman publications, London.

Tenkouano, A., Crouch, J.H., Crouch, H.K., Vuylsteke, D. and Ortiz, R. 1999: Comparison of DNA marker and pedigree-based methods of genetic analysis in plantain and banana (*Musa* spp.) clones. I. Estimation of genetic relationships. *Theoretical and Applied Genetics*, 98:62-68.

Tenkouano, A., Crouch, J.H., Crouch, H.K., Vuylsteke, D. and Ortiz, R. 1999: Comparison of DNA marker and pedigree-based methods of genetic analysis of plantain and banana (*Musa* spp.) clones. II. Predicting hybrid performance. *Theoretical and Applied Genetics*, 98:69-75.

Valsala Kumari, P.K. and Nair, P.C.S. 1990. Cytotaxonomical studies on banana cultivators. *South Indian Horticulture* 38(4):177-182.

Vuylsteke, D., Swennen, R., Wilson, G.F. and De Langhe. 1988. Phenotypic variation among in vitro propagated plantain (*Musa* spp. cv. AAB) *Scientia Horticulture*. 36:79-88.
Zambrano, A.Y., G. Martinez, Z. Gutierrez, E., Manzanilla, L., Ruiz, L., Diaz, Contreras, A., Marquez, Demey, J.R. 2005. Identification of RAPD marker linked to resistance of *Musa* to *Fusarium oxysporum*. *Acta Horticulutrae*. 738:535-542.

13

Pomegranate Breeding

S.S.Hiwale

Pomegranate is emerging as one of the important fruit crops owing to its hardiness and ability to withstand adverse soil and climatic conditions. India contributes 40-45 percent of total 10 lakh tons production in the world. Currently the export of pomegranate is around 10,000 tons, which is just 5 percent of the total export. With the increase in population, the domestic demand of fruits has also increased substantially. As on date, out of total area of 1, 25,000 ha. under pomegranate, major area under the crop is found in Maharashtra. The state has 85,000 ha. (68 % area) under the crop, of which only 51,000 ha is under production, followed by Karnataka 12,727 ha. Gujarat 3,787 ha, up to 2003 –2004 (Anon, 2005). Earlier orchards were raised mainly with variety Ganesh, which was the leading variety. However, the variety is fast being replaced with the new cultivars like Bhagawa and Arakta, which have attractive skin and aril colour and are suitable for export.

The pomegranate is an ancient and favorite table fruit of tropical countries. It is commercially grown for its sweet-acidic fruits, which provide a cool refreshing juice, and is valued for its medicinal properties. The fruit rind, juice, leaf and roots are used in the preparation of various ayurvedic medicines. Its popularity is also due to ornamental nature of the plant especially due to bearing bright red flowers throughout the year. The fruit juice is refreshing and contains both glucose and fructose. The juice and seed contain large quantities of tannin and agolic acid, which is essential in curing several diseases. The fruit can be successfully grown even under purely rain fed conditions. In the recent years pomegranate cultivation has become an economically viable proposition.

Pomegranate (*Punica granatum* L.) belongs to the natural order Punicaceae; Punica perhaps is the only known genus of this family, which includes large shrubs or small trees with two species, *Punica protopunica* Balf. found wild in Socotra Island and the other *Punica granatum* cultivated in the sub-tropical parts of the world. *Punica granatum* is classified into two sub-species chlorocarpa and porphyrocarpa, each having two varieties. These sub-species have been established on the basis of colour of ovary, a stable feature that is retained even when they are reproduced by seeds. Sub-species chlorocarpa is mainly found in the *Transcaucasus*, whereas, the second Sub-species *porphyrocarpa* is mainly distributed in central Asia.

Types and varieties

Several types of pomegranates with varying shapes, sizes and color are cultivated in India. Fruits are round, obovate in shape and vary in diameter from 8-12 cm. The rind may be thick or thin and the color ranges from pale yellow to crimson. The seed pulp in superior types is thick, fleshy and very juicy, while in inferior types it is thin. The seed coat varies in hardiness, some of the softer seeded types being known as seedless (Bedana). Lack of lignifications of the testa is the main cause of so called seedlessness in pomegranate. Among the numerous types grown, the Bedana and Kandhari are considered the best.

The varieties Ganesh, Mridula, Arakta, G-123, P-23, P-26, Mastani, and Bhagwa are popular in Maharashtra, Bassein Seedless, Ganesh, Mridula and Arakta in Karnataka, Dholka in Gujarat, Jalore Seedless, Jodhpur Red and Jodhpur White in Rajasthan, Kabul Red, Vellodu, Yercaud-1, CO-1 in Tamil Nadu.

Though we know a great deal about pomegranate and its origin there are a number of seedling varieties that are being grown. Selecting a variety with known qualities is the better choice. Most horticulturists divide pomegranate varieties into three categories Sweet, Sweet tart and Sour, we prefer to include the seed hardiness in this and divide them into six groups.

1. Soft seeded sweet
2. Soft seed sweet tart
3. Early variety (mostly sweet)
4. Normal (harder) seeded sweet tart
5. Normal (harder) seeded sweet
6. Sour (Nearly always normal seeded).

The pomegranate varieties grown in different countries are:

Spain

1. Molar: The best seedling pomegranate in Spain, sweet good red fruit and productive.
2. Valencia: An early ripening sweet variety with moderate production.

Malta

1. Blanca: Medium whitish fruit, seeds are undeveloped (nearly seedless), aril pink in color, productive.
2. Dulce Colorada: Large red color, red arils, soft seeded, sweet, productive.
3. Gorda de Jativa: Large red fruit, red aril, sweet, productive.
4. Pignoncea: Large red fruit, small seed, sweet.
5. Cagin: Large red fruit, very hard small seed, sour.
6. Quiseppe: Large, red sweet fruits with high production.

Italy

Dolce, Alappi, Dolce Nostrale.

Cyprus

Zaitiki: Large, sweet fruit with soft seed.

Greece

1. Patras acide: Extra large, red fruit with very sour taste, good for syrup.
2. Patras Douce: large red, sweet fruit, productive.

Yogoslavia

1. Slatki Barshi Nar: Early ripening, large, yellowish green fruit with sweet taste.

Bulgaria

1. Uzbekskii Saldikii (Uzbek sweet)-sweet fruit.
2. Nikiskii Rannii- sweet tart fruit, early ripening and early to produce.

Israel

1. Wonderful: The sweet tart variety from USA. High production with medium hard seeds.
2. Asmar (Black): Dark purple almost black fruit, white, hard seeds.
3. Ras-el-Bghil: Very large fruits with normal seed.
4. Red Lufani: Red, large with normal seed, good production.
5. Malissi: Nearly seedless (called seedless) early ripening.

Arabian Peninsula

1. Roman chouall: Medium dark, almost dark fruit, sweet tart flavour, grown in Iraq.
2. Mellassi: Large red fruit with soft seeds and sweet taste.
3. Cherabani: Medium red fruit, sour taste and the syrup is used for preparing beverages.
4. Selimi: Large red fruit with sweet tart taste and small seed.
5. Aswar (black), Ahmar (red) and Halwa also grown.

Turkey

1. Hicaznar: A large fruited, having good productivity.
2. A K-anar: yellowish white juice and light seeds.
3. Kizil anar: Large red fruit with sweet tart flavour, high production.

Iran

1. Shahvar (Saveh) - Red skin, very large fruit.

Afghanistan

1. Kandhari large white: Extra large yellowish sweet fruit early ripening.
2. Kandhari large red: Extra large fruit with sweet flavour, mid season ripening. This is one of the best varieties in the world of Pomegranate.
3. Kandhari large black: Dark purple (looks black), extra large fruit, sweet tart fruit.

India

1. Ganesh: It was one of the most popular pomegranate varieties in India, large yellowish red, with soft seeds.
2. Mridula: A hybrid of Ganesh and a red Russian variety, pink red skin with soft seed, sweet taste, early in mature (150 days).
3. Bhagwa: Sweet, soft seeded, red fruit, suitable for export.
4. Muscat: The rind is yellowish pink, grains are medium hard.
5. Bassein seedless: The outer skin color is red, seeds soft.
6. Jodhpur Red: Grown in Rajasthan, rind is yellowish in color with whitish pink arils.
7. Jalore seedless: Grown in Rajasthan, the rind is reddish. Arils are soft, pink colored.
8. Dholka: Grown in Gujarat, arils pink red colored.
9. Jyoti: Released from UAS, Bangalore, the variety red coloured fruit with soft arils.
10. RCR-1: Released from UAS, Dharwad. Fruit red coloured, soft arils.
11. G-137: Open pollinated variety from Ganesh, released by MPKV, Rahuri.
12. Yercaud Selection: CO-1- Released by TNAU, Coimbatore having soft seeds.
13. Ruby Red: It is a multiple hybrid developed at IIHR, Bangalore, rind red, seeds soft. Fruit size small.

China

1. Cin- Kwene-liu-Cin- pehine.
2. Echen- Large thin skinned (paper shell) fruit, sweet taste.

Tunisia

1. Gabsi: Large yellowish fruit, with sweet taste.
2. Tunsi: Medium red fruit, dark red aril with sweet tart taste.
3. Chelfi: Medium red, seedless variety.
4. Djebeli: Very large red fruit, late ripening sweet variety with very small seed.

Mexico

1. Grande de china and Granada Agaria are grown.

United States of America

1. Granada: Very early, red medium a bud sprout of wonderful.
2. Early foothill: Large fruit similar to wonderful ripens after Granada, low and higher sugar.
3. Early wonderful: Large, early.
4. Wonderful: Bright red variety with sweet tart. This is the main commercial variety of USA. Very productive in San Jaquin Valley of California. It is a frost sensitive variety.

Commercial varieties

Ganesh

It is a selection from cultivar Alandi. The growth habit of tree is spreading type with evergreen nature, the leaf is elliptic in shape. The tree flowers throughout the year with three main flushes. Fruit is round and smooth and pink in color with reddish tinge. Fruit weight per fruit was maximum in Mrig bahar (255g). Arils are sweet in taste with light pink in color. The juice percent is in the range of 51.5 to 52.76 percent. T.S.S. of the juice was 16.6^{o} Brix to 17.4 ^{o}Brix. The acidity is about 0.4, seeds are soft with a range of 1.15 kg/cm^2 to pressure 1.59 kg cm^2. The variety is susceptible to fruit borer and fruit spot.

Jalor seedless

The growth habit of tree is spreading type with evergreen nature. The leaf is entire and elliptic in shape. The tree flowers throughout the year with three main flushes. It is a good yielder. Fruit is round and smooth and pink in color with reddish tinge. Fruit weight per fruit is is about 154 to 175g. Arils are sweet in taste with light pink in color. The juice percent varies between 50.25 to 51.72

percent. TSS of the juice is 13.2° - 15.2 °Brix. The acidity is about 0.46 to 0.59 percent. Seeds are soft (2.24 kg/cm^2 pressure). The variety is susceptible to fruit borer, leaf spot and fruit spot.

Dholka

The variety is grown in Dholka area of Gujarat. The growth habit of tree is spreading type with evergreen nature. The leaf is entire and elliptic in shape. The tree flowers throughout the year with three main flushes. Fruit is round and smooth and light green in color with reddish tinge. The average fruit weight is about 298g. Arils are sweet in taste with light pink in color. The juice percent varies from 51.95 to 52.53. T.S.S. of the juice is about 16.2 °Brix. The acidity is between 0.51 and 0.47 percent. Seeds are soft with soft (1.31 kg/cm^2 to1.25 kg cm^2), the variety is susceptible to fruit borer, leaf spot and fruit spot.

P-13

The growth habit of tree is spreading type with evergreen nature. The leaf is entire and elliptic in shape. The tree flowers throughout the year with three main flushes. Fruit is round and smooth and yellow in color with reddish tinge. Fruit weight is about (217 to 233g). Arils are sweet to taste with light pink color. The juice percent varies from 49.52 to 53.5. The T.S.S. of the juice is about 15.4 °Brix. Seeds were soft (1.64 kg/cm^2 pressure). The variety is susceptible to fruit borer, leaf spot and fruit spot.

P-16

The growth habit of tree is spreading type with evergreen nature. The leaf is entire and elliptic in shape. The tree flowers throughout the year with three main flushes. Fruit is round and smooth and pink in color with reddish tinge. Fruit weight is about 245 g, the arils are sweet in taste with light pink in color. The juice percentage varies between 48.92 to 52.69. The T.S.S. of the juice is 16.2° Brix. The acidity is about 0.47 percent. Seeds are soft (1.63 kg/cm^2 pressure). The variety is susceptible to fruit borer, leaf spot and fruit spot.

P-23

It is a selection from Muscat. The growth habit of tree is spreading type with evergreen nature. The tree flowers throughout the year with three main flushes. Fruit is round and smooth and pink in color with reddish tinge. Fruit weight is about 260.5g, the juice percent varies from 49.65 to 51.65. The T.S.S. of the juice is 17.2 to 16.4 °Brix. The acidity is in the range of 0.49 to 0.58 percent. Seeds are soft (1.92 kg/cm^2 pressure). The variety is susceptible to fruit borer, leaf spot and fruit spot.

P-26

This is also a selection from Muscat. The growth habit of tree is spreading type with evergreen nature. The tree flowers throughout the year with three main flushes. Fruit is round and smooth and yellow in color with reddish tinge. On an average, fruit weighs about 268g. Arils are sweet in taste with light pink in color. The juice percent varies between 51.45 in Mrig bahar and 52.69. The T.S.S. of the juice ranges from 14.62 to 15.2 °Brix. The acidity varies from 0.43 and 0.36 percent. Seeds are soft (1.46 kg/cm^2 pressure), the variety is susceptible to fruit borer, leaf spot and fruit spot.

G-137

The growth habit of tree is spreading type with evergreen nature. The leaf is entire and elliptic in shape. The tree flowers throughout the year with three main flushes. Fruit is round and smooth and pink in color with reddish tinge. Average fruit weight is about 270g. Arils are sweet in taste with light pink in color. The juice percent varies from 55.23 to 54.87 and the T.S.S. of the juice is from 17.0 to 17.4 °Brix. The acidity varies from 0.42 to 0.49. Seeds were soft (1.04 kg/cm^2 pressure). The variety is susceptible to fruit borer, leaf spot and fruit spot.

Musket

The growth habit of tree is spreading type with evergreen nature. The leaf is entire and elliptic in shape. The tree flowers throughout the year with three main flushes. Fruit is round and smooth and pink in color with reddish tinge. Ave rage fruit weight is 310.27g; arils are sweet in taste with light pink in color. The juice percent varies from 50.33 to 49.74. The T.S.S. of the juice is from 14.4 to 14.8 °Brix. The acidity varies from 0.46 to 0.39 percent. Seeds are soft (1.64 kg/cm^2). The variety is susceptible to fruit borer, leaf spot and fruit spot.

Kandhari

The growth habit of tree is spreading type with evergreen nature. The leaf is entire and elliptic in shape. The tree flowers throughout the year with three main flushes. Fruit is round and smooth and pink in color with reddish tinge. Average fruit weight is 317.63g, arils are sweet in taste with light pink in color. The juice percent varies from 52.33 to 50.15. The T.S.S. of the juice ranges from 14.4 to 14.8 °Brix, the acidity ranges from 0.41 to 0.40 percent. Seeds are soft (1.26 kg/cm^2 pressure), the variety is susceptible to fruit borer, leaf spot and fruit spot.

Kabul

The growth habit of tree is spreading type with evergreen nature. The leaf is entire and elliptic in shape. The tree flowers throughout the year with three main flushes. Fruit is round and smooth and yellow in color with reddish tinge. Fruit weight per fruit was maximum in Mrig bahar (298.18g), whereas, in Ambe bahar it was 193.67 g / fruit. The length of the fruit was found to vary, in mrig bahar it was 8.17cm and in Ambe bahar it was 6.56 cm. The breadth also showed similar trend, it was 8.3 cm in Mrig bahar and 7.03 cm in Ambe bahar. Arils are sweet in taste with light pink in color. The juice percent showed variation which was 48.09 in Mrig bahar and 53.20 in Ambe bahar. T.S.S. of the juice observed to be 16.2 °Brix in Mrig bahar and 17.4 °Brix in Ambe bahar. The acidity in Ambe bahar was 0.55 percent and in Mrig bahar it was 0.47 percent. Seeds were soft with softest in Ambe bahar (1.66 kg/cm^2 pressure) compared to Mrig bahar seeds (1.78 kg cm^2 pressure). The variety is susceptible to fruit borer, leaf spot and fruit spot.

Mridula

It is a hybrid developed at MPKV, Rahuri from the cross Ganesh x Gul-E-Shah Red. The growth habit of tree is spreading type with evergreen nature. The tree flowers throughout the year with three main flushes. Fruit is round and smooth and pink in color with reddish tinge. The fruit weighs on an average about 252g. Arils are sweet to taste with light pink in color. The juice content varies from 51.79 to 50.15 percent. The T.S.S. of the juice ranged from 14.92 °Brix to 14.8 °Brix. The acidity is about 0.45 percent. Seeds are soft (1.09 kg/cm^2). The variety is susceptible to fruit borer, leaf spot and fruit spot.

Bhagwa

The variety is known by various names viz. Shendri, Asthagandh, Mastani, Jai Maharashtra, and red Diana in various parts of Maharashtra. The variety is as on date is the leading variety of Maharashtra state. It is a heavy yielder and has desirable fruit characters like attractive fruit colour, dark red arils, no cracking for domestic as well as export market. The fruits mature late in 180-190 days after fruit set. Average yield is 30-40 kg fruits / tree. The fruit is having attractive saffron colored thick, glossy skin, arils dark red and bold. The variety is less susceptible to fruit cracking, fruit spot and thrips.

Phule Arakta

The variety is released by Mahatma Phule Krishi Vidyapeeth. It is a heavy yielder and suitable for export due to attractive skin and aril color; fruits are ready for harvest within 120-130 days.

Alandi

It is a local variety found growing in Maharashtra.Spreading type variety, produces medium sized fruits with hard seeds, arils are red in color and sweet in taste.

Jyoti

It is highly promising type earlier known, as GKVK-1 .It is a selection from basin seedless. It produces fruits with soft seeds and red aril color.

Ruby

It is released from Indian institute of Horticulture research Bangalore. The fruit size is small having deep red arils. The variety is a medium yielder.

Amlidana

It is a F_1 hybrid from the cross Ganesh x Nana. It grows well under tropical climate with quality fruit attributes. Amlidana is superior to sour variety Daru, whose trees come up naturally in temperate region of North India. Its fruits provide more acidic (16-18 %) anardana and higher fruit yield /tree. In addition, short statured trees are suitable for high density planting, giving increased fruit yield /unit area (Jalikop *et al*, 2002).

Goma Khatta

It is a F_1 hybrid from the cross Ganesh x Nana cultivar identified for high acidity, high yield and anardana, released from Central Horticultural Experiment Station, Godhra, Gujarat. The cultivar is suitable for making anardana. It has high acid content, an ideal character for making anardana. It is suitable for high density planting at 5x4 m spacing and 500 plants can be accommodated in a hectare. It requires 150 to 160 days to mature with yield per plant being 11.08 kg under rainfed conditions of Gujarat. The hybrid also yielded anardana to the tune of 1.55 kg/plant (Hiwale *et al.,* 2009).

Germplasm evaluation

The world's largest collection comprising 1117 entries is maintained at Turkmenistan (CIS). They exhibit great variability for fruit size and quality, skin color, resistance to abiotic and biotic stresses but most of them are temperate in nature and do not grow well in the Indian tropics.

Most of the varieties have one or the other superior traits but are susceptible to Bacterial Nodal Blight, a disease that has threatened the pomegranate cultivation. The quality parameters are highly influenced by the soil and climatic conditions. There is a need to develop a variety with desirable traits like bold arils with red color with soft seeds, resistant to blight, drought, cracking and aril browning.

Evaluation of pomegranate germplasm under semiarid rainfed conditions indicated that seeded varieties bear fruits with higher fruit weight, where as their aril-seed ratio was the least and 100 seed weight the highest. Seasonal variation affected all the physico- chemical characters studied. In 'Hasta bahar' the quality of the fruit was the best compared to 'Mrig bahar' fruits. Acidity declined in Hasta bahar. The seeds of hasta bahar were harder than Mrig bahar fruits but the juice color was attractive in Hasta bahar. Fruit weight in Mrig bahar ranged from 95 g to 200g whereas; in Hasta bahar it ranged from 141g to 261g. Fruit length varied from 75.5 to 50.0 mm and 75.5 to 66.5mm in Mrig and Hasta bahar respectively. Fruit diameter varied from 73.5 to 50.05cm and 78 to 70.2cm in both Mrig and hasta bahar respectively. Aril seed ratio varied from 1.42 to 0.63 in Mrig and 3.02 to 0.69 in Hasta bahar. Juice percent varied from 77.3 to 58.3 in Mrig bahar and 78.25 to 62 29 in Hasta bahar. Acidity of the fruit ranged from 1.45 to 0.35 in Mrig and 0.90 to 0.28 in Hasta bahar Hiwale (2009). Purohit (1985) in his studies on seed softness of commercial varieties reported that thickness of the testa, weight of whole seed, density of the seed, and weight of testa as % of seed weight can be used to determine softness of seed of pomegranate. The scale fixed by him indicated that variety having thickness of seed testa <0.5mm, the density of whole seed< 0.6g/ml, the density of seed testa <0.4g/ml, weight of the seed testa <50 % of the weight of whole seed. Bedana, Bassein seedless, Ganesh and Dholka were soft seeded. Though Bedana had soft seeds it cannot be grown commercially owing to its deciduous nature and sparce flowering.

Evaluation of 24 genotypes carried out by Meena *et al.,* (2003) indicated that all the characters except total soluble solids and total sugar had less pronounced differences between the value of phenotypic and genotypic coefficient of variation. Genotypic coefficients of variation were higher for seed content, wt. of 100 aril, wt. of 100 seeds, acidity and average fruit weight. The results indicated that higher magnitude of variation for above traits offer better opportunities for improvement through selection. These results are in consonance with those of Chaturvedi *et al.,* (1980) and Pandey and Bist (1998). Higher estimates of heritability were obtained for all the characters except total soluble solids, ascorbic acid and total sugar content, which may be due to environmental effect on these traits. Acidity and wt. of 100 seeds had high heritability in spite of low genetic advance, which may be due to introduction of replication in the system (Burton and Devmore, 1953).While low genetic variability and low genetic advance were noted in total soluble solids. High heritability estimates coupled with high-expected genetic advance as percent of mean were observed for mean seed content, wt. of 100 seeds, acidity content and average fruit weight indicating additive genes effect on these traits. These characters can be considered as reliable selection indices. High to moderate heritability along with low genetic advance were recorded for total sugar and ascorbic acid content indicated that these characters might be controlled by non additive gene action and improvement through selection for these characters would be rather ineffective.

Breeding

Pomegranate is a genetically heterozygous material, thus there exists lot of variability in the nature. There is scope for crossing two selecting genotypes. The main objectives in the breeding are;

- Develop varieties resistant to cracking and splitting of fruits.
- Varieties resistant to fruit borer and stem borer.
- Develop varieties having high sugar (16%) and less acid (1.5 %)
- Varieties bearing high percentage of fertile flowers should be achieved.
- To develop better varieties for Western India by hybridization between the locally adopted *Alandi* and *Dholka* types. *Alandi* types have medium sized fruits, deep red to deep pink flesh, hard seeds and quite sweet juice but are slightly more acidic than *Dholka* types. *Dholka* types bear larger fruits with greenish white rind, whitish flesh, soft seeds and sweet juice.
- To develop better varieties for Southern India by hybridization between prolific bearing Paper Shell or Spanish Ruby, and the soft seeded Musket Red and the large-sized fruit bearing Vellodu types.
- Breeding varieties having dark red aril, skin color irrespective of the season
- Breeding anardana type varieties with dark red and bold arils, high acidity, bigger fruit size and high yield.

Breeding methods

a) Selection

Cheema *et al.*, (1954) at Fruit Research Station, Ganesh khind, Pune was the first to indentify a selection from Alandi, which became very famous amongst the farmers of Maharashtra and was named "Ganesh". A critical evaluation of Ganesh resulted in identification of five promising clones viz. G-107, G-132, G-

133, G-134, G-137(Sawant, 1993). The performances of these selections were further studied (Keskar *et al.*, 1990), which revealed that clone G-137 performed better than Ganesh. Seedling selection work initiated in cv. Muscat at MPKV, Rahuri resulted in selection of over 37 large fruited types with soft seeds and high TSS (Chaudhari and Shirsat, 1976). The survey in seedling gardens resulted in the identification of four promising types P-13, P-16, P-23, P-26 (Naik, 1975). In Karnataka seedling selection from Bassein seedless was identified at UAS, Bangalore and was named "Jyoti" for its softseededness and bigger fruit size. In Rajasthan "Jalore Seedless" is being grown as the traditional variety.

b) Hybridization

The commercial cultivars Ganesh and Musket though prolific bearers and soft seeded lacked the dark red colour of the Russian types, which were sour in taste. The hybridization programme initiated at MPKVV, Rahuri resulted in the F1 progeny screening but none of the hybrids could be used directly because of the defects observed in them (Kale, 1986). The F2 progeny developed from open pollinated F1 resulted in the identification of superior type combining all the desirable attributes and was named Mridula (Keskar *et al.*, 1994). Hybridization work carried out at IIHR Bangalore resulted in identification of multiple cross hybrid "Arka Ruby", released at institute level (Jalikop *et al.,*).

Breeding work initiated at CHES, Vejalpur using various cross combinations along with chemical mutagen colchicine @ 0.5 % resulted in identification of good F_1 hybrids. Mutation, however, resulted in smaller size fruits which are not a desirable character.

Characters of hybrids developed at CHES Godhra

Hybridization work carried out at CHES, Godhra under semiarid rain-fed conditions resulted in identification of some hybrids with big arils and varying aril colour, with superior performance to existing hydrids. Some of the features of identified hybrids are enlisted.

Hybrid -1

It is from the cross Ganesh x Jalore Seedless. It is superior in fruit size and aril size to both the parents. The mean fruit weight is 249.66 g and aril size is 10.5 x 8.5 mm (bold aril) with a fruit to aril ratio of 1.33 and fruit to skin ratio of 4.18.The number of days to maturity is 110-115 in Hasta bahar crop under semiarid rainfed conditions. The hybrid is superior to its parents in respect of fruit set (71.69 %) , fruit retention (51.88%) and ultimately yield per plant (13.73 kg) in four year old plant.

Hybrid -2

It is from the cross Ganesh x Sindhuri. The hybrid has bold arils (10.0 x 7.87 mm) and bigger fruit (216 g.). The arils are light pink in colour with a fruit to aril ratio of 1.31 and fruit to skin ratio of 4.15.The number of days to maturity is 120-125 in Hasta bahar crop under semiarid rainfed conditions. The hybrid is superior to its parents in respect of fruit set (70.85 %), fruit retention (59.95 %) and ultimately yield per plant (8.86 kg) in four year old plant.

Hybrid -3

It is a hybrid between Ganesh x Line I of anardana type fruit. The fruit size is superior than the male parent and simlar to female parent. The hybrid has dark red aril colour. The arils are of medium size (10.75x6.5 mm) with a fruit to aril ratio of 1.32 and fruit to skin ratio of 4.42. The percent fruit set was 67.64 and fruit retention was 60.29 %. The Total soluble solids was 15.1° Brix with an acidity of 0.48 %. The yield was 8.21 kg per plant under semiarid rainfed conditions.

Methodology

Selfing is done by bagging the bud a day prior to anthesis. Select appropriate young bud, which will open the next morning and by using the forceps emasculation is done. This can be done easily as stamens emerge from within upper half or more of hypanthium (Calyx tube) and are distant from ovary. Crossing is carried out the next day morning, by collecting the desired anthers and smearing them with the help of a brush on to the receptive stigma of the emasculated bud, the next morning.

Hybridization is done in the morning hours before 11am. Fruits require about 120-130 days for maturity. Crossed fruits are harvested and the seeds extracted. They are sown in plastic trays. Seeds germinate within 10 days of sowing. The seedlings are transplanted to polythene bags and subsequently to field for further evaluation, Bhandari (1979).

Inheritance

Inheritance studies carried out indicated that high acidity was always dominant to low acidity, pink aril color and hard seed nature was dominant to soft (Jalikop *et al.*, 2005). Recessive markers for yellow foliage color and rosette forming habit originated from a mutant of Kabul Yellow were used for breeding and studying the mode of pollination of pomegranate (Jalikop *et al.*, 2003). Hard seeded pomegranate had higher fruit weight and volume than soft seeded ones (Jalikop and Kumar 1998). Karale and Desai (2000) found that heterosis values were the highest for juice weight and aril weight percent. Levin (2006) in his work on breeding for frost resistance developed some hybrids with good resistance to frost.

In Israel, pomegranate breeding resulted in identification of early ripening, good colour and soft seeds from open pollinated seedling selections of known cultivars. The cultivar 'Emek' was identified with pink red skin and bright red arils .The seeds are soft, sweet .The average fruit weight is higher than other commercial cultivars. Another new Israeli cultivar 'Kamel' is superior to cv. Wonderful with respect to fruit quality and tree growth .Its distinguishing characters are dark red skin color.

Mutation breeding

Chemical mutagenesis was carried out by Shao *et al*., (2003), by treating the shoots with Colchicine. The resultant tetraploids had short roots, wider and shorter leaves and flowers with enlarged diameter as compared to diploid Matuskovic and Micudova (2006). Kerakadze (1987) created mutation by gamma irradiation. Terakami *et al*., (2007) developed genetic transformation in pomegranate. Development of transformation in 'Nana' was useful as a model system to study genetic manipulation of pomegranate.

Molecular characterization

Talebi Baddaf et al., (2003) reported low variability among Iranian genotypes using random amplification of polymorphic DNA (RAPD). However, RAPD analysis of 24 Iranian cultivars reported by Zemani *et al*., (2007) indicated a high level of polymorphism among the genotypes. In China, RAPD analysis of 25 pomegranate accessions showed that the genetic background is complex and difficult to classify. The RAPD data disagree with the traditional taxonomy based on flavor, petal color and skin and aril colour (Feng et.al.2006, Yang *et al*., 2007). There is a narrow variation and lack of significant divergence in central Asian pomegranate varieties. Amplified Fragment length Polymorphism (AFLP) analysis of 65 accessions representing Central and West Asia region as well as Russia and United States, (Aradhya *et al*., 2006). Yilmaz *et al*., (2006) also found narrow genetic base in Turkey using RAPD analysis. However, RAPD and AFLP analysis of Tunisian collection reported high polymorphism (Hasnoni *et al*., 2006). The Israeli germplasm has been evaluated for antioxidant content like ellagic acid, punicalgin, punicalin and galagic acid (Tzulker *et al*., 2007). Polymorphism was noticed in 36 genotypes by Inter-small Sequence repeats (ISSR) technique.

Gene extraction work also initiated and a few genes were isolated. Most of the genes isolated involved in production of unsaturated fatty acids, genes that encoded for parts of riboA and RNA, the mitochondria – R gene and rbcl genes, about 10 genes involved in anthocyanin biosynthesis pathway have been isolated from several pomegranate cultivars (Jbir *et al.,* 2006 and Zemani *et al*., 2007).

Future line of work

1. Most of the breeding is done through traditional method of crossing. Thers is scope for use of advance breeding methods such as genetic transformation system.
2. Breeding to resistance/tolerance to biotic stresses, particularly Bacterial blight, Fusarium wilt, Fruit rot and fruit borer.
3. Identification of genes for increasing storage life through different breeding methods.
4. Confirmation of marker assisted polymorphism needs to be undertaken.
5. Database of expressed sequences tags, Ests sequences and genetic maps need to be developed.

References

Anonymous., 2005. Pomegranate Smaranika, Published by Maharashtra Pomegranate Growers Association, Pune.

Aradhya, M., 2006. Low genetic diversity and lack of genetic differences in Punica as evidenced amplified fragment length polymorphism P.64. In ISHS 1st Symp. Pomegranate, Minor and Mediterranean fruits. 16-19 Octo. Adana, Turkey.

Bhandari, M.M. 1979. Practical's in Plant Breeding, Oxford & IBH Publishing Co. New Delhi. 23:52-79.

Burton, G.W. and De Van .W. 1953 Estimating heritability in tall fescue (*Festuca aruninancea*) from replicated clonal material. *Agronomy Journal* 45:478-81.

Chaudhari, K.G. and Shirsat, N.S. 1976. Improvement of pomegranate (*Punica granatum*) by selection. *South Indian Hort*. 24 (2):56-59.

Chaturvedi, K.N., Rai,·J.N. and Cupta, D.B. 1980. Variability in grape. *Ind J. of Hort*. 37:262-4

Cheema, G.S., Bhatt, S.S. and Naik, K.C. 1954. Commercial Fruits of India, MacMillan & Co.

Feng, Y.Z., Song, M.T. and Han D.B. 2006. The general status of pomegranate germplasm resources in China. *China fruits* 4:57-58.

Hiwale, S.S., Bagle, B.G. and More, T.A. 2009. Root distribution pattern in pomegranate cv. Ganesh. Paper presents in paster session of International symposium on pomegranate including mediterranian fruits held at UAS Dharwad from 23rd June to 27th June.

Hasnoui. N., Mars, M., Chibani Ben, J., Marrakhi, M. and Trifi, M. 2006. Molecular Polymorphism in Tunisian pomegranate (*Punica granatum* L.) germplasm as related by RAPD finger prints.P.70. In ISHS.1st Symp. Pomegranate, Minor and Mediterranean fruits.16-19- Octo. Adana, Turkey.

Jalikop, S.H., Tiwari, R.B. and Sampath Kumar. 2002. Amlidana: a new pomegranate hybrid. *Indian Hort.* July Sept. 2002 PP: 22-23.

Jalikop, S.H., 2003. Rossetted siblings in F2 of a cross in pomegranate (*Punica granatum* L) can be useful model for resetting investigations. *Euphytica,*131:333-342.

Jalikop, S.H., Rawal, R.D. and Kumar, R. 2005. Exploitation of sub-temperate pomegranate Daru in breeding tropical varieties. *Acta Hort.* 696:107-112.

Jalikop, S.H. and Kumar, P.S.1998. Use of soft ,semisoft and hard seeded types of pomegranate (*Punica granatum*) for improvement of fruit attributes. *Indian J. Agri. Sci.*68:87-91.

Jbir, R, Hasnoui, N., Trifi, M., Marrakchi, M. M.M. and Mars, M. 2006. Molecular polymorphism in Tunisian pomegranate (*Punica granatum* L.) cultivar as related by AFLP: pp71. (In) ISHS.1st Symp. Pomegranate, Minor and Mediterranean fruits. 16-19- Oct. Adana, Turkey.

Kale, R.H. 1986. M.Sc. (Agri) Thesis MPKV, Rahuri.

Karale, A.R. and Desai, U.T. 2000. Study of hetrosis for fruit characters in inter cultivar crosses of pomegranate (*Punica granatum* L.) *Genet. Plant Breed.* 60:191-196.

Kerakadze, I.G. 1987. Radiation mutagenesis in subtropical crops. p. 231-254. (In) Radiationnyi mutagenez I ego rol' v evoliutsii Nauka, Moscow, Russia.

Keskar, B.G., Karale, A.R., Dhawale, B.C. and Kaulgud, S.N. 1994. Variety release Proposal, Dept. of Hort., MPKV, Rahuri.

Keskar, B.G., Karale A.R. and Kale, P.N. 1990. Improvement of pomegranate. (In) K.L. Chadha and O.P.Pareek (eds.) Advances in fruit crops. pp.399-405Vol.1. Malhotra Publishing House.

Levin, G.M. 2006. Pomegranate roads: A soviet botanist exile from Eden pp.15-183 edits. B.L. Baer (ed) Florest press, Forestville, CA.

Matuskovic, J, and Micudova, O. 2006. Practices with chemical mutagens (Nitrium azide) on growth habit (*Punica granatum* L.) P.41 (In) ISHS.1st Symp. Pomegranate, Minor and Mediterranean fruits. 16-19 Octo. Adana, Turkey.

Mehranna, K.D. Badraldin, E.S.T. and Masoud, B. 2006. Characterization of microsatellites and development of SSR markers in pomegranate P.66. (In) ISHS.1st Symp. Pomegranate, Minor and Mediterranean fruits. 16-19- Octo. Adana, Turkey.

Meena, K.K. Room Singh and Singh S.K. 2003. Genetic variability and genetic advance to average fruit weight and its components traits in pomegranate (*Punica granatum*) genotypes. *Indian J. Agril. Sciences,* 73(11):630-632.

Naik, V.B. 1975. M.Sc. (Agri) Thesis, MPKV, Rahuri.

Pandey, G. and Bist H.S. 1998. Variability, correlation and path analysis in pomegranate germplasm. *The Hort. J.*, 11:7-12.

Purohit, A.G. 1985. Soft seediness of commercial pomegranate varieties. *Indian J. of Agril. Sciences*, 55(5): 367-368.

Sawant, B.A. 1993. Biochemical investigations of pomegranate arils in relation to blackening. M.Sc. Agri. Thesis submitted to MPKV Rahuri.

Shao, J, Chen, C. and Deng, X. 2003.*In vitro* production of tetraploids in pomegranate (*Punica granatum*) .*Plant Cell Tiss. Organ Cult.*75:241-246.

Talebi Baddaf, M. Sharifi Neia, B., and Bahar, M. 2003. Analysis of genetic diversity in pomegranate cultivars of Iran RAPD markers. *Proc. 3rd National Cong. Biotech. Mahad*, Iran 2:343-345.

Terakami, S., Matsuta, N., Yamamoto, Yamamoto, T., Gemma, H. and Soejima, J. 2007. Agrobacterium mediated transformation of the dwarf pomegranate (*Punica granatum* L. cv. Nana) *Plant cell Rep.* 26:1243-1245.

Tzulker, R., Galzer, I., Bar-Illan, I., Holland, D., Aviram, M. and Amir, R. 2007. Antioxidant activity, polyphenol content and related compounds in different fruit juices and homogenates prepared from 29 different pomegranate accessions. *J. Agric, Food Chem.*

http:/'Pubs .acs.org. cgibin. /abstract.

Yang R.P. Long, W.H. Zhang, H, Xu, B. and Li, W.X. 2007. RAPD analysis of 25 *Punica granatum* germplasm resources collected from Yunnan province. *J fruit Sci*.24: 226-229.

Yilmaz, M., Akka Kacer, Y., Yalcin-Mendi., Y. and Ozguven, A.I. 2006. Molecular characterization of some Pomegranate (*Punica granatum* L.) genotypes by RAPD markers pp72. (In) ISHS 1st Symp. Pomegranate and Minor Mediterranean fruits. 16-19 Octo. Adana, Turkey.

Zamani, Z., Sarkhos, A., Fatahi, R. and Ebadi, A. 2007. Genetic relationship among pomegranate genotypes studied by fruit characteristics and RAPD markers. *J. Hort. Sci. Biotech*. 82:11-18.

14

Papaya Breeding

M.R. Dinesh and G.L. Veena

Papaya is one of the fruit crops grown both commercially as well as a backyard crop. Papaya was mainly a backyard crop till recently. It emerged as a main commercial crop during the last decade, because of the high nutritive value. It is a rich source of Vitamin A (2020 I.U), Vitamin B1 (40 mg), Vitamin C (46mg), protein (0.5%) and mineral matters (0.4%). Ripe papaya is a favourite breakfast and dessert fruit available year-round. It is used for making fruit salads, refreshing drinks, jams, jelly, marmalade, candies and crystallized fruits. Green fruits are pickled or cooked as vegetable. Carpine, an alkaloid present in papaya, can be used as a heart depressant, amoebicide and diuretic. It is also grown for the extraction of papain, a proteolytic enzyme present in the latex, collected mainly from the green fruit. In 1875, T. P. Lucas, a British physician discovered the medicinal value of papaya and started a hospital in Brisbane, to treat patients solely with papaya. The people in Papua New Guinea use papaya for the skin and to treat rashes or sunburn or, with repeated applications, to remove the brown spots of aging. They also say that if you put papaya into the compost pile, they grow especially large! Natives in the Pacific Islands use all parts of the papaya tree for medicinal purposes — leaves, skin, seeds, bark, roots, flowers and of course, the fruit. South American women massage their breasts with thin slices of green papaya to stimulate the milk glands. Eating the fruit provides energy and nutrients for both herself and the baby. Papaya bark is used as a toothache remedy and the flowers in teas, to treat bronchial infections. For centuries, teas from the roots were used to expel parasites and to alleviate bleeding, kidney colic, and jaundice.

Papaya is widely grown in the tropical and sub-tropical regions of the world. It is being grown in about 57 countries of the world. Brazil, India, Indonesia, Mexico and Nigeria are the leading countries. Other countries which grow papaya are Indonesia, Mexico, China, Thailand, Peru and Philippines.

Origin, distribution and species status

Papaya (*Carica papaya* L.) is native to Tropical America. It is believed to have originated in Southern Mexico (De Candolle, 1884). It is presumed that the genus *Carica* L., must have originated from natural hybridization involving *C. peltata* Hook. & Arn. It was introduced to India in the early part of 16th century from Philippines through Malaysia (Schroeder, 1958). It is being grown for fresh fruit and for papain extraction. Opinions differ on the origin of *C. papaya* in tropical America. It is likely that papaya originated in the lowlands of eastern Central America, from Mexico to Panama (Nakasone and Paull, 1998). The seeds were distributed to the Caribbean and Southeast Asia during the 16th century, and to India and Africa.

Caricaceae family

Papaya, *Carica papaya* L. (2n =18), belongs to family Caricaceae. Caricaceae contains herbaceous, shrubby or arborescent plants with a well-developed system of articulated laticifers. The usually sparsely branched plants are mostly dioecious and bear berry-like fruits (Badillo, 1971, 1993, 2000). Caricaceae consists of six, relatively small genera. Now *Carica* is monotypic containing the best-known and economically most important species of this family *Carica papaya*.

i. *Cyclicomorpha:* It is the only genus of the family that is restricted to humid habitats, occurs in montane forests of equatorial Africa.

ii. *Horovitzia*: It is a monotypic genus of hairy herbaceous plants endemic in Mexico.

iii. *Jarilla:* This genus consists of three herbaceous species; all restricted to Southern Mexico and Guatemala.

iv. *Jacaratia:* It contains seven species of trees with compound leaves, distributed from Mexico to the north of Argentina.

Vasconcellea, by far the largest genus of the family with 21 of the 35 Caricaceae species, has only recently been resurrected on generic level by based on morphological and genetic characteristics (Aradhya *et al.,* 1999 and Badillo 2000). Earlier the genus *Carica* L. was divided into two sections, Carica and Vasconcellea, based on the number of locules in the ovary (1-locular and 5-locular, respectively). Now Carica is monospecific containing the best-known and economically most important species of this family, *Carica papaya*. It is opined that evolutionary process of recent differentiation and speciation most likely lies at the origin of the high incidence of interspecific intercompatibility in Vasconcellea. Since Vasconcellea seems to be a young complex network of closely related species, it might therefore be useful to introduce the concept of hybridizing species complexes or "syngameons" into future taxonomical revisions of the genus Vasconcellea. Such complexes consist of numerous genetically distinct species or lineages, which periodically split and/or fuse as they extend through time. During splitting and fusion, morphologic intermediates form and species overlap. Based on all available evidence, the following species complexes with different subspecies might be recognized within Vasconcellea:

- *Complex 1: V. stipulata, V. × heilbornii, V. weberbaueri, V. Parviflora*
- *Complex 2: V. quercifolia, V. chilensis, V. candicans (+V. crassipetala?)*
- *Complex 3:* all other analyzed taxa (+ *V. crassipetala*?). The genetic relationships among *C. papaya* cultivars and related species have been established using AFLP markers by Kim *et al.*, (2002). They observed that self-pollinated hermaphrodite cultivars were as variable as open pollinated dioecious cultivars and supported the notion that *C. papaya* diverged from the rest of Carica species early in the evolution of this genus.

The morphological resemblance with the common papaya and their being mainly found in higher altitudes, *Vasconcellea* spp. are often referred to as highland papayas. They are particularly common in upland valleys of Ecuador and Columbia but they can be found from Venezuela down to the north of Argentina. In comparison to papaya, most of these plants tend to be smaller, less succulent and quite different in fruit taste. *Vasconcellea* as a species based on morphological and genetic characteristics has been reviewed and the information revised at generic level and now the genus consists of 21 species (Badillo, 2000). Its biodiversity center is located in Ecuador, where 16 species have been described to grow in the Andean valleys up to 3,500 m above sea level (Romeijn-Peeters, 2004). Five species of this genus have been placed on the IUCN Red List of Threatened Species: *V. horovitziana, V. omnilingua, V. palandensis, V. pulchra* and *V. sprucei.*

Interesting traits, for instance cold tolerance and disease resistance, might be valuable for papaya breeding programmes, to extend the range of papaya cultivation and improve its yield. *V. candicans, V. cundinamarcensis, V. stipulata, V. × heilbornii and V. cauliflora* are reported resistant to "PRSV'

(Papaya Ring Spot Virus), a major threat to papaya cultivation. The pleasant fragrance of *V. stipulata* and 'Babaco' fruits, the monoecious habit of *V. monoica*, the cold tolerance of *V. cundinamarcensis* and *V. stipulata*, and the ornamental qualities of pink-flowered *V. parviflora* can also be used in the breeding programme as they have desirable traits, which will be of interest to breeders. Intergeneric hybridization has been attempted between *C. papaya* and *Vasconcellea spp*., but mostly without success confirming the recent resurrection of Vasconcellea, which separated *Carica papaya* from the highland papayas. Magdalita *et al.,* (1997) showed that genome incompatibility barriers prevented intergeneric hybridization, leading to abortion of hybrid seeds soon after their formation.

Botany, reproductive mechanism and floral biology

Papaya being a polygamous species has male, female and hermaphrodite forms of inflorescence (Storey, 1937). Papaya, though dioecious in nature, the sex expression has been one of the most perplexing and lacking in explanation. Only in recent years there has been an understanding of the basic facts concerning flowering by several workers of the world. Ovideo (1526) whose account is believed to be the earliest published on papaya noted that the trees were male and female. Linnaeus (1753) classified papaya as a dioecous species. Storey (1937) reported the existence of four sex forms, namely, typical females or pistillate, typical males or staminate, hermaphrodite and intermediate. Hofmeyr (1938) reported the extent of nine different sex forms, the male, female, elongata sterile, hermaphrodite and monoecious, pentandria, coexistence of elongata, pentandria and pistillate. He further classified the hermaphrodite based on the nature of structural modifications into three forms, namely; pentandria, intermediate and elongata. Storey (1958) classified papaya flowers into eight broad categories for classifying the trees according to the modifications of sex. He categorized them as staminate, reduced elongata, elongata, carpelloid elongata, pentandria, carpelloid pentandria and pistillate. He further classified hermaphrodite trees into fifteen classes on the basis of seasonal shifts in female sterility, carpelloidy of stamens and both in combination. Hofmeyr (1939) reported that there are two forms of female trees, a typical stable form and sex reversing type. According to Singh (1964) the papaya has only 3 basic types; the staminate or male, hermaphrodite and pistillate.

Staminate flowers or male flowers

The flowers are bracteorate, sessile in cluster or recema with incomplete and actinomorphic flowers. The calyx consists of five lobes which are united (gamosepalous) small. The corolla consists of five petals which is gemopetalous and elongated and yellow in colour. The androecium consists of 10 stamens in whorls and epipetalous. The anthers are bilocular and intiose.

Pistillate flowers

The pistillate flowers are sub-sessile and bracteate. The calyx consists of five sepals and gamosepalous which is light green in colour. The corolla has 5 petals which is linear, polypetalous with twisted activation. Androecium is absent in pistillate flowers and gynoecium is sessile. The style is very short with a five lobed stigma. Ovary is superior, monocarpellary, point at placentation with numerous seeds. The fruit is a berry which is pulpy in nature. Seeds are blackish to light brown in colour, straight embryo and fleshy endosperm. The cotyledons are oblong and flat.

The hermaphrodite or bisexual flowers are similar to pistilate flowers but the inflorescence is multiflowered i.e. 5-6 flowers corymb. Corolla is gamopetalous. Stamens are 10 and 5 in two whorls, sessile at the base of petals. The ovary is functional, with five locules with parental placementation.

Flower types

The flowers of the papaya fall into different graded series of types from one extreme to the other with regard to morphology and sexual functionality. The unisexual staminate flower which is trumpet shaped having gamopetalous corolla with only free short petal limbs, 10 epipetolous stamens, and a non-functional rudimentary pistil is at the beginning. The unisexual pistillate flower having a corolla, which is gamopetalous in nature but superficially, appears to be polypetalous. It has a large functional pistil, but is entirely devoid of stamens. Numerous modifications of flowers, mostly bisexual, which differ in numbers of functional carpels and stamens, in carpellody of stamens, in degree of gamopetaly, and in other morphological features occur in between the pistillate and staminate flowers. Storey (1937) designated the several normal flower types and grouped the all irregular types into a few inclusive categories between the normal types. Hence, the flower classification becomes feasible and practical.

Storey (1958) has classified papaya flowers into eight broad categories for classifying trees according to modifications of sex expression. The categories are:

1. Staminate,
2. Teratological staminate,
3. Reduced elongata,
4. Elongata,
5. Carpelloid elongata,
6. Pentandria,
7. Carpelloid pentandria and
8. Pistillate.

In all, fifteen classes of sex variation were proposed for hermaphrodite trees on the basis of seasonal shifts in female sterility, carpellody of stamens and both in combinations. Likewise fifteen comparable classes are found among male trees. Two forms of female trees are known, the typical stable form and a sex reversing type reported by Hofmeyr. According to Storey (1958) there are in all 32 heritable sex forms in papaya.

Staminate

This is the typical flower of the male tree. The corolla is gamopetalous, forming a long slender tube for two-thirds to three-fourths of its length, but terminating in a short five-parted limb. There are ten stamens inserted at the throat of the corolla tube. These are in two series of five each, an inner series lying opposite the corolla lobes and an outer series lying opposite the sinuses between the corolla lobes. The flower is unisexual in that it has no functional pistil although a rudimentarey structure is present in the position which a pistil would be expected to occupy. This is generally referred to as a rudimentary or vestigial pistil, but anatomical evidence suggests that it is instead, an extension of the floral axis from which carpels no longer differentiate.

Teratological staminate

This term is used to designate certain staminate flowers in which there is some degree of carpel initiation and development. In these flowers, a number of hair like processes, doubtless vestigial carpels, arise around the base of the pistil-like structure. The number of these processes may lie between one and ten, with five being most common. Frequently, they are quite free from each other except at their base, but sometimes, they are connate for part of their length. At times, flowers may be found in which some of the hairlike processes show evidence of carpel morphology, sometimes having well differentiated stigmatic tissue at the apex.

Reduced elongata

The reduced elongata is in effect, a staminate flower. In so far as is known, it occurs only on hermaphrodite trees. It resembles the staminate flower morphologically, but differs from it histologically by having a shorter, stiffer, thicker walled corolla tube. These flowers are larger than but otherwise identical to the staminate flowers of staminate trees'. It fails to develop a functional pistil under certain environmental conditions, suggesting that it may be a transitional form.

Elongata

It is regarded as the typical hermaphrodite flower of the papaya, wherein the petals are fused together for one-fourth to three-fourths of their length, forming a tube. As in the staminate flower, there are 10 stamens arranged in two series. It possesses a fully functional elongate pistil which develops into a long, more or less cylindrical fruit. Typical flowers of this type contain five laterally fused carpels. However, one or more carpels fail to develop ontogenetically, so it is not unusual to find fruits consisting of one to four carpels. Also, frequently, pistils are produced having supernumerary carpels, so that fruits having six to nine carpels are commonly found. In a typical fruit, all carpels open into a central ovarial cavity. In many fruits, however, some carpels may be closed, and isolated from the central cavity. In nature more variations of the 'elongata' types have been identified.

Carpelloid elongata

The teratological forms of the elongata flower, is grouped under this. The main characteristic is the tendency of the stamens of the inner series to become carpelloid to some degree. The flowers in this category were described as different forms by different workers. In this flower type, numerous variations have been recorded. The degree of 'carpellody' may vary from slight to the point where the carpelloid stamen is fully developed with an ovule-containing locule and a stigma. Carpelloid stamens may remain affixed to the corolla or be partly or wholly free. Coincidental with stamen carpellody is a tendency towards freedom of the petals from each other, fusion of the carpelloid structure to the original elongata ovary, and, sometimes, abortion of one or more carpels of the elongata ovary.

Pentandria

The pentandria flower is generally regarded as a normal hermaphroditic type. It is characterized chiefly by its very short corolla tube, long corolla limbs, the presence of only five stamens with long filaments, and a more or less globose but furrowed functional pistil. Several workers defined it in different ways owing to the variations seen in this. The pistil of the pentandria flower consists of five carpels opening into a central ovarial cavity; number of carpels may vary from five to ten. Depending on the locules of individual carpels opening into the central cavity, or remaining closed and isolated from the central cavity, several forms can be described.

Carpelloid Pentandria

In this category, teratological forms are seen wherein stamens of the pentandria flower, i.e., the stamens of the outer series, become carpelloid. As in the carpelloid elongata flower, various degrees of carpellody are seen. Flowers of this type resemble pistillate flowers when all five stamens are fully carpelloid and fused laterally, especially when accompanied by abortion of some or all of the underlying original carpels, flowers of this type may resemble pistillate flowers superficially. They are conveniently referred to as 'Pseudopistillate flowers'. In certain regions they have been found only in certain hermaphrodite trees which, without close inspection, might be mistaken for pistillate trees.

Pistillate

This is the typical flower of the pistillate or female tree. It is unisexual in that it has a large functional pistil but lacks stamens. The petals appear to be completely free from one another; actually they are inconspicuously fused at the base. This type has also been described as different forms by different workers as in nature we find unstable hermaphrodite types, which are confused for female plants.

Floral biology

Knowledge of flowering, pollen characteristics pollination biology, fruit set and fruit development is pre-requisite for any planned hybridization programme. Besides, it is helpful in selecting cultivars having a good set and to adjust cultural practices in relation to flower emergence and fruit development. Information on pollen characteristics and stigma receptivity would facilitate right interpretation, on hybridization studies so that lack of obtaining hybrid seed is not merely ascribed to cross ability basis between them. Many workers have studied the season of flowering, flowering height, period of anthesis and dehiscence, pollen viability, stigma receptivity and seed set. It is noticed that there is variation in all these parameters depending on the climate existing in different regions.

The different *Carica* species, viz., *C. papaya, C. monoica, C. cauliflora* and *C. goudotiana* flower through out the year in India. The period of flower at anthesis after bud initiation varies in all the three sex forms. The female flowers develop within 32 days while the male flowers 42 days, after bud initiation. However, variation ranging from 45 to 47 days was observed in some regions. However, It is to be presumed here that the variation is because of the change in the climate. The anthesis in three sex forms viz., staminate, hermaphrodite and female takes place between 6 a.m. and 11 a. m.

The dehiscence of anther also shows variation depending on the existing climate. It takes place 24 hours before anthesis in all the species of Carica. However, there have been reports that dehiscence in staminate and hermaphrodite flowers take place between 10 to 12 hours before anthesis. High temperature and low humidity hasten the time of anther dehiscence.

The pollen grain in papaya remains viable 2-3 days before to 2 days after anthesis with maximum on the same day of anthesis. Allen (1963) was first to germinate papaya pollen *in vitro* by utilizing Boric acid. Later, it was found that the 100 percent pollen could be germinated only with the help of 5 percent sucrose solution (Subramanyam and Iyer, 1986). The amount of pollen produced in a single anther in different varieties of papaya showed the number of pollen grains to be in the range of 7,500 to 23,750 pcr anther among papaya cultivars. Papaya pollen can be stored up to 5 years without loosing the viability if proper conditions are created for storing. Pollen stored at room temperature in desiccation with 50% relative humidity could remain viable for 48 days. Papaya pollen could be stored successfully at sub-zero temperature at various relative humidity levels for upto 5 years. The receptivity of stigma as evidenced by percentage of fruit set was found maximum on the day of anthesis. The stigma receptivity of female and hermaphrodite flowers remains two days before and two days after anthesis.

Cytology, cytogenetics and genetics

Cytology

The chromosome number for all the cultivated varieties of papaya and species of Carica was 2n=18 indicating that it is diploid species. Cytology and cytogenetics of papaya have been studied by several worker Heilborn (1921) was first to report that the chromosome number for the species, *C.chrysopetala, C. pentandria, C.candamarcensis* and *C.papapya* to be 2n=18. Storey (1941) reported that *C.cauliflora* and *C.quercifolia* were diploids having the chromosome under 2n=18. Hofmeyr (1941) reported that the haploid number of chromosome is 9 for each of the known sex forms. All the species of the genus *Carica* has more or less similar size and similar looking chromosome (Hofmeyr, 1941). None of the workers who made cytological investigations in the genus *Carica* could detect any aneuploidy distinguishable of heteromorphic chromosome viz., sex chromosomes.

Genetic balance hypothesis of sex determination

The genetics of sex is very intriguing in papaya. Although several hypotheses were floated, the one suggested by Hofmeyr (1941) who put forward a genetic balance hypothesis to explain the basis of sex determination in papaya is widely accepted. According to this hypothesis, it is assumed that the factors determining femaleness are present in the sex chromosome and that of maleness in the sum total of autosomes. It is further assumed that M_1 and M_2 respectively represent the inert segments of sex chromosomes. The M_1 is slightly larger in dimensions than M_2 segment. It is presumed that the genes responsible for life are eliminated in these inactive region and causes lethality of the genotypes M_1M_1, M_1M_2, and M_2M_2. However, the genotypes M_1m and M_2m will be viable due to presence of an active sex chromosome.

This hypothesis has been further explained by assigning arbitrary value to sex chromosomes and the autosomes. For example, the balance value in favour of femaleness in the sex chromosomes may be represented as m=25, M_1=21, M_2=22 and in the sum total of the autosomal chromosome compliment in favour of maleness as 47. Hence, the value obtained for mm=50-47=3, which is in favour of femaleness; M_1m=46-47=1 in favour of maleness; M_2m= 47-47=0. The genotype mm has a value of 3 in favour of femaleness, M_1m has a value of 1 in favour of maleness and M_1m and M_2m have the value of 0. The females are stable because the value of 3 is more than the threshold value 0, whereas the male value and hermaphrodite are quite instable because of the value of 1 which is lower than the threshold value 0. Hence, the sex expression of males and hermaphrodites are determined by the modifying factors which push the balance closer are far away from the threshold value. The work on colchicines induced polyploidy by Hofmyr (1945) supported this hypothesis.

Sex linked inheritance

Male and hermaphrodite plants in papaya are heterozygous sexes and females as homozygous sex and the sex determination in papaya are of XY types by some workers (Singh, 1990). Hofmeyr (1941) with the help of three point test involving the genes, the flower colour, Y (yellow), y (white), stem Colour P (Purple), p(non-purple) and sex of the plant suggested the order of genes on the sex chromosomes with their approximate cross over values M-23.7, Y17-3 P.

Determination of sex in papaya at seedling stage poses a problem in papaya cultivation since the plants flower six months after planting. Attempts have been made to discover such sex linked character but none have succeeded in finding the immediate practical values.

Sex problem

Papaya though normally dioecious in nature, the sex expression has been one of the most perplexing problem to papaya growers. It is very difficult to identify the sex at the nursery stage. Although there are three basic types of sex i.e, male, female and hermaphrodite several variation in the sex forms are observed. Many workers suggested several ways and means to overcome the problems of sex in papaya cultivation. Singh (1990) based on studies on controlled pollinations over years, established that in the progenies of a cross whose parents are known (pure line), the males and bisexual or female follows a definite ratio. He also established that if flowers of female plants are pollinated with the flower of male plants, the progenies obtained will have approximately 50% males and 50% females. Similarly if bisexual plants are selfed or cross pollinated by bisexual flowers, the progenies obtained will be 25% females and 75% bisexuals.

Till date no definite criteria are found to identify the sex of papaya before flowering. However, a few attempts were made to identify the productive plants in the nursery stage. Kumar (1951) while studying the correlation between seed colour and sex of the plant observed that seeds having deep brown colour produce more number of female plants while light coloured seeds produced less of female plants. He also observed that 12 weeks old male platns had maximum stem elongation while the hermaphrodite or female plants had minimum. Studies on internodal length before flowering in different sex indicated that male plants had short internode while the female or bisexual had longer internodal length (Singh and Sirohi, 1977).

Studies on chemical analysis to identify the plant sex at seedling stage indicated that alkaline water extract of dried leave of undifferentiated seedling could be effectively used to obtain desired result. When the extract of the dried leaves of undifferentiated seedling is added to a mixture of Semon reagent and saturated solution of Copper sulphate the colour of the solution varied for different samples. This variation of colour was matched with colour of the leaf extracts after adding the reagent along with Copper sulphate to differentiate the sex (Singh, Majumder and Sharma, 1961). Different protein patterns were obtained in leaf and flower extracts of vegetative, male and female plants in papaya. As the flower developed, new bands were observed and others disappeared, and these changes were specific to each (Jindal and Singh, 1976). In the chemical analysis study of papaya parts, it was concluded that it may be possible to predict the sex of papaya plants by a study of their physiological metabolism. Physical mapping of sex determination gene in papaya was attempted by Zhiyong Liu*et al.,* (2003), their classical genetic analyses led to the conclusion that sex determination in papaya is the result of a single gene with three alleles, they used three hermaphrodite sex-linked markers, W11, T12 and CPBE55 for fine mapping of the sex locus on 991 F2 and 755 F3 individuals derived from the two cultivars Kapoho and SunUp. No recombination was found among these three markers in either the F2 or F3 populations. BAC end cloning and chromosome walking have allowed construction of a 900 kb BAC DNA containing the W11, CPBE55, and T12 markers. The physical distance between W11 and CPBE55, CPBE55 and T12, W11 and T12 is 500 kb, 400 bp and 900 kb, respectively. The large distance spanned suggests severe suppression of recombination in the genomic region that harbors the sex determination gene. They also cloned and sequenced 62 hermaphrodite sex-linked AFLP fragments based on a mapping population of 54 F_2 plants. Forty-three SCAR markers were developed and 27 of them were used to screen a papaya BAC library. Southern hybridization of three markers on BAC contigs revealed three duplicated segments in this genomic region. Southern hybridization of selected markers to total genomic DNA of the three sex types indicated that these markers are present only in male and hermaphrodite plants but not in female plants.

Sex reversal

Much has been written about the sex behaviour of papaya and particularly about the change of sex. The exact reasons of sex reversal in papaya are still not clear. Storey (1953) while studying the sex reversal phenomenon in papaya indicated that different sets of modifying genes may be responsible for the sex reversal in male and female trees. Although, the basic sex types are determined by genes, occasionally certain males and hermaphrodite plants undergo sex reversal. To overcome sex reversal problem in papaya, physical, chemical and cultural treatments have been suggested. Chowdhury (1957) reported that sex in papaya is a physiological phenomenon which is amenable to the changes by variation in environment. It has also been mentioned that season plays a major role on shift from female sterility to female fertility. He also reported that female sterility was higher during cool nights and short days. The explanation of environmental variation responsible for seasonal stifts in sex form in papaya seems to be more justified and the same has been reported by many workers. Lange (1961) found that during winter the young vigorous trees flower early, produce more number of flowers which are females and more fruits. However, in less vigorous tree, not much effect was noticed. Teaotia and Singh (1967) reported an increase in female sterility during winter months and restoration of female fertility in the same plant during summer months. Seemanthini (1967), while studying the effect of climate on sex differentiation in the cultivar Coorg Honey Dew indicated that the climate has some effect on the above phenomena. She observed high percentage of female in cv. Coorg Honey Dew during January planting than July planting under Coimbatore condition.

So far not a single method was found to be useful for solving the sex reversal problem. Several reports with regard to the use of growth regulators in changing the sex have not yielded productive results. In the nursery stage the taller seedlings have the tendency to become males and dwarf ones to be females (Seemanthini, 1968, Singh and Sirohi, 1977). Thinning of the tall plants in a population leads to increase in the female population thereby increasing the production per unit area. The production of gynodioecious varieties of papaya helps to a great extent in having the unproductive male plants removed in a plantation since this variety gives all productive plants i.e., females and bisexuals.

Modifiers of sex expression

In papaya, as mentioned above there are three basic sex types. However, more often than not sex expression gets changed in males and hermaphrodite flowers. Higgins and Holt (1914) compiled the first enumeration of sex forms which were described previously in the literature. To this they added forms observed originally by them. Their enumeration described thirteen forms. Additional forms were described later by Kulkarni, (1970) raising the number to fifteen. For the most part, these forms were described from single observations, the observers not realizing, apparently, that a given tree might represent a certain form at one time of the year and a different form at another time of the year. That seasonal transitions from one form to another did occur were observed by many workers.

Hofmeyr (1938) in reporting his hypothesis for sex determination classified all forms into three groups, namely;

1. Staminate and variations from it,
2. Pistillate, and
3. Hermaphrodite and variations from it.

He suggested that the variations within each group may be due to the interaction of modifying factors of a genetic nature with the environment producing various forms. Storey (1953) recognizing that some of Higgins' and Holt's forms represented seasonal variations of a single form, reduced all variations to the following four basic sex forms:

1. Form A, Pistillate or female tree;
2. Form B, Continuously fertile hermaphrodite tree;
3. Form C, Summer-sterile hermaphrodite tree;
4. Form D, Staminate or male tree.

He mentioned that sex reversal, could be the best interpreted as the response of the male (form D) and the seasonally barren hermaphrodite (Form C) to physiological changes. Under natural conditions, this is the most evident as a response to seasonal change, but comparable effects can sometimes be produced by beheading these forms of trees. Continuing studies since 1940 have indicated that the concept of three primary sex types based on both primary and secondary sex characters continue to be most reliable. However, Storey's classification of all types of trees into four all-inclusive classes has proven to be a case of oversimplification. Hofmeyr's concept of modifying factors more closely fits the facts as they are now known, although he did not hypothesize as to the genetical nature of the factors. Horvitz (1955) postulated the existence of specific male and female florigenic substances which are genetically controlled by a single gene with two alleles. The dominant allele promotes male structure, and the recessive allele promotes female structure in the flowers, which they do through the production of enzymes which catalyze compounds promoting maleness and femaleness respectively. The degree of sexuality of a given flower depends upon the proportion of male and female forming substances present in the initial primordial. Within a given genotype, temperature is the decisive factor in the production of a particular type of flower at a given time.

Storey, postulated a hypothesis that there are two sets of genetic factors, which modify sex expression in males and hermaphrodites under proper physiological stimulus. One set is responsible for seasonal shifts from female fertility to sterility and back to fertility by affecting the genetic development of the carpels. The other set is responsible for causing stamens to become carpelloid, often with their fusion to the pistil, and for their return to normal morphology. Female sterility behaves as a sex-limited character in that it is expressed only in male and hermaphrodite trees. The genetic factors, however, may be carried by female trees without exerting any influence on development and functional ability of the carpels of the pistillate flower's and passed on to the offspring. Carpellody of stamens is, perforce, limited to expression in males and hermaphrodites, but the causative factors, like those of female sterility, maybe carried by females and transmitted to the offspring.

The genetics of female sterility and stamen carpellody have not yet been fully worked out, so the number of genes responsible for each of these character modifications have not been determined. Studies to date suggest, however, that at least tow pairs of genes, but possibly more, are concerned in each. Furthermore, all gene pairs are independent of one another and none is sex-linker. When factors for female sterility and stamen carpellody occur together in the same tree, those for female sterility tend to be epistatic. A point of particular interest is that carpellody of stamens apparently can occur only in female fertile flower's and never in purely staminate flowers. The phenotype of a given individual tree, as judged by the modifications it undergoes in response to environmental changes, presumably is determined by the proportionality of the factors of each set that it carriers.

Male and hermaphrodite trees can be catergorized into five readily recognizable classes with regard to female sterility and fertility. These classes are:

1. Continuously female sterile throughout the year;
2. Female sterile during most of the year, but female fertile for a short period, usually three months or less;
3. Female sterile for about half the year, and female fertile the rest of the year;
4. Female sterile for a short period, usually less than three months;
5. Continuously female fertile throughout the year.

It has been observed in Hawaii, and elsewhere, that female sterility in sex reversing trees is enhanced by warm weather and long days, and female fertility by cool weather and short days. Trees which are female sterile in the warm months may become female fertile in the cool months. Trees, which are female sterile in the cool months, remain sterile for the rest of the year and, trees which are female fertile in the warm months, remain fertile for the rest of the year. No deviation from this pattern has ever been observed.

The periods of transition from female sterility to female fertility correspond closely to changes in the seasons of the year. Female sterility is the greatest from the end of June well into September. Some trees which are female sterile during this period become female fertile toward the end of September and remain sterile from June well into December when they become fertile, remaining so until June when they revert to sterility. Another class of trees remains sterile from June to middle of March when they become fertile for a short time, then revert to sterility. The times of transitions correspond so closely to the yearly equinoxes and solstices that the period of sex expression may be defined conveniently as summer, fall, spring and winter.

The expression of stamen carpellody follows a pattern similar to that of female sterility in that transitions from normal stamens to carpelloid and back to normal occur around the yearly equinoxes and solstices. Similarly, trees may be classified into five categories, as follows:

1. Those producing normal flowers the year around;
2. Those producing flowers with normal stamens during the summer, fall, and winter months, and carpelloid stamens during the spring months;
3. Those producing flowers with normal stamens during summer and fall months, and carpelloid stamens during winter and spring months;
4. Those producing flowers with normal stamens only during summer months, and carpelloid stamens the rest of year; and
5. Those producing flowers with carpelloid stamens throughout the year.

It may be seen that, in forms subject to seasonal variation, the cool weather promotes tendency towards stamen carpellody and normalcy is restored towards warm weather manifestation. The pattern of behaviour is analogous to that of female sterility in that a continuously fruitful tree producing flowers with carpelloid stamens during summer continues to produce such flowers during rest of the year, while a tree producing normal flowers during winter and spring continues to produce such flowers during the rest of year. It should be noted, also that a tree which produces carpelloid stamens only in the spring, reverts to production of normal flowers in summer; a tree which produces carpelloid stamens initially in winter, continues to do so in spring before reverting to normal; and a tree which produces carpelloid stamens initially in the fall, continues to do so through winter and spring before reverting to normal.

In trees having lesser degrees of stamen carpellody, i.e., carpellody in winter and spring only, the greater percentage of flowers is of the carpelloid elongata type, but pentandria flowers are commonly seen. In trees having higher degree of stamen carpellody, especially those which produce flowers with carpelloid stamens throughout the year, a high percentage of the flowers are of the pentandria and carpelloid pentandria types. Sometimes nearly all of the flowers are of the pseudopistillate type. When this is so, the trees resemble female trees superficially, and can be mistaken for them if not inspected closely.

Varieties

The genus *Carica* although has about 48 species only the species *Carica papaya* is grown its edible fruits. A large number of papaya varieties are found in cultivation. The number of varieties which is quite large has been found because of out crossing in papaya and these cannot be called as varieties.

Some of the varieties, which are popularly grown in different states, are given below:

States	Varieties
1. Bihar	Pusa Majesty, Pusa Delecious, Pusa Dwarf, Pusa Nanha and Ranchi
2. Karnataka	Coorg Honey Dew, Solo, CO 2 and Washington
3. Maharashtra	Washington, CO 2, Pusa Delicious, Pusa Majesty, Local, Ranchi and Taiwan hybrid.
4. Odisha	CO 2, Coorg Honey Dew, Washington, Local, Ranchi, Pusa Dwarf and Pusa Delicious.
5. Tamil Nadu	CO 1, CO 2, CO 3, CO 4, CO 5, CO 6 and Coorg Honey Dew
6. Uttar Pradesh	Coorg Honey Dew, Pusa Dwarf, Pusa Delicious CO 3, Local and Barwani Red.

Some of the important varieties and their description are given below:

1. **Washington**: The tree is hardy, petioles are purple pigmented and fruits are round to ovate in shape. The fruits are medium sized weighting 1 kg to 1.2 kg on an average. Flesh is yellow in colour and cavity is large. Keeping quality is moderate. Fruits are sweet to taste with 12° Brix T. S. S. It is dioecious in nature. It yields about 60 kg/ plant.

2. **Barwani Red**: It is similar to Washington, but is dwarf in height. It is a dioecious plan type. The fruits are larger and weigh on an average about ½ kg to 2 kg. It is devoid of purple colour. The fruits are sweet to taste and keeping quality is good. The TSS is 11.0° Brix. Fruit cavity is large. It yields about 40 kg per plant.

3. **Coorg Honey Dew:** It is a selection from Honey Dew or popularly called as Madhy Bindu variety. It is gynodioecious in nature and semi dwarf. Fruits are big sized, weighing on an average about 1.75 kg to 2 kg. Fruits are dark green in colour and have slight ridging skin surface. Fruits from hermaphrodite trees are elongated and oval whereas from female trees it is ovoid and not as long as bisexual ones. Pulp is yellow in colour, cavity large and keeping quality is poor. TSS is around 13.5° Brix. It can be used for both table and papain extraction.

4. **CO1:** It is a selection from the variety Ranchi. It was evolved by sibmating the cultivar Ranchi over a period of eight years at the Tamil Nadu Agricultural Univeristy, Coimbatore. The plant is semi-vigorous and the first fruiting takes place at the height of 60-75 cm from the ground level. It is a dioecious variety. Fruit is medium to big weighing around 1.5 kg spherical in shape. Pulp is orange yellow in colour, medium firm, moderately

juicy and keeping quality is good with 12° Brix TSS. It produces about 70 kg fruits per tree.

5. **CO 2:** It is a selection purified from a local type and released from Tamil Nadu Agricultural University, Coimbatore. It is, dioecious in nature. Fruits are medium sized weighing about 1.5 kg. Fruits obovate and skin is greenish yellow and ridged at the apex. Flesh is dark yellow in colour, soft to firm and moderately juicy. Fruit cavity is large and TSS is around 11° Brix. It produces about 70 to 80 kg fruits per tree. It has been mainly developed for papain extraction. It gives about 4 to 6 kg of dried papain per. It produces about 400-500 kg of dried papain per hectare.

6. **CO 3:** This is tree/year hybrid from the cross CO2 x Sunrise Solo. This was released from Tamil Nadu Agricultural University, Coimbatore. It is a gynodioecious variety. Fruits are obovate, smooth, weighing on an average about 800g. Fruit cavity is medium, pulp colour is red, firm and sweet to taste. TSS is around 13.5° Brix. Keeping quality is good. It is bred for table purpose. It yields about 100 to 120 fruits weighing about 60 to 65 kg during the bearing period of two years.

7. **CO 4:** This is an improvement over CO 1. This was released by Tamil Nadu Agricultural Univeristy. It is a hybrid progeny from the cross CO1 x Washington. The fruit is medium in size weighing on an average about 1.3 to 1.5 kg. Fruits are round in shape. The tree has purple colouration in all parts of the plant. Flesh is yellow in colour, firm and cavity is medium. It is quite sweet to taste with a TSS of 13° Brix. Trees give 80 fruits over a period of two years. It produces about 80 kg of fruits per tree. It is dioecious in nature.

8. **CO 5:** It is a selection from the variety Washington, isolated by Tamil Nadu Agricultural University, Coimbatore. It is a dioecious variety producing fruits of about 1.5 kg. Flesh is yellow in colour and moderately soft. It has been isolated mainly for high papain content. It gives 75 – 80 fruits per tree in two years.

9. **CO 6:** It is a selection from the variety Pusa Majesty developed by TNAU, Coimbatore. It is dioecious in nature and dwarf in stature. It produces large sized fruits of 2 kg. It is a dual purpose variety, being useful for table and papain extraction. Flesh is yellow in colour moderately firm and cavity is large. The TSS is about 12° Brix. It yields about 45 fruits / tree during bearing period and 745 kg papain/ ha. It yields about 60 kg fruits per tree.

8. **CO 7:** It is a hybrid from the cross Coorg Honey Dew x CP. 85 developed by Tamil Nadu Agricultural University, Coimbatore. It is gynodioecious in

nature. Fruits on average weigh about 1.15 kg. Pulp is red in colour and cavity is small. TSS is reported 16.7% under Coimbatore conditions. It yields about 98 fruits per tree and about 340 tons per hectare.

9. **Pusa Majesty:** This is a gynodioecious line developed from the variety Ranchi by sibmating. It starts bearing at the height of 48 cm within 245 days after planting. The average fruit weight is around 1 to 1.5 kg. It is a good yielder. The fruit has firm flesh of 3.5 cm thickness with orange colour fruit and cavity is 17x9 cm. The TSS ranges 7 to 10° Brix. Its shelf life is good hence it is suitable for long distance transport. It yields about 38 kg per plant. The Pusa varieties were developed at Regional Research Station of IARI, Pusa, Bihar.

10. **Pusa Giant:** This promising selection was developed by sibmating and selection from the variety Ranchi. It is dioecious in nature. Plants are vigorous and bear fruits of 2 to 3 kg. Plant being highly vigorous it can stand well for storm and windy conditions. It bears first fruit at one meter height. Flesh is yellow in colour and cavity is 18 x 10 cm, flesh is moderately firm and 5 cm thick. TSS is 7 to 8.5° Brix. It yields about 40 kg per plant.

11. **Pusa Delicious:** This is a gynodioecious variety. The plants of this variety are medium sized. It is a high yielding variety developed by sibmating from the variety Ranchi. The first fruiting height is 80 cm. In 253 days of planting, it starts bearing. The average fruit size is 1 to 2 kg. Fruits have distinct flavor. Flesh is deep orange in colour, 4 cm in thickness and TSS varies between 10 to 13° Brix. Keeping quality is moderate and seed cavity is 14 x 8 cm. It yields about 41 kg per plant.

12. **Pusa Dwarf :** This is a selection from the variety Ranchi developed by sibmating. Plants are dwarf in stature. It is dioecious in nature and is a medium yielder. It is the most suitable for high density planting and kitchen garden. Fruits are medium sized, weighing on an average about ½ to 1 kg and bears oval round fruits. It bears fruit at the height of 40 cm above the ground level. Flesh is yellow in colour, moderate to firm 3.5 cm in thickness and cavity is 12x8 cm. TSS is between 6.5 to 8° Brix. It yields about 40 kg / plant.

13. **Pusa Nanha:** It is a dwarf mutant of 106 cm height. It was evolved by treating the seeds with gamma irradiation and obtaining mutants. It bears fruit at a height of 30 cm above the ground. It is dioecious in nature. Fruits are medium sized, round to ovate in shape. Pulp is thin and yellow in colour. TSS is about 8° Brix and cavity is low. It is well suited for high density planting and is also tolerant to water logging. About 6,400 plants

can be accommodated per hectare. It is also useful for pot cultivation. It yields about 63 tons per hectare and about 10.1 kg per plant.

14. **Sunrise Solo:** It is an improved variety from Solo developed in Hawaii. It is an inbred strain resulting from a cross between two inbred strains 'Line 9' Solo strain with red flesh Kariya Solo and selected after three successive generations. The flesh is red in colour, with mild flavour with an average TSS of 15.5 per cent under good growing conditions. The fruits on an average weigh about 400 to 500g, pear shaped and smooth in appearance. It yields about 20 kg per plant.

15. **Waimanalo:** It is progeny of the cross between Betty variety of Florida and the line 5 and line 8 Solo strains developed in Hawaii. The fruits are round in shape. The fruits weigh on an average about One kg. Flesh is orange in colour, firm with small cavity. Fruits are sweet, with a TSS 13.5 to 14° Brix. Keeping quality is good. It yields about 35 kg per plant.

16. **Pant papaya-1**: This variety is a dwarf plant type (125 to 135 cm) bears first flower at the height of 45 to 60 cm. Fruits are oblong and medium size (1 to 1.5 kg), with good quality.

17. **Pant papaya – 2:** Plants are vigorous, medium height (180 to 220 cm in the first year), and bear first flower at the height of 80-100 cm from the ground level. It produces medium to large sized fruit (1 to 2 kg) of good quality, with higher yield potential of 35 to 40 kg per plant.

18. **Pant Papaya -3:** Plants are vigorous with medium height (225 to 250 cm). It bears first flower at a height of 115-130 cm from the ground level. It produces small to medium sized fruits (0.5 to 0.9 kg) of excellent quality with medium yields (20 to 30 kg per plant). It was found resistant/ tolerant to water logging.

19. **Punjab Sweet:** This variety was developed at Punjab Agricultural University, by sibmating the lines obtained from Kenya. This line is dioecious in nature and bears fruits at a height of 103 cm from the ground level. Fruits weight more than 1 kg and pulp is deep yellow and melting. It has an average yield of 50 kg fruits per tree. Its TSS ranges 9 to 10.5° Brix. Fruit cavity is medium. The average yield per plant is 50 kg.

20. **Solo:** A small-fruited papaya from Barbados was named 'Solo' in 1919. It is a gynodioecious variety. The fruits weigh 500 to 600g and are of excellent quality. When the fruit is fully ripe, the thin skin is orange-yellow and the pulp light pink and very sweet.

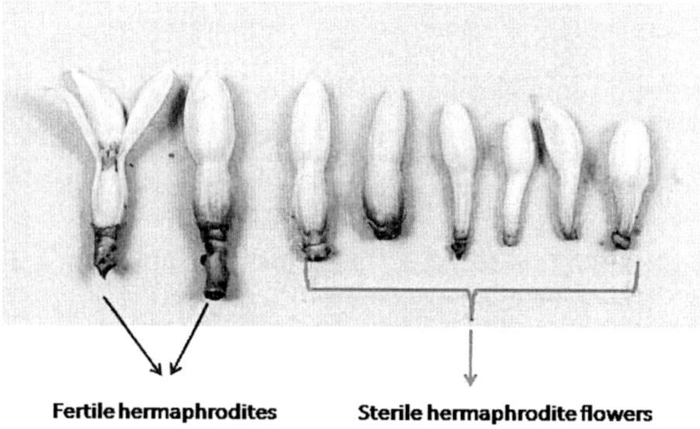

Different flowertypes

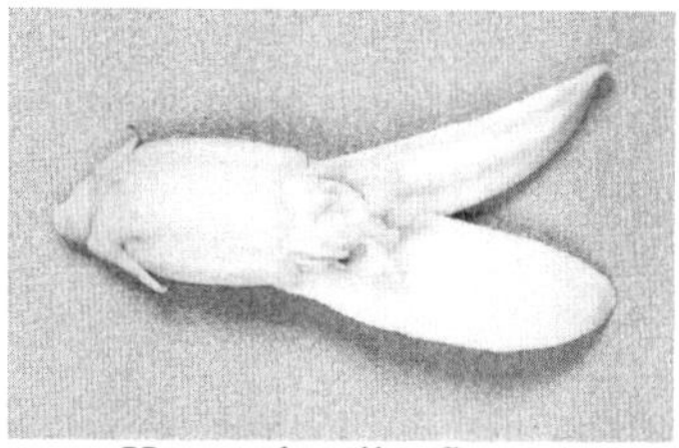
Hermaphrodite flower

Male flower

Maleflower

Ringspotplant

Fruit from a female tree

Arka Prabhath

Fruit from bisexual tree

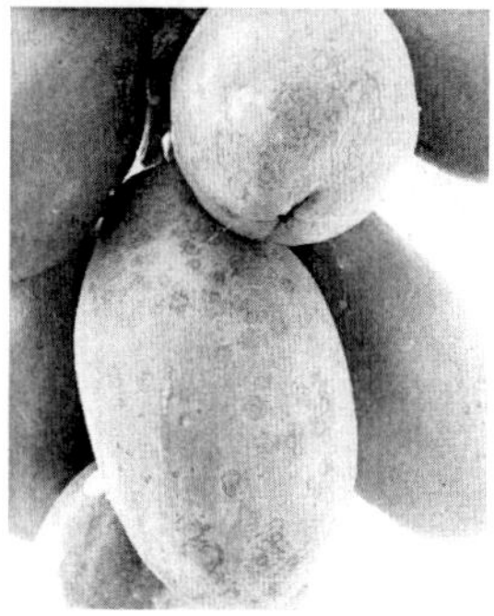
Ringspot

Arka Surya

Breeding

The yield of papaya which was quite low about 13 to 26 tones per acre has now gone up to 80 tons, due to systematic cultivation. The development of new gynodioecious varieties, which produce productive female and hermaphrodite plants within a variety, has also contributed to higher yields. However, due to the following hindrances, improvement on the expected lines has not been seen in papaya. This is due to:

i. Polygamous nature of crop
ii. Lack of genetical information
iii. Highly sensitive to climatic variations
iv. Susceptibility to fungal and viral diseases
v. Lack of systematic characterization
vi. Local prejudices and misconceptions.

Papaya, being cross pollinated and seed propagated has high degree of variability. Breeding programme based on artificial hybridization was started in Hawai in 1915 (Storey, 1969), in South Africa in 1931 (Hofmeyr, 1933) and in India (Rao, 1974).

Improvement of papaya by developing new varieties through breeding following selection has been the most successful method. Hofmeyr (1938) noted that a strain can definitely be improved by selection inspite of the plant being dioecious. Agnew (1941, 52) developed the varieties "Peterson" and "Bethina" by inbreeding in Australia. Hofmeyr (1953) pointed out that dioecious nature of papaya limits the inbreeding process to sibmating which is a long process. He stated, however, that employment of sex reversal may speed up the in breeding programme very considerably since this will allow self pollination. He further pointed out that when male trees inbred, their suitability as parents will have to be judged indirectly from the performance of their sisters as regards yield, shape of fruit etc. Storey (1953, 1969) reviewed the status of commercial varieties of papaya with special reference to Hawaii. According to him the only bonafide varieties were "Solo" of Hawaii and "Hortus Gold" of Sourth Africa. These have originated as inter varietals hybrids between distinct strains. He also suggested that any desirable dioecious variety can now be converted to gynodioecious with complete elimination of male by pollinating the females with hermaphrodites and by back-crossing until the desired varietal characteristics have been restored.

In earlier days the breeding work was aimed at increasing the yield and to evolve varieties which were gynodioecious. The genetic divergence studies in papaya have shown that there is greater variability among the varieties of papaya. Hence, there is a greater chance for selection (Subramanyam, 1978). The varieties CO1, CO2, CO5, CO6, Pusa Delicious, Pusa Majesty, Coorg Honey Dew, Pant Papaya 1, Pant Papaya 2 and Pant Papaya 3, Solo varieties of Hawaii, Bettina, Peterson and Hortus Gold of Queensland and many African varieties have been developed through selection and inbreeding (Agnew 1954, Hamilton 1954, Aiyappa and Nanjappa., 1959, Rao, 1974).

Objectives in breeding

Hybridization work has been taken up in different places with the following objectives:

i. To evolve varieties, which are gynodioecious

ii. To evolve varieties with high TSS, small cavity and good keeping quality

iii. To evolve varieties, which are resistant to ring spot and mosaic virus.

As a result of the work carried out using different varieties as parents several hybrids have come out, although they have come out as a result of hybridization and followed by sib mating and advanced generation selection like CO3, CO4, CO7 and Surya. The introduction has played a great role in the development of papaya cultivation around the world. Among the introduction, the Solo lines of Hawaii have made great impact. Ploidy manipulations in papaya have not been far successful (Hofmeyr, 1945, Singh 1953, 1963; Zerpa, 1957). With regard to mutation breeding, new dwarf variety has come out from Pusa known as Pusa Nanha (Manshra Ram, 1983).

In papaya F_1 seed production seems to be gaining importance, as by crossing two stable gynodioecious genotypes the segregation in F1 progenies can be reduced. Heterosis in papaya has been observed by many workers as mentioned above in this chapter for various quantitative characters. The main advantage is that fruitset and seed setting are not problematic as in other perennial crops.

Approaches to breed new varieties

a) Recurrent selection

Recurrent selection in papaya can be used for development of superior genotypes with desirable characters. The genetical studies conducted in papaya have shown that certain varieties like Waimanalo, Coorg Honey Dew, Pink Flesh Sweet and Sunrise Solo are good gene donors for quality parameters. The varieties Coorg

Honey Dew, Pink Flesh Sweet and Washington are good gene donors for yield (Dinesh, *et al.*, 1991). Hence, these varieties can be used in the breeding programme to develop new varieties having good yield as well as quality.

Production of F_1 hybrids is a distinct possibility as heterosis has been observed by several workers (Subramanyam, 1978; Dinesh *et al.*, 1991). The following steps are involved in the development of a new variety.

i. Identify superior genotypes for intercrossing

Identify the varieties for crossing by taking up the evaluation for different plant and fruit characteristics. The general and specific combining ability effects can be worked out using biometrical methods like diallel, although, the plant materials chosen sometimes do not fulfil all the assumptions.

ii. Select for characters needed

There should be clear cut idea about the characters for which selection needs to be carried out. The estimates of heritability and heterosis may be estimated to select the parents having the desirable characters in question.

iii. Sibmating may be done after selecting the progenies

Selection of progenies once carried out in the F_1 generation should be followed up with sibmating so that the characters can be fixed in the further generation. This continuous selection needs to be carried out after every generation of sibmating. Recurrent selection can be successfully employed in papaya for converting a dioecious variety into gynodioecious variety and recurrent parent has to be gynodioecious in nature.

b) Back Cross Method

The main aim of backcrossing method is to transfer desirable alleles from a non-recurrent parent to genetic background of a recurrent parent.

Dioecious Variety

i. Identify suitable male parents from same progeny which have resemblance to female parents of desired types.

ii. Select females with the most desirable features and cross with 3 male parents identified earlier.

iii. Progenies raised from BC1 and BC2 inbred lines are screened and desired females are selected for further sib-mating. This process is continued for a minimum of 7-8 generations. On back crossing the ratio of male: female is 1:1.

Gynodioecious Variety

i. Self a desirable hermaphrodite and sibmate the female with hermaphrodite

ii. Select hermaphrodites whose sex does not change with climatic variations. Cross desired female with at least 3 selected hermaphrodites as male parent. Sibmating in females and cross with hermaphrodites may go simultaneously. Homozygosity can be achieved in 7-8 generations.

c) Mutation breeding

Mutation breeding helps in creating more variability. In papaya, one mutant Pusa Nanha has been developed at Pusa (Mansha Ram, 1983) by using gamma irradiation. This approach can also be used for developing varieties with dwarf habit in case of gynodioecious varieties as it does not bring about too many changes in other characters. At present there are no gynodioecious dwarf varieties.

Breeding for different traits

Genetical studies conducted in papaya have paved the way for breeding programmes aimed at improvement of specific characters. Zimmerman and Kowalski (2002) through breeding and selection have developed three pawpaw lines, which initiated fruit set between 40 and 60 cm from the soil surface. These lower height bearing pawpaw lines produce fruits that are marketable one month earlier than other cultivars. Breeding for red pulp varieties is one of the requirements as consumer preference is there only for this type of fruits. In the case of papaya, it has been noticed that yellow pulp colour is dominant over red pulp colour. Cynthia *et al.*, (2003) have observed that red pulp colour is governed by recessive gene, the segregation was observed to be 2:1 for yellow to red pulp. Hence, choosing the parents with red pulp colour will help in improvement of pulp colour and other fruit characteristics.

Resistance breeding

Papaya plants are highly susceptible to extremes of temperature and water logging. Papaya is highly susceptible to frost. Few attempts have been made to breed frost resistant varieties. However, some of the wild species, like *C. candamarcensis* Hook, *C. quercifolia* Benth and Hook and *C. pentandra* are tolerant to cooler temperatures. Three selections PP1-4, PP1-9 and PP-3 were made in the Nainital area of Uttaranchal by surveying for frost resistant types.

There are about seven major fungal diseases reported so far under field conditions and about nine post harvest diseases. Mosqueda Vazquez and Nakasone (1982) in a 5x5 diallel cross of inbred lines ranging from susceptible to resistant to Phytophthroa root rot obtained high heritability in narrow sense indicating that selection should be effective in breeding for root rot resistance. More than the fungal diseases, it is the viral diseases which cause heavy damage to papaya crop. It is still not clear as to whether the Papaya Ring Spot Virus is same as the 'Papaya Mosaic Virus'. This viral disease was first described by Lindner *et al.,* (1945). In India breeding for resistance to Papaya Ring Spot Virus was started at Pune. The varieties screened were found susceptible to PRSV, the wild species *Vasconcellea microcarpa, Vasconcellea candamarcensis, Vasconcellea monoica*, and *Vasconcellea cauliflora* were resistant. However, hybridization was not successful due to embryo abortion at 70-90 days and through *in vitro* culture, hybrid plants were raised. In the F_7 generation progenies, some degree of resistance to PRSV was recorded (Phadnis *et al.,* 1970). According to Conover (1976) tolerance to Distortion Ring Spot Virus is an inheritable trait and tolerant types can be developed by inbreeding selected plants or by hybridization of inbred lines.

The hybrid progeny from the cross of *C. papaya* and *C. cauliflora* was found tolerant to PRSV (Iyer *et al.,* 1987). Conover *et al.,* (1986) reported that the variety Cariflora developed from dioecious seed line accessions as tolerant to PRSV. Dinesh *et al.*, (2013) proposed the methodology for developing tolerant progenies for 'PRSV'by intergeneric crossing of *C. papaya* var. Arka Surya with *V. cauliflora.*The methodology involves the using of gynodioecious varieties as female parent and then sibmating the hermaphrodite progenies, screening them and simultaneously validating them through markers. This would ensure fixing of characters in 5 to 6 generations.The schematic diagram for the procedure adopted for developing progenies from the intergeneric crosses is given below:

Arka Surya x ***V. cauliflora***

(Sunrise Solo x Pink Flesh Sweet)

↓

F_1 progenies from the cross

(Segregation ratio of 3:1 for papaya to castor)

↓

Screening in the laboratory for PRSV

↓

Field planting of the resistant progenies—2 sets one with castor leaf and another with papaya leaf Sibmate the progenies and raise the next generation (S_1) Screen the progenies for

↓

PRSV and evaluate them for fruit characteristics

Sibmate the progenies and raise the next generation (S_2) (Single plant selections)

↓

Repeat the operation for another five cycles with two sets of progenies

Selection of progenies with desirable fruit characteristics and PRSV tolerance

Arka Surya (♀)

X

↓

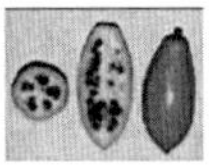

V. cauliflora (♂)

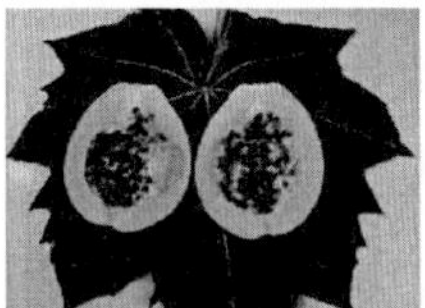

Yellow pulp

Single plant of intergeneric hybrid

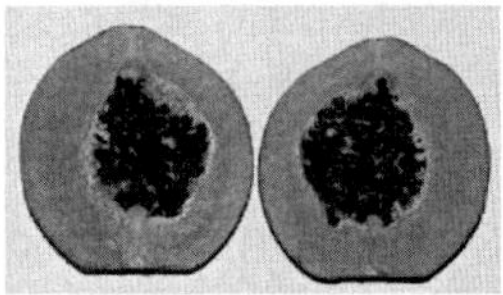

Pink pulp

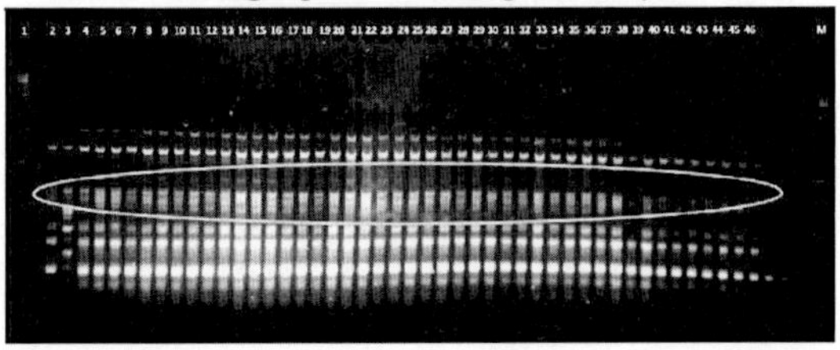

PCR amplified products showing polymorphism using ISSR primer UBC 834:1& M- 1kb ladder, 2- *A. Surya*, 3- *V. cauliflora*, 4 to 7- F1s & 8 to 46 Intergeneric hybrids

Cryo-preserved Pollen in Breeding

Storage of pollen in papaya was initially investigated for breeding and conservation of nuclear genetic diversity (Ganeshan, 1986). Cryopreserved pollen for intra and inter-specific crossing in Carica was investigated by Shashikumar *et al.* (2007). In *C. papaya* cv Washington and *cauliflora*, pollen cryostored after six and eight years showed germination of pollen tubes comparable to controls, they indicated the possibility of establishing a pollen bank for *Carica* species to improve breeding efficiency, heterosis breeding and hybrid seed production as well as conserving nuclear genetic diversity.

Genetical studies

Papaya being polygamous, large variability is seen. The amount of variability and heritability is of great importance in carrying out breeding programme aimed at improving the crop. The flowering and fruiting habit of papaya vary greatly as it is a highly cross pollinated crop. As a result of this, innumerable land races are available in the country and the seeds are sold by nurserymen under different names. However, these do not breed true.

Information with regard to the methodology of breeding and development of hybrids depend on the magnitude of heterosis. Heterosis for yield and yield components have been reported in papaya by Hamilton (1954). Storey (1953) observed manifestation of heterosis in intervarietal hybrids. In the crosses of Philippine × Solo, heterosis was observed by Dai (1960) for vegetative characters, advanced flowering data and resistance to waterlogging. Jayarama Reddy (1962) observed vigorous plants in open pollinated ones than in inbred progenies of CO1 variety. Desikan (1972) found that F1's of CO1 × Coorg Honey Dew, its reciprocal and Washington × CO1 had higher TSS compared to their parents. In the progenies of CO1 and Coorg Honey Dew \, total sugars and reducing sugars were higher. Ahmed Shah and Shanmugavelu (1975) found higher leaf area, thicker petiole, large sized fruit, higher TSS and ascorbic acid in first generation hybrids than their parents. Subramanyam and Iyer (1984) in a study of intervarietal hybrids involving the varieties Coorg Honey Dew, Thailand, Washington, Pink Flesh Sweet and Solo Yellow Sweet, noticed heterosis for vegetative characters and yield and its components. Dinesh (1989) found heterosis in the progenies involving crosses of the varieties Washington, Coorg Honey Dew, Thailand, Pink Flesh Sweet, Waimanalo and Sunrise Solo for vegetative characters and yield. Aquilizan (1983) observed absence of in breeding depression in papaya. Hamilton (1954) also observed absence of inbreeding depression in Solo Strains and found inbred strains to be more vigorous than commercial strains. Genetic variability and character association studies involving several genotypes have shown wide range of variability for many of the quantitative characteristics. The phenotypic coefficient of variation was higher than genotypic coefficient of variation for all characters. Path analysis revealed that total sugar content, fruit width and number of fruits per plant had high positive direct effect on yield per plant at the genotypic level, which suggests significance of these characters during selection in a pawpaw breeding programme. Chan (1992) found that heterosis can be exploited in increasing the yield of F1 papaya hybrids. The three inbred lines L7, L19 and Eksotika were crossed *inter se* and the hybrids were evaluated. It was found that there was no difference in plant vigour between the hybrids and the inbred parents but marked heterosis was observed for yield.

In papaya, the gene for yellow flower colour was dominant to that for white and purple stem and petiole dominant to green stem and petiole colour (Hofmeyr, 1938). The studies carried out in mutant genes so far have shown that albino plant is recessive to normal green, dwarf recessive to tall, crippled leaf recessive to normal flat, rugose leaf recessive smooth, wavy leaf recessive to normal flat bladed, red flesh of fruit recessive to yellow and grey seed coat dominant to black.

The studies with regard to inheritance of quantitative characters are limited in papaya. The fruit shape depends on sex of plant and the type of flower from which it develops. Storey (1953) reported that the number of nodes do not determine the plant height, but only the length of internode determines the height. Hofmeyr (1940) reported occurrence of dwarf plants governed by a recessive gene'd' and in Storey's studies were governed by a recessive gene'dp' Nakasone (1952) reported that two independent sets of multiple factors control the height of first flowering and time from seed germination to flowering as highly correlated with number of nodes. Wasee *et al.,* (1983) found that additive gene effects controlled fruit weight, fruit shape, flesh thickness and total soluble solids. Dinesh *et al.,* (1991) reported that the variety Coorg Honey Dew is a good combiner for yield, fruit length, fruit breadth and fruit volume, Sunrise Solo, Waimanalo and Pink Flesh as good gene donors for quality characters. Subramanyam and Iyer (1981) observed high heritability in broad sense for morphological characters and yield. They also reported high phenotypic and genotypic coefficient of variability for stem thickness and moderately high for fruit weight and volume. Ghanta and Mandal (1992) observed high heritability and genetic advance for plant height, number of leaves per plant and fruit yield per plant. Mosqueda Vazquez (1973) also obtained high heritability values for number of fruits, Wasee *et al.,* (1983) reported high narrow sense heritability for fruit weight, flesh thickness and total soluble solids. Khadi and Singh (1980) found that genotypic variability and heritability (broad sense) were quite high for yield per plant, number of fruits per plant, fresh and dry weight per fruit and additive genetic effects were recorded for leaf area, number of fruits per fruiting length, number of seeds per fruit and yield per plant. Dinesh (1989) also found that additive effects controlled the characters plant height, plant spread, number of fruits, fruit length, fruit breadth, pulp thickness, TSS, fruit cavity index, reducing sugars, total sugars and total carotenoids. Ghanta and Mondal (1992) found that fruits per plant and number of leaves per plant showed high genetic coefficient of variation.

In papaya, simple correlation of characters was studied by different workers, fruit weight was correlated with number of normal seeds per fruit by Allan (1969). Valsche and Mitidieri (1954) found positive and significant correlation between TSS and total sugars in papaya. Subramanyam (1978) also bserved positively significant association of fruit yield with plant height, stem girth, leaf area, number of leaves at first flowering, fruit volume, fruit weight, fruit length and fruit diameter. Mansha Ram and Majumder (1984) observed that fruit yield was highly correlated with fruit weight, leaf length and number of fruits per plant. Magdalita *et al.,* (1984) found that fruit weight was positively and highly correlated with fruit length, width, volume, flesh thickenss and cavity volume in papaya. Valsche and Mitidieri (1954) and Murashige and Abuzied (1964) found a positive and significant correlation between the TSS and total sugars. Phenotypic correlation co-efficient of fruit yield with number of fruits per plant, fruit volume, weight per fruit, fruit bearing length, leaf area per plant and seed weight per fruit were significant. Dinesh (1989) observed significant, positive phenotypic and genotypic correlation of yield with fruit length, fruit weight and number of fruits. Cynthia *et al.,* (2000) observed that fruit characteristics had higher correlation with yield than the number of fruits, indicating that selection based on these characters would enhance indirectly. Heterosis was observed by Mansha Ram *et al.,*(1999) for different fruit characteristics. Heterobeltiosis has also been observed.

Path coefficient analysis in papaya indicated number of fruits per plant and weight per fruit had positive direct effects on yield per plant, fruit volume and fresh seed weight on TSS and fruit bearing length and number of fruits plant on plant stature (Khadi and Singh, 1981). The selection index for fruit yield was observed to be the combination of plant height, number of fruits per plant, weight per fruit, fruit volume, cavity of fruit, weight of fresh seed per unit and number of seeds per fruit. The selection index for fruit yield was observed to be the combination of plant height, number of fruits per plant, weight per fruit, fruit volume, cavity of fruit, weight of fresh seed per unit and number of seeds per fruit. The selection index for plant stature measured a plant height comprised the height at first fruiting, leaf area per plant, stem girth, number of fruits per plant, fruit bearing length and fruit yield per plant (Khadi and Singh, 1981).

Biotechnological applications in papaya

Use of RAPD technique to polymorphism is important in papaya as genetic variability is limited (Saraswathi *et al.,* 2007). RAPD was used for the construction of genetic linkage map, which is the first step towards the understanding of complete genome (Sondur, *et al.*, 1996). Several workers have used RAPD and SCAR markers for confirming the hybridity and sex determination in papaya. Devitt *et al.,*(2006) identified several cDNAs and determined the genes to be homologous to other plant carotenogenic cDNAs, while attempting to determine the genetic difference between red and yellow-fleshed *C. papaya*.

Tissue culture techniques are extremely important for producing suitable plant material. In papaya totipotent embryogenic cultures have been used. High frequency somatic embryogenesis and plantlet regeneration in papaya were developed by Fitch (1993) and Monmarson *et al.,* (1995).

Papaya has become a model for other fruit crops with respect to biotechnology applications. This is largely due to excellent control of *in vitro* development gained over two decades of research. Micro propagation has been highly developed in subtropical regions to stabilize selected dioecious genotypes. Embryogenic regeneration, obtained efficiently from several kinds of immature tissues in the presence of auxins, is often difficult to achieve on selective media employed in transformation procedures; however, improved methods of preparing embryogenic cultures are likely to overcome this bottleneck. Anther culture for production of dihaploid inbred lines is an under-researched objective that merits more attention. The relative ease with which *Carica* interspecific hybrids can be produced by embryo or ovule rescue has directed research activity away from protoplast fusion. Papaya has been hybridized with 6 other *Carica* species, and some hybrids inherit useful disease resistance genes, but serious fertility barriers limit breeding efforts. Phylogenetic analyses based on isozyme and DNA polymorphisms indicate that papaya is a distant relative of other members in the genus. The greatest economic impact of biotechnology on papaya industries in the near term will result from creation of virus-resistant cultivars through genetic transformation with coat protein (cp) genes. Multi-year field tests in Hawaii have demonstrated the effectiveness of translatable cp genes against local strains of Papaya Ringspot Potyvirus (PRSV). Greenhouse inoculations in New York and Taiwan revealed narrow specificity and broad range resistance in different transformants, with gene dosage being a major determining factor. Untranslatable cp genes and replicase genes have also been introduced into papaya and provide protection against PRSV in greenhouse tests. At University of Hawaii, transgenic papaya against PSRV was developed using coat-protein mediated resistance (Manshardt, *et al.*, 1998). Two varieties

Sun Up (from Sunset) and UH Rainbow (from Kapoho Solo) were found to be resistant to 'PRSV'.

Chan-Tai, *et al.,* (2003) in a very interesting observation noticed that hermaphroditic papaya mutant (SR*) obtained was derived from the 'Sunrise' papaya cultivar mutant. Self-pollination of the mutant resulted in all progenies being hermaphroditic. The genotype of the female was lethal, as a result of a lethal gene being linked to the mm female gene complex in this case. However, a 3:1 segregation ratio was obtained from the progeny of the hermaphroditic cultivar 'Thailand' crossed with SR*, indicating that all genotypes survived. Homozygous genotypes (M_2M_2) would be lethal according to Storey's model. Randomly selected F_1 plants of the 'Thailand' SR* combination were self-pollinated to obtain an F_2 generation. The F_2 segregation ratio suggested that the SR* mutant had a different form of the M2 allele, now designated as $M_{@}$, which allowed the dominant $M_{@}M_2$ to survive in cross combinations. Genetic study has proved that SR* has the $M_{@}m_1$ genotype, a new mutant. It is capable of producing all hermaphroditic papaya progenies.

Future line of work

Papaya, although was varied uses, it is under exploited. However, of late papaya growing for processing and papain production has become a profitable proposition. Looking back at the progress made, the future needs of varietal improvement in papaya are:

1. Exploration, conservation and characterization of genetic variability.
2. Development of stable gynodioecious genotypes for fruit and papain production.
3. Exploitation of heterosis by producing F1 hybrids.
4. Production of genetically pure seeds by controlled pollinated.
5. Development of dwarf gynodioecious lines, and
6. Breeding varieties resistant to "Ring Spot Virus".

References

Agnew G.W.J.1948. Papaya culture in Queensland. *Queensland Agric. J.* 66: 208-226.

Agnew G.W.J. 1954. Papaya. *Queens land Agric. J.* 79: 77-90.

Ahmed Shah, H and Shanmugavelu K.G. 1975. Studies on the first generation hybrids in papaya (*Carica papaya* L) I. Morphological, floral and fruit characters. *S. Indian Hort.* 23(2 & 4): 100-08.

Aiyappa, K.M and Nanjappa, P.P. 1959. Coorg Honey, a new find in papaya. Indian Hort., 3:3

Allan, P. 1969. Effect of seeds on fruit weight in *Carica papaya. Agroplantae (S. Afr.)* 1: 163-170.

Allan, P., 1963. Pollination of papaws. *Fmg. S. AFr.* 39: 35-40.

Aquilizan, F.A., 1983. A breeding system for establishing stable dioecious papaw varieties. In Proceedings Australian Plant Breeding conference, Adelaide, South Australia, 14-18 February

Aradhya, K.M., Richard, M. Manshardt, Francis Zee and Morden, Clifford W. 1999. A phylogenetic analysis of the genus Carica L. (Caricaceae) based on restriction fragment length variation in a cpDNA intergenic spacer region. *Genetic Resources and Crop Evolution* 46: 579–586.

Badillo, V. M. 2000. Carica L. vs. Vasconcella St.-Hil. (Caricaceae) con la rehabilitacion de este ultimo. *Ernstia* 10(2): 74-79.

Badillo, V. M. 1971. Monographia de la familia Caricaceae. Maracay, Venezuela.

Badillo, V. M. 1993. Caricaceae, Segundo Esquema. Publicada por la Asociacion de Profesores, Alcance 43, Universidad Central de Venezuela, Maracay, 111 pp.

Chan, Y.K., 1992. Progress in breeding of F1 papaya hybrids in Malaysia. *Acta Hort.* 292: 41-49.

Chan-Tai, C., Yen, C.R., Chang, LS., Hsiao, C.H. and Ko,T.S. 2003. All hermaphrodite progeny are derived by self-pollinating the sunrise papaya mutant. *Plant-Breeding* 122(5):431-434

Chowdhury, R.S., Garg, O.K. and Borah, P. C. 1957. Physiological changes in relation to sex in papaya (*Carica papaya* L.). *Phyton*, 9:137-41.

Conover, R.A., 1976. A program fro development of papaya tolerant to the 'Distortion Ring Spot Virus'. *Proc. Fla. State Hort. Soc.*, 89:229-231.

Conover, R.A., R.E. Litz and Malo, S.E. 1986. 'Cariflora' – a papaya ring spot virus tolerant papaya for South Florida and the Caribbean. *Hort. Science*, 21:1072.

Cynthia. B., Kumar. N. and Soorianathasundaram. K. 2003. Breeding for red pulped fruits in dioecious lines of papaya (*Carica papaya* L.). *Indian Journal of Horticulture*: 60(1):10-15.

Cynthia. B., Kumar, N. and Soorianathasundaram, K. 2000. Genetic variability and association of economic characters in red fleshed dioecious lines in papaya (*Carica papaya* L.). *South Indian Horticulture* 48(1/6):11-17.

Dai, B.B. 1960. Experiences on the utilization of F_1 papaya hybrids. *Mungychen-Chiu/Agric Res. Taipei.*, 8(1):17-29.

Desikan. K.R. 1972. Studies on first generation hybrids in papaya (*Carica papaya* Linn). M.Sc. Thesis submitted to the Tamil Nadu Agricultural University, Coimbatore.

De Candolle, A., 1884. Origin of cultivated plants. Londo, 486 pp.

Devitt., L.C., Sawbridge. T., Holton, T.A., Mitchelson K. and Dietzgen, R.G. 2006. Discovery of genes associated with fruit ripening in *Carica papaya* using expressed sequence tags. *Plant Science.*170:356-363.

Dinesh, M.R., C.P.A. Iyer and Subramanyam, M.D. 1989. Hybridisation in papaya-Paper presented at the All India Co-ordinated Fruit Improvement Project Workshop held at Tirupathi.

Dinesh. M.R., Veena, G.L., Vasugi, C., Krishna Reddy, M. and Ravishankar, K.V. 2013. Intergeneric hybridization in papaya for 'PRSV' tolerance. *Scientia Horticulturae.* 161(2013): 357–360.

Dinesh. M.R., Iyer, C.P.A. and Subramanyam, M.D. 1991. Combining ability studies in *Carica papaya* L. with respect to yield and quality characters. *Gartenbauwissenchaft*, 56(3): 81-83.

Dinesh, M.R., Iyer, C.P.A. and Subramanyam, M.D. 1991. Genetical studies in *Caricapapaya* L. *Acta Horticulturae*, 321(1):152-15581-83.

Ganeshan, S., 1986. Cryogenic preservation of papaya pollen. *Scientia Hort.* 28:65-70.

Ghanta, P.K and Mondal, S.K. 1992. Genotypic variability and correlation coefficients relating to fruit yield and a few other quantitative characters in papaya (*Carica papaya* L.) cultivars. *South-Indian-Horticulture* 40(5): 242-247.

Hamilton R.A., 1954. Fertilizer guide for tropical and subtropical farming. 64:111-116.

Hamilton, R.A. 1954. A quantitative study of growth and fruiting in inbred and crossbred progenies from two Solo papaya strains. *Hawaii Agr. Exp. Sta. Tech. Bull.* 20.

Higgins, J.E. and Holts, 1914. The papaya in Hawali. *Hawaii Agric. Exp. STa. Bull.* 32: 44 pp.

Hofmeyr, J.D.J. 1933. Pawpaw selection. *Farming in S.A.*, 8:159-160.

Hofmeyr, J.D.J. 1938. Genetical studies in *Carica papaya* L. I. The inheritance of sex and certain plant characteristics. II. Sex reversal and Sex forms. *Bull. Dept. Agric. Far. Sci. S. Afr.*, 187.

Hofmeyr, J.D.J. 1939. Sex reversal inc. papaya L. *Chronica Botanica*, 6:245-47.

Hofmeyr, J.D.J. 1940. The genetics of C.papaya L. *Chronics Botanica*. 6:245-47.

Hofmeyr, J.D.J. 1941. The use of colchicines in Horticulture with special reference to *Carica papaya L. Fmg. S.Afr.*, 16:311.

Hofmeyr, J.D.J. 1953. Sex reversal as a means of solving breeding problems of *Carica papaya L.S.AFr. J. Sci.*, 49: 228-32.

Hofmeyr, J.D.S., 1938. Genetical studies in *Carica papaya* L. I. The Inheritance of sex and certain plant characteristics. II. Sex reversal and Sex forms. *Bull. Dept. Agric. Far. Sci. S. Afri.* 187.

Hofmeyr, J.D.J. 1945. Further studies on tetraploidy in *Carica papaya* L. *S. Afr. J. Sci.*, 41:225-30.

Horovitz, S.,1955. Sex in *Carica papaya* L. Proc. 14th Inst. Hort. Congr. Schemeneingen.

Iyer, C.P.A., M.D. Subramanyam and Jagdeesh Chandra, K. 1987. Inter specific hybridization in the genus Carica. Paper presented at the Workshop of All India Co-ordinated Research Project on Tropical Fruits'. College of Agriculture, PKV, Nagpur.

Jayaram Reddy, P.A., 1962. Studies on papaya-A comparison of inbred and open pollinated progenies of CO1 papaya and certain observations on seed fruit characters. M.Sc(Ag) Thesis submitted to Faculty of Horticulture Regional PG Training Centre, AC & RI, Coimbatore.

Jindal, K.K and Singh, R.N. 1976. Modification of flowering pattern and sex expression in *Carica papaya* L. by morphactin ethephon and TIBA. *Zeitschriftfur pflenzenphysiologic*, 78:403-10.

Khadi, B.M and Singh, I.D. 1980. Estimates of variability, heritability and genetic advance in papaya (*Carica papaya* L). *Pantnagar J. Res.* 5(2).

Khadi, B.M and Singh, I.D. 1981. Selection indices in papaya (*Carica papaya* L), *Pantnagar. J. Res.* 6 (1).

Kim, M.S., P.H. Moore., F. Zee., M.M.M. Fitch, D.L. Steiger, R.M. Manshardt., R.E. Paull., R.A. Drew., T. Sekioka and Ming, R. 2002. Genetic diversity of *Carica papaya* as revealed by AFLP markers. *Genome.* 45(3): 503-512.

Kulkarni, H.Y., 1970. Decline viruses of papaw (*Carica papaw* L.) in East Africa. *Ann. Appl. Biol.*, 16:1-9.

Kumar. V., 1951. Studies in *Carica papaya* L. I. Preliminary observations on the relation of sex to the pre-flowering growth of papaya seedlings and external characters of seeds. *Indian J. Hort.*, 8:26-34.

Lange, A., 1961. The effect of temperature and photoperiod on the growth of *Carica papaya* L. *Ecology*, 42:481-86.

Linder, R. C., Jensen, D.D. and Ikeda, W. 1945. Ringspot: New papaya plundered. *Hawali Farm and Home* 8(10:10-14).

Linnaeus, Carl Von., 1753. Species plantarum. Laurentii Salvi Holminae, P. 1200.

Magdalita, P.M., Drew, R.A., Adkins, S.W. and Godwin, I.D., 1997. Morphological, molecular and cytological analyses of *Carica papaya* x *C. cauliflora* interspecific hybrids. *Theoretical and Applied Genetics* 95: 224-229.

Magdalita, P.M., Pimentel, R.B., Rosario, R.B., Del, E.E., Sotto, R.C., Rivera. F.N. and Espino, R.R.G. 1984. Phenotypic variability in some characters of papaya (*Carica papaya* L) *Philippine Agriculturist.* 67(3): 289-294.

Mansha Ram, 1983. Papaya Pusa Nanha. *Int. Agri.*, 21:18-19.

Mansha Ram., Majumder, P.K. Singh, B.N. and Akhtar, S.A.1999. Studies on hybrid vigour in papaya (*Carica papaya* L.) varieties. *Indian Journal of Horticulture*, 56(4): 295-298.

Mansha Ram and Majumder, P.K. 1984. Character association in papaya. Proce. of National Seminar on papaya and papain production, March 26-27, TNAU, Coimbatore.

Manshardt R.M. and Drew R.A. 1998. Biotechnology of papaya. *Acta Horticulturae*, 461:65-73.

Maureen M.M. Fitch, Richard M. Manshardt, Dennis Gonsalves and Jerry Slightom, L. 1993. Transgenic papaya plants from*Agrobacterium*-mediated transformation of somatic embryos. *Plant Cell Reports*, 12 (5): 245-249.

Monmarson S., Michaux-Ferriere N. and Teisson C. 1995. Production of high-frequency embryogenic calli from integuments of immature seeds of Carica Papaya, L. *Journal of Horticultural Science* 70(1):57–64.

Mosqueda Vazquez, R. and Molina Galan, J., 1973. A study of correlated factors and an analysis of yield components using path coefficients in Carica papaya. *Agrocienica*. 11: 3-14.

Mosqueda Vasquez, R. and Nakasone, H.Y. 1982. Diallel analysis of Root rot resistance in Papaya. *Hort. Science*, 17(3): 384-385.

Murashige, T. and Abuzied, Y. 1964. Refractometric dry solids as an indicator of the sugar content of papaya fruit. *J. Agric. Food Chem.*, 12(6):520–522.

Nakasone, H.Y. and Paull, R.E.,1998 . Tropical fruits. CAB International, Wallingford.

Nakasone, Henry Yoshiki., 1952. Studies on the Inheritance of Fruiting Height of *Carica papaya* L. M.Sc Thesis submitted to University of Hawaii. 36 p.

Ovideo, Y.G.F., 1526. Historia Genreally Natural de las Inlian Impredela RealAcad. De la Historia, Madrial. 1851-1855.

Phadnis, N.A., Budrukkar, N.D. and Kaulgud, S.N. 1970. Embryo culture technique in papaya *Carica papaya. Poona Agric. College Magazine* 60:101-104.

Rao, V.N.M., 1974. Papaya in India. Farm Information Unit, Directorate of Extension, *Bull.* No. 9.

Romeijn-Peeters E. 2004. Biodiversity of the genus Vasconcellea (Caricaceae) in Ecuador: a morphological approach. PhD dissertation, Faculty of Sciences, Ghent University.

Saraswathi, M.S., Siva, S.A., Uma, S. Sathiamoorthy, S. Vadivel, E. 2007. Rapd Markers in flesh colour determination of papayas (*Carica papaya* L.)., *Acta Horticulturae* 740.

Schroeder, C.A., 1958. The origin, spread, and improvement of the Avacado, Sapodila and Papaya. *Indian J. Hort.*, 15(3):263-26.

Seemanthini, B., 1967. Study of influence of fruit size on yield of papain in certain inbreed selections of papaya (*Carica papaya* L.). *Indian J. Hort.*, 24: 41-55.

Seemanthini, B. 1968. A Preliminary note on the heigh tat which the first flower is produced and its relation to the sex of the plant in papaya (*Carica papaya* L.). *Sl Indian Hort.*, 7:5-10.

Singh, R.N., 1953. Colchicine induced polyploidy in papaya (*Carica papaya* L). *Sci. Cult.* 19:44.

Singh, R.N., 1963. The problematic papaya. *World Crops*. 15:82-85.

Shashikumar, S., Ganeshan, S. and Tejavathi, D.H. 2007. Fertilizing ability of cryopreserved pollen in inter-intra specific crosses in *Carica papaya* L. *ISHS Acta Horticulturae* 740

Singh, I.D. and Sirohi, S.C. 1977. Sex expression studie sin papaya (*Carica papaya* L.). *Pant J. Res.*, 2:150-52.

Singh, I.D. 1990. Papaya. Oxford & IBIT Pub., Co. PVt. Ltd., New Delhi.

Singh, R.N., 1964. Papaya Breeding : A Review. *Indian J. Hort.*, 21:148-54.

Singh, R.N., Majumdar P.K. and Sharma, D.K. 1961. Sex reversal in Papaya (*Carica papaya* L.). *Indian J. Hort.*, 18:148-57.

Sondur, S.N., Manshardt R.M. and Stiles, J.I. 1996. A genetic linkage map of papaya based on randomly amplified polymorphic DNA markers. *Theoretical and Applied Genetics*. 93(4):547-553.

Storey, W.B. 1937. Segregation of sex types in Solo papaya and Their application to the selection of seeds. *Proc. Amer. Soc. Hort. Sci.*, 35:80-82.

Storey, W.B. 1941. papaya production in Hawaiian Inslands. Part-I. Botany and Sex relationships of the papaya. *Hawai Agric. Expt. St. Bull.* 87: 5-22.

Storey, W.B. 1953. Genetics of the papaya. *J. Hered.*, 44:70-78.

Storey, W.B. 1958. Modification of sex expression in papaya. *Horticultural Advance*, 2:49-60.

Storey, W.B. 1969. Papaya- in: Outlines of perennial crop breeding in the tropics. Edited by F.D., Ferwerda and F. wit. Misc. papers 4. Landbow Whogeschol. Wegeningen, Netherlands, pp. 389-407.

Subramanyam, M.D. 1978. Cytogenetical studies in the genus Carica. PhD Thesis submitted to the Faculty of Science (Biology) Guru Nanak Dev University, Amritsar.

Subramanyam, M.D. and Iyer, C.P.A. 1981. Studies on varia-bility, heritability and genetic advance in *Carica papaya* Cvs. *S. Indian Horticulture*, 29:167-74.

Subramanyam, M.D and Iyer, C.P.A. 1984. Exploitation of heterosis in papaya. *Indian Journal of Horticulture* 41(1&2):40-46.

Subramanyam. M.D. and Iyer, C.P.A. 1986. Flowering behavior and floral biologyin different species of Carica L. Genus. Haryana *J. Hort. Sci.*, 15(3-4):179-187.

Teaotia, S.S. and Singh, R.N. 1967. S Esonal variation in sex expression of papaya (*Carica papaya* L.). *Indian Agric.* 11:45-49.

Valsche, O. and Mitidieri, J.1954. Briex in the determination of the sugars content of papaws as an aid in breeding. *Ann. Esc. Symp. Agric. Queiroz.* 11:85-92.

Wasee, S., Subhadrabandhu, S., Srinivas, P., and Dhanasobhan, C., 1983. A study of the inheritance of some characters in two cultivars of papaya (*Carica papaya* L.). *Kasetsart Journal* 17(1):16-25.

Zerpa, S.M., 1957. Triploides de *Carica papaya* L. *Agron. Trop.*,7:83-86.

Zhiyong Liu, Paul H. Moore, Hao Ma, Minna S. Kim, Qingyi Yu, John I. Stiles, Maureen M.M. Fitch , Andrew H. Paterson and Ray Ming, 2003. Physical mapping of sex determination gene in papaya, *Plant Animal and Microbe Genomes Conference*. 541-210.

Zimmerman, T.W. and Kowalski, J.A. 2002. Breeding and selection for early bearing papayas. *Acta Horticulturae*. (632):53-55.

15

Guava Breeding

M.R. Dinesh and C. Vasugi

Guava (*Psidium guajava* L) is an important fruit crop of India. It has gained considerable prominence on account of its high nutritive value, availability at moderate prices, pleasant aroma and good flavour. It is one of the commonest fruits liked by the rich and the poor and is popularly known as the 'apple of tropics'. It is one of the hardiest fruit trees adaptable to a variety of soil and climatic conditions. It comes up well even under neglected conditions and in fact it is even considered sometime as a weed in Fiji and Hawaii. Guava fruit is relished when mature or ripe or when freshly plucked from the tree. It is also used for many commercial products like jelly, fruit butter, juice, etc.The guava is said to have originated from tropical America (Hayes, 1953). De Candolle (1904) stated that it originated in Mexico, while Purseglove (1968) opined that it originated in Brazil. It is widely distributed all over the equatorial regions of tropical and sub-tropical climate. It was introduced to India during the 17th Century. In Spanish, the tree is known as *guayabo*, or *guayavo*, the fruit *guayaba* or *guyava*. The French call it *goyave* or *goyavier*; the Dutch, *guyaba*, *goeajaaba*; the Surinamese, *guave* or *goejaba;* and the Portuguese, *goiaba* or *goaibeira*. Hawaiians call it guava or *kuawa*. In Guam it is *abas*. In Malaya, it is generally known either as guava or *jambu batu*, but has also numerous dialectal names as it does in India, tropical Africa and the Philippines where the name, *bayabas,* is often applied. Various tribal names–*pichi, posh, enandi, etc*., are employed among the Indians of Mexico and Central and South America.

Species status

The genus *Psidium* belongs to the family Myrtaceae and has a basic chromosome number of X= 11. All the cultivars of Indian guava belong to the single species *Psidium guajava* L. Hayes (1953), reported that the genus contains about 150 spp., though only a few have been studied in detail. Bailey (1919) reported that the two species pyriferum and pomiferum mentioned by Linnaeus are nothing but trees with pear shaped and round shaped fruits. Subsequently other species were recognized and documented. The wild species are of considerable importance in breeding programmes. The species that are important are:

P. cattleianumvar lucidum

It is a small tree or shrub with smooth bark. Leaves are obovate elliptic and glabrous. Fruits are obovate to round with light yellow colour skin. It is also known as lemon guava. Skin is thin; flesh is soft with numerous seeds.

P. cattleianum var longipes

It has sweet flavour and good aroma. It is also known as "Strawberry guava" because of the sweet aroma reminiscent of strawberry. Since it lacks muskiness of common guava, it is preferred among certain tribals. It is cold tolerant type bearing red skinned fruits with round shape and about 2.5 cm diameter (Normand 2001).

P. molle Bertol

It is a shrub or small tree. The leaves are obovate. The lower surface of leaves is reddish velvety. Fruits are pale yellow in colour when fully ripe. Flesh is white with many seeds. The flavour is acidic. It is also known as *P. guineense* Sw, *P schiedeanum* Berg and *P araca* Raddi. It is also called as Brazilian guava (Morton, 1987).

P. friedrichsthalianum *niedenzu*

It is called as 'Chinese guava' or 'Costa Rican guava'. It is a tall tree of about 7-10m high. The branches are slender and smooth. Leaves are oval or oblong/ oval, smooth almost glossy above and pubescent below. Fruits are globose, small and sour. Fruits are good for jelly making because of high acidity (Villota and Agudelo, 1998).

***P. montanum* Swartz**: It is found generally in mountains of Jamaica. The branchlets are 4-angled; leaves oblong to oval, glabrous, fruits are round, flesh white with more number of seeds. It produces fruits of poor quality.

P. acutangulum DC

It is a tree, which grows to a height of 8 to 13 ft. Its branchlets are quadrangular and winged near the leaf base. It occurs wild and and cultivated at low and medium elevations throughout Amazonia and from Peru to Colombia, Bolivia, Venezuela and Guianas. The fruits are round, pear shaped or ellipsoid, pale yellow in colour, very acidic but well flavoured. The fruits contain a few hard triangular seeds.

Breeding

Guava is mainly a self-pollinated crop but cross-pollination also does occur. This results in large variability in the seedling population from which promising genotypes have been selected in different agro-climatic regions of the country. In India, during early days guava plants were generally propagated by seeds of limited varieties available with nurserymen and pomologists. The seedling population obtained by open pollination has given rise to considerable variation in the form and size of fruit, the nature and flavour of pulp, seediness and other morphological characters such as spreading or erect growth habit of trees (Naik, 1949). Cheema *et al.,* (1954) observed all the cultivars of guava to be highly heterozygous. The commercial producers utilized the variation thus obtained for selection of desirable genotypes and by propagating them vegetatively. The assessment of genetic diversity and relationship among *Psidium* spp was carried out by Sharma *et al.*, (2007). They observed high genetic similarity between Chinese guavas grouped with *Psidium guajava* cultivars.

It is observed that only a few named varieties are under cultivation. Most of these varieties suffer from one defect or the other. Hence, guava improvement by breeding was started mainly with the following objectives of developing new cultivars with most of the following characteristics:

i. dwarf plant habit suitable for high density planting,
ii. fruits with uniform shape, size, good colour, firm and thick pulp, good aroma, a few and soft seeds, high TSS and high pectin,
iii. long shelf life,
iv. resistance to *Fusarium* wilt.

Introduction

Most of the guava varieties have evolved as selection from seedling variants. Varietal variability has come about because of open pollination from highly heterozygous parents. Although, introduction is a potential tool in any crop improvement programme, as it considerably cuts down on the time, there is also a danger of new diseases getting introduced. It has been our observation that some of the exotic acidic types like Beaumont, which were introduced, are prone for 'Stylar End Rot' caused by *Phomopsis psidii* and this characteristic was even inherited in hybrids. Hence, utmost care needs to be taken while introducing the varieties and unless they are observed under plant quarantine; further multiplication or usage on a field scale may not be made.

Selection

Guava improvement work for the first time was initiated during 1907 at Ganeshkhind Fruit Experimental Station, Pune, India primarily with the collection of seeds of varieties grown in different places to isolate superior strains. One strain from open pollinated seedlings of 'Allahabad Safeda' collected from Lucknow was selected and released as 'Lucknow-49' (Cheema *et al.,* 1954) which became very popular and has now been renamed as 'Sardar'. One superior selection viz., S-I having good fruit shape and quality, a few seeds, sweet taste and high yield was isolated at Saharanpur (Singh, 1959). A guava selection, CISH-G 4, has been selected at CISH Lucknow. It has high TSS (14°Brix) and attractive pink blush (Anon, 1999). At Faizabad, seedling selections were made out of 'Allahabad Safeda' and many selections under the 'Faizabad Selection' (Pathak and Dwivedi, 1988). At CISH, Lucknow three seedling selections of coloured guava from Allahabad namely; CISH-G-1, CISH-G-4 and 'Lalit' were released. The variety 'Lalit' has higher yield as compared to 'Allahabad Safeda' besides the fruits having attractive saffron yellow colour and medium sized fruits having pink pulp and good sugar-acid blend. The fruits of CISH-G-1 are deep red coloured having attractive shape, a few soft seeds and longer shelf life. It has export potential. At the Indian Institute of Horticultural Research, Bangalore, from 200 open pollinated seedlings of variety 'Allahabad Safeda' collected from Uttar Pradesh, one seedling selection 'Selection-8' was promising and released as 'Arka Mridula'. This variety also performed well with respect to yield and quality under rainfed conditions of Bihar (Ramkumar, 1998). At 'Central Institute of Sub-tropical Horticulture', Lucknow, open pollinated seedlings of the red coloured guava collected from guava belt of Allahabad were evaluated for various characteristics; out of these progenies six were selected. One progeny selection was named as 'Lalit' and released. The performance of open pollinated seedling progenies of guava cv. 'Apple Colour' was studied by Marak and Mukunda (2007).

Hybridization technique

In guava, the flowers chosen for crossing are emasculated when they are at the 'calyx break stage' a day before opening. The pollen from the pollen parent is brought from an unopened flower, the generally preferred one being at calyx break stage. The stigmatic surface is gently smeared using the pollen and the flowers are bagged. It is our experience at Bangalore that pollination when carried out during the morning hours between 10 AM to 12 noon has given better results.

Intervarietal hybridization

In general, intervarietal crosses in guava are successful without any crossability barriers. However, varietal crossing incompatibility in guava has been reported, when crosses were made between 'Behat Coconut' and 'Sardar', 'Behat Coconut' and 'Apple colour'. In India, breeding work for guava improvement has been going on at several research institutions. At HETC, Basti a number of cross combination of 'Seedless' x 'Allahabad Safeda', 'Seedless' x 'L-49', 'Allahabad Safeda' x 'Patillo', 'Apple colour' x 'Red fleshed' and 'Apple colour' x 'Kot Prud' were made. None of the fifty five hybrids obtained from these crosses were found promising. From Fruit Research Station, Sangareddy, A.P., two hybrids 'Safed Jam' and 'Kohir Safeda' were selected out of the crosses of 'Allahabad Safeda' x 'Kohir' and 'Kohir' x 'Allahabad Safeda' respectively and released. These hybrids have been particularly recommended for semi arid tropical areas. These have also been found suitable for the preparation of guava juice. The other hybrid released was 'Arka Amulya' (Allahabad Safeda x Triploid) from Indian Institute of Horticultural Research, Bangalore. Parents in intervarietal hybridization should also take into account the reaction of the cultivars to guava wilt. Negi *et al.,* (2001) summarized the varietal reactions. Under natural conditions Lucknow-49 was free from wilt and 'Allahabad Safeda' had the incidence of only 4%. Relative field tolerance of 20 guava cultivars studied at CISH, Lucknow showed that 'Chittidar', 'Portugal', 'Seedless', 'Spear Acid' and 'Smooth Green' were tolerant while 'Apple Colour' and 'Red Flesh' were moderately tolerant.

Characteristics of some of the guava hybrids released recently are given below:

Safed Jam: This is a hybrid 'Allahabad Safeda' x 'Kohir' (a local collection from Hyderabad Karnataka region). It is similar to 'Allahabad Safeda' in growth habit and fruit quality. Fruits are bigger in size with good quality and a few soft seeds.

Kohir Safeda: It is a heavy yielding cross of selected line of 'Kohir' x 'Allahabad Safeda'. Tree is vigorous; fruits are larger with a few soft seeds and white flesh.

Arka Amulya: It is from the cross 'Allahabad Safeda' x 'Triploid'. Fruits are medium sized, flesh is white, TSS is high (12.5° Brix) with good keeping quality. Plants are semi - vigorous and spreading.

Arka Kiran: It is from the cross 'Kamsari' x 'Purple Local'. It produces medium sized fruits of 180-200 g, has soft seeds, deep pink pulp and high TSS (12°Brix) and high Lycopene content (7.45 mg 100 g^{-1}).

Phylogenetic affinity, usefulness of wild species and interspecific hybridization

Utilization of wild species for crop improvement has been one of the ways to introduce certain gene(s) for specific purposes like hardiness, disease and pest resistance. In perennial crops, because of the long time gap, efforts have not been to their full potential. Exploitation of wild species requires extensive knowledge of taxonomy, reproductive biology and cytogenetics, genetics, crossability barriers and fertility of hybrids. Although, the success obtained in fruit crops is less, many desirable traits having great potential for crop improvement are found in wild species. Interspecific crosses in many of the fruit crops between cultivars and species have resulted in hybrids, which are partially sterile due to ploidy level differences, genomic incompatibilities and cytoplasmic imbalances. In order to develop rootstock tolerant/resistant to guava wilt and to test possible role of different species, studies have been undertaken on interspecific hybridization, in the genus *Psidium.* Phylogenetic studies carried out in *Psidium* species by utilizing the difference in flavonoid patterns showed that there is close affinity between *P. guajava* and *P. molle*. The species *P. molle* and *P. guineese* were found similar morphologically with minute differences in chromatographic pattern (Das and Prakash, 1981). They also found close affinity between *P. guineense, P. pumilum* and *P. chinensis*. It was observed that *P. guajava* and *P. chinensis* are crossable. However, *P. guajava* and *P. molle* are cross incompatible, when *P. guajava* is used as female parent (Subramanyam and Iyer, 1982). Leslie *et al.,* (1995) and Edward and Shankar (1964), reported that *Psidium friedrichsthalianum* Niedenzu is resistant to guava wilt. The other species reported to be resistant to the wilt are *P. cumuni, P. cattleianum* var. lucidum and *P. molle, P. guineense.* However, Singh *et al.,* (1977) reported that *P. cattleianum* var *lucidum, P. corecium, P. cajuvalis, P. guineense* and *P. friedrichsthalianum* developed wilt infection. Hence, intensive work is needed to develop good interspecific hybrids resistant to wilt by using these species in the breeding programme.

The contribution of wild species in crop improvement and management programmes mainly involves their utilization as rootstocks for regulation of vigour, yield fruit quality and disease and pest resistance. Pathak and Ojha (1993) enumerated their potential uses. *P. cujavalis, P. molle, P. cattleianum* and *P. guineense* can be used as rootstock. Chinese Guava (*P. friedrichsthalianum)* and Philippine guava are compatible rootstocks and are reported resistant to wilt disease. 'Allahabad Safeda' trees grafted on *P. pumilum* had a dwarfing influence. *P. cujavallis* produced the largest trees but with non-uniform and rough skinned fruits. Singh *et al.,* (1976) observed that fruits of 'Allahabad Safeda' contained higher sugar content on *P. pumilum* while higher ascorbic acid content was recorded in these grafted on *P. cujavallis.* High yields were obtained while using *P. cattleianum* rootstock. The other species reported resistant to the wilt are *P. cumunii, P. cattleianum* var. lucidum, *P. molle, P. guineense*. Singh *et al.,*(1977) reported that *P. araca, P. cattleianum* var lucidum, *P. corecium, P. cujavalis, P. guineense* and *P. friedrichsthalianum* developed infection. Interspecific hybridization carried out at CISH, Lucknow between *P. chinensis* and *P. molle* resulted in the development of wilt resistant rootstock.

Inheritance studies

Genetical studies conducted in guava at IIHR, Bangalore, indicated that red pulp colour is dominant over white and that this character is governed monogenically. Many cultivated red fleshed varieties were found to be heterozygous for this character. Bold seeds were dominant over soft seeds and this was also determined monogenically. A linkage was also found between flesh colour and seed size i.e., red flesh with bold seeds (Subramanyam and Iyer, 1982). Mitra and Bose (1985) reported heterosis in guava. Studies conducted at Coimbatore by Raman *et al.,* (1969 and 1971), have shown that triploidy and genetic factor(s) are responsible for female sterility and the variation among triploids is due to their independent origin from a distinct diploid variety. Iyer and Subramanyam (1971), opined that the seedless varieties of guava are triploids, grow vigorously and bear fruits, which are irregular in shape with ridges because of the irregular distribution of seeds of various sizes. Dinesh and Yadav (1998) carried out half-sib analysis in the progenies of the variety 'Apple Colour'. They observed that the genotypic variance was lower than the phenotypic variance and heritability to be moderately high for all the characters implying that selection can be practiced for the improvement of fruit characteristics. Hence, hybridization among less seeded diploids can be adopted in the improvement programme. It is our observation at Indian Institute of Horticultural Research on the inheritance pattern using 'Purple guava' and Arka Mridula as parents that the hybrids segregated in a ratio of 3:1 for green leaf types and purple leaf types.

Utilisation of germplasm

The accessions viz., Allahabad Safeda and seedless were used in the breeding programme for developing varieties like Arka Amulya (Allahabad Safeda x Seedless) and Arka Mridula (selection from Allahabad safeda). Interspecific hybridization was carried out using the wild species *Psidium chinensis* x *P.guajava* cv.Beaumont to produce root stocks resistant to wilt disease. Apple colour and Sardar were also used in various combinations. Red flesh and Philippine guava are being used in the breeding programme for imparting red/ purple colour to the progenies.

Future needs

Priority needs to be given for good fruit quality since there is little merit in increasing yields and disease resistance, if not accompanied by high quality. High quality should include high TSS, good sugar-acid blend, good aroma, attractive skin and flesh colour and soft seeds, processing quality, which includes juice colour, high Vitamin C, higher lycopene content, good pectin content and good flavour. While selecting new varieties, keeping quality may be given adequate importance. In this connection the flavour and firmness of the pulp (that could contribute towards better keeping quality) of the 'apple coloured guava' as gene donors to improve other commercial cultivars. Hence, attempts to hybridize these genotypes with 'Allahabad Safeda' and other commercial cultivars are to be intensified and selections may be made of those even without apple colour provided they have all other desirable characteristics.

In the recent past, efforts have been intensified to develop apple coloured cultivars to make them attractive for local as well as export markets. However, deep red colour has been found to be a very unstable character, the skin colour changing from deep red to yellowish white from season to season as well as within the same tree. Hence, 'stable' types are to be identified and great care has to be taken while selecting hybrids from large population.

The cultivar 'Allahabad Safeda' possibly represents a population of guava trees grown extensively in Uttar Pradesh (India) than the descendents of single clone. Hence, enormous variation has been observed in this so-called cultivar. It is for the horticulturists to make rigorous screening of the population to identify superior genotypes keeping specific objectives in mind. A survey to locate such genotypes is certainly required especially in Uttar Pradesh. While breeding or selecting superior types for suitable for high density planting characters like early bearing, compact plant type, favourable response to pruning, good branch angle to minimize branch breakage even under heavy bearing and with a high fruit-shoot ratio need to be given due attention.

Allahabad safeda

Apple Colour

Arka Kiran

Variability for fruit shape in guava

Guava flower

Mridula 4

Since, guava cultivation in many localities is threatened by wilt (*Fusarium* spp), resistant rootstocks and (or) cultivars need to be identified or developed. However, where self-rooted plants are used for planting, development of scions combining good fruit quality and resistance to wilt becomes essential. In this connection, the Philippine Guava (Purple type) which has shown some promise as a wilt-resistant rootstock needs extensive experimentation. Since, this species segregates into purple and white types on crosses with *P. guajava* cultivars, there is need to look for any possible linkages between purple leaf types and resistance to wilt. Extensive screening of other related *Psidium* species need to be made for their resistance to wilt. In this connection, reliable pathological screening techniques need to be standardized to hasten the process of disease resistance breeding in guava.

Varietal introduction, though an essential part of any crop improvement programme, needs to be made with great caution and with strict customary plant quarantine measures. To quote some examples of caution, the 'Beaumont' variety of guava when introduced to India was found to be severely infected with 'Stylar End Rot' (*Phomopsis psidii),* although, it is not a severe problem in its original habitat. Similarly, the 'Giant Thailand Guava' when introduced to Bangladesh was found to be severely susceptible to many insect pests. Such examples are many and should be borne in mind.

Molecular characterization of germplasm needs to be taken up with a view to work out the genetical distance so that good recombinants can be arrived by crossing suitable parents. Biotechnological tools need to be employed extensively. DNA finger printing and similar tools may be used to study the extent of variation even among the 'Allahabad Safeda' cultivar. Attempts to spot genetic markers for wilt resistance may be attempted to improve efficiency in developing wilt resistant clones and rootstocks.

References

Anonymous., 1999, Report of National Horticulture Board. pp 42-44.

Anonymous., 1999, DARE/ICAR Annual Report. pp 45.

Balasubramanyam, V.R., 1959. Studies on blossom biology of guava (*Psidium guajava* L.), *Ind. J. Hort.*, 16(2): 69-75.

Braganza, M.A., 1990. "Floral Biology studies and varietal evaluation in genus *Psidium*". M.Sc. Ag. Thesis submitted to University of Agricultural Sciences, Bangalore.

Chadha, K.L., Singh Harmail and Tandon, D.K. 1981. A varietal trial of guava. National Symposium Tropical Sub-tropical Fruit Crops, Bangalore, pp.17.

Chandra., R, A Bajpai, Soni Gupta and Tiwari, R.K. 2004. Embryogenesis and plant regeneration from mesocarp of *Psidium guajava* L. (guava) *Indian Journal of Biotechnology* 3:246-248.

Dass, H.C. and Prakash, D. 1981. Phylogenetic affinities in Psidium spp as studied by Flavonoid patterns. National Symposium on Tropical and Sub-tropical Fruit Crops, Bangalore, pp15.

De Candolle, A.P. 1904. Origin of cultivated plants. Kogan Paul, London.

Dinesh, M.R. and Reddy, B.M.C. 2001. Evaluation of *Psidium guajava* accessions and some other *Psidium* species for fruit characters. *J. Appl. Hort.*, 3(1):41-43.

Dinesh, M.R. and Yadav, I.S. 1998. Half-sib analysis in guava (*Psidium guajava*). *Indian J. Hort.*, 55(1):20-22.

Du Preez., R.J., 1995. Guava cultivars. *Inligtingsbulletin-Instituut-vir-Tropiese-en-Subtropiese-Gewasse*. 272:10-16.

Edward, J.C. and Shankar, G. 1964. Rootstock trial for guava (*Psidium guajava* L.). *Allahabad Farmer*, 38:249-50.

Golberg, L and Levy, L. 1941. The vitamin C content of fresh, canned and dried guava. *Nature*, 148-286.

Hayes, W.B. 1953. Fruit Growing in India. Kitabistan, Allahabad.

Iyer, C.P.A. and Subramanyam, M.D. 1971. Problems with triploidy in guava. *Sabrao Newsletter* 3(1):31-33.

Jaiswal,V.S and Amin, M.N. 1992. Guava and jackfruit. Biotechnology-of-perennial-fruit-crops. Hammerschlag, F.A; Litz, R.E. (Ed) 421-431. Wallingford, UK.

Kumar, L.S.S. and Ranade, S.G. 1952. Autotriploid in guava (*Psidium guajava* L.). *Curr. Sci.*, 21:75-76.

Leslie, R.W. Landrum., Dennis Clark, P. William and Brendecke, Jeff. 1995. Hybridization between *Psidium guajava* and *P. guineense* (Myrtaceae), *Economic Botany.* 49(2):153-161.

Leu, L.S. and Kao, C.W. 1979. Artificial inoculation of guava with *Myxosporium psidii. Plant Disease Report*. 63:1077-79.

Mitra, S.K. and Bose, T.K. 1985, Guava. Fruits of India-Tropical and Sub-tropical (ed) by Bose Naya Prokash, Calcutta.

Morton, J. 1987. Guava. (In) Fruits of warm climates. Julia F. Morton and Miami, FL. (eds) p.356-363.

Negi, S.S., Mishra, A.K. and Rajan, S. 2001. *Guava wilt. Ind. J. Hort.*, 58(1&2):145-151.

Normand, F. 1994. Strawberry guava, relevance for Reunion. *Fruits* 49: 217-27.

Normand, F. and Habib, R. 2001. Phenology of strawberry guava (*Psidium cattleianum*) in Reunion Island. *J. Hort. Sci. and Biotech*. 76: 541-545.

Ojha, A.P., Tiwari, J.P. and Mishra, K.K. 1986. Studies on growth, flowering and yield of guava (*Psidium guajava* L.) under terai condition of U.P. *Prog. Hort.*, 8:205-06.

Pandey, S.D., 1968. The guava of Uttar Pradesh: A classification. *Hort. Adv.*, 7:72-98.

Pathak, R.A. and Dwivedi, R. 1988, Report of Fruit Research Workshop Sub-tropical and temperate fruits. Rajendra Agricultural University, Pusa, 76-77.

Pathak, R.K. and Ojha, C.M. 1993. Genetic resources of guava. In Advances in Horticulture (Vol I) Eds. Chadha, K.L and O.P. Pareek, Malhotra Publication House, New Delhi

Purseglove, J.W. 1968. Troical Crops: Dicotyledons. John Wiley and Sons, Inc., New York.

Raman, W.M., Manimekalai, G. and Ramalingam, R.S. 1969. Observation on seedlessness, fruit development and cytology of varieties of guava. *Mad. Agri. J.*, 56: 255-61.

Raman, V.S., S. Sri Rangaswamy and Manimekalai, F. 1971. Triploidy and Seedlessness in guava (*Psidium guajava* L). *Cytologia*, 36:392-399.

Ram Kumar., 1973. Inducing ploypoidy and cytological studies in guava (*Psidium guajava* L). *Ind. J. Hort.*, 32(3-4):128-130.

Ram Kumar., 1998. Performance of guava under rainfed condition of Bihar. *Haryana J. of Hort. Sci.* 27(3):145-47.

Risterucci, A.M., Duval, M.F., Rohde, W. and Billotte, N. 2005. Isolation and characterization of microsatellite loci from *Psidium guajava* L. Molecular Ecology Notes doi:10.1111/j.1471-8286.

Sachan, B.P., Pandey, D. and Shankar, G. 1969. Influence of weather on chemical composition of guava fruits (*Psidium guajava* L) var. Allahabad Safeda. *Punjab Hort. J.*, 9: 119-23.

Sehgal, O.P. and Singh, R. 1967. Studies on blossom biology of guava (*Psidium guajava* L) I. Flowering season, flowering habit, floral bud development, anthesis and dehiscence. *Ind J. Hort.*, 24:118-126.

Shafaat Mohammed,1975. Investigations on the breeding behaviour of aneuploids of guava. Thesis submitted to the Division of Fruits and Horticultural Technology, IARI, New Delhi.

Singh, L.B., 1959, S1, a new promising selection of guava (*Psidium guajava* L). Annual Report, Fruit Research Station, Saharanpur, pp. 58-60.

Singh, V.R., L. Dhar and Singh, G. 1977. Note on the performance of guava cultivars and *Psidium* species against wilt disease under natural field conditions. *Haryana J. Hort. Sci.* 149-150.

Singh, U.R., Pandey, I.C., Upadhyaya, N.P. and Tripathi, B. M. 1976. Effect of different rootstocks on the growth yield and quality of guava. *Punjab Hort. J.* 16:121-128.

Singh, R. and Sehgal, O.P., 1968, Studies on blossom biology of *Psidium guajava* L (guava). II. Pollen studies, stigma receptivity, pollen and fruit set. *Indian J. Hort.*, 25:52-59.

Singh, I.S., Singh, H.K. and Gupta, A.K. 1979. Effect of post harvest application of ethephon on quality of guava (*Psidium guajava*) cultivar Lucknow - 49. *Haryana J. Hort. Sci.* 8(1/2): 12-16.

Srivastava, O.P. 1974, Studies on the flowering habit, blooming period, anthesis, dehiscence and pollen grain of *Psidium guajava* L varieties Apple Colour, Chittidar and Red Flesh. *Prog. Hort.*, 6(3):71-77.

Srivastava, H.C.1977. Cytological studies in *Psidium friedrichsthalianum N. Cytologia*, 42:395-400.

Srivastava, R.P and Srivastava, R.K. 1965. Physico-chemical studies on Safeda Allahabad and Red fleshed guavas. *Punjab Hort. J.*, 5:12-15.

Subramanyam, M.D. and Iyer, C.P.A. 1982. Improvement of guava by breeding. Report, Fruit Workshop, Nagpur. pp. 117-118.

Syamal, M.M., Singh R.K. and Chhlonkar, V.S. 1980. Studies on growth and flowering in guava. *Psidium friedrichsthalianum*. 37:243-45.

Teaotia S.S., Pandey, I.C. and Agnihotri, B.N. 1962. Study of some guava varieties (*Psidium guajava* L.) of Uttar Pradesh. *Indian Agriculturist*. 6:47-53.

16

Sapota Breeding

A. Rekha

Sapota [*Manilkara zapota* (L.) P. Royen, (*Manilkara achras* (Mill.) Fosber] is a member of the family sapotaceae. It is a fairly slow growing and long-lived tree. It is valued for its delicious fruit in India, while in Guatemala, south East Mexico, British Honduras, chickle is commercially produced which is the main ingredient in preparation of chewing gum. Chickle is extracted from trunk and unripe fruits of the tree as white latex exudate. Although synthetic gums are primarily used, countries like Mexico, Venezula and Guatemala still grow sapota for chickle.

It is native to South Mexico as well as North East Guatemala. *Manilkara* species is found in forests throughout tropical America, which was introduced long ago followed by West Indies, Bahamas, Bermuda and Southern part of Florida mainland. Early in colonial times, it was carried to Phillippines, and later distributed to Malaysia and other countries in the tropics (Purseglove, 1974). It is not known when it was first introduced to India. Commercial cultivation was first taken up in Maharashtra during 1898 (Cheema *et al.,* 1954). Future of sapota appears to be promising, since attention in this crop is enthusiastic from growers and consumers in many countries. Sapota is grown on commercial basis in India, Philippines, Sri Lanka, Malaysia, Mexico and Venezuela.

India is one of the largest producers of sapota in the world. Major states growing this fruit crop are Maharashtra, Gujarat, Andhra Pradesh, Karnataka, Tamil Nadu, West Bengal, Kerala, Uttar Pradesh, Punjab, Assam and Haryana. India's production of sapota continues to grow up and there is an active research program in India with specific goals towards improving storage, transport and marketing strategies.

Composition and uses

Fully ripened fruits are eaten along with skin, as skin is more nutritive than pulp (Gopalan *et al.,* 1981). Fruits contain sugars ranging from 12 to 18%. Ripe fruits contain carbohydrate (21.48g), protein (0.7 g), fat (1.1 g), calcium (28 mg), phosphorus (27 mg), iron (2 mg) and ascorbic acid (6 mg) per 100 g.Ripe fruits are used for making mixed jam, jelly, sherbet and halwa. It is also used for manufacture of industrial glucose and pectin. 'Chickle' made from milky white latex of bark and immature fruits is the base material in chewing gum and statutes. Bark is used as a tonic, antipyretic and astringent. Decoction is given in diarrhea and in paludism. Tender leaves are used as a vegetable, the wood for making agricultural implements, building construction and furniture industry.

Botany and varieties

Botanical name of sapota is *Manilkara zapota* (L.) P.Royen, other synonyms are *Manilkara achras* and *Achras sapota*. It is a slow growing evergreen tree with a height of about 9 m. Main stem is strong, stout and produces radial whorls of branches at more or less regular intervals. Main branches bear a number of short branch lets, which produce flowers and fruits terminally in leaf axils. They also give out vegetative shoots from their sides. Leaves are elliptic to obovate, flowers small, clustered in leaf axils, produced on past season shoots, sepals 6, tomentulose, petals 6, staminodes 6 and petaloid stamens 6. Fruit is a berry, oval/ round shaped. The skin of the fruit is thin, rough and brownish. Seeds are black, shiny and 3 to 6 in numbers. The basic chromosome number (x) = 7, but ranges 9 to 13 in the Sapotaceae family.

Varieties

There are about 40 to 50 varieties grown all over India. These varieties are classified into four groups based on branching habit, colour and shape of foliage, rind character and pulp consistency.

Trees with erect growing habit

Branches appear in whorls, foliage deep green, broad and oval. Fruits are large in size with a smooth and yellow skin. Pulp is butter like and sweet.

Trees with drooping habit

Branches appear in whorls; foliage light green, narrow and elliptical. Fruits are small with rough brownish skin and pulp of inferior quality.

Trees with spreading habit

Branching irregular, foliage light green, narrow and elliptical. Fruits with rough skin and pulp inferior in quality.

Each of these groups is further divided into two sections based on shape of fruits, either round or oval.

A good table purpose sapota has a few seeds with melting sweet pulp. Thick skinned hard fleshed varieties with sandy texture are inferior in quality.

Important varieties in sapota are Kalipathi, Chaatri, Dwarapudi, Guthi, Gavarayya, Cricket Ball, Oval and kirthibarti. Improved varieties like CO 1, CO 2, CO3, PKM-1, PKM-2, PKM-3, PKM-4, DHS-1, DHS-2, are also evolved in this crop (Table 1)

Crop Improvement

In sapota, crop improvement works were reported from Agricultural college and Research Institute, Coimbatore and Periyakulam, Tamil Nadu (Sambamurthy and Ramalingam, 1954) and Agricultural College, Dharwad, Karnataka (Farooqui and Rao, 1976). Main objectives are dwarf stature, higher production, large sized fruits, good dessert and keeping quality. Improvement works were through selection and hybridization methods. Many other hybrids are also under evaluation (Table 2).

Table 1: Characteristics of sapota varieties

Name of the variety	Characters	Fruit characters	Pulp characters	Other features
Calcutta Round	Foliage light green	Large and round	Flesh gritty, medium quality	Commercial variety in West Bengal and Karnataka. Susceptible to leaf spot.
Baramasi	Foliage light green	Medium and round	Medium quality	Popular in West Bengal, Bihar, U.P. Bears though out the year.
Bangalore	Foliage green	Large and oval, Ridges on the fruit	Golden coloured, medium sweet	Popular in A.P.
Chaatri	(Dropping branches, appear in all directions, have the shape of umbrella) Foliage light	Medium and oval	Sweet, medium in quality	Common in Maharashtra, medium bearer, trees have the shape of an umbrella
Cricket Ball (Calcutta large)	Foliage light green	Large and round	Gritty, granular not very sweet	Shy bearer, suitable for arid region, grown in Tamil Nadu, Andhra Pradesh, West Bengal, Karnataka,Maharashtra
Dhola Dowani	Foliage light green	Medium and oval	Quality superior	Grown in Maharashtra
Bhuripatti	Foliage thick, light green to brown	Medium, large, oval	Quality good, moderately gritty	Medium bearer
Jhumaki	Dark green foliage	Small and oval	Moderately gritty and medium in quality	Medium in bearing fruits borne in cluster.
Jonnavalasa 1	Light green foliage	Medium and oval. Fruit skin thin, rough, without ridges.	Pulp cream coloured, sweet	Grown in Andhra pradesh
Jonnavalasa	Light green foliage	Medium and oval, fruit skin buff coloured with eight marked ridges, prominent depression at stalk end	Musk melon like flesh, centre with golden tinge	Grown in Andhra pradesh

Contd.

Name of the variety	Characters	Fruit characters	Pulp characters	Other features
Kalipathi	Dark green broad and thick leaves	Medium and oval	Mellow flesh,very sweet	Leading variety of Maharashtra, Gujarat, N. Karnataka. Excellent in quality, single fruited.
Kirthibarthi	Foliage light green	Medium and egg shaped. Fruit skin rough, medium thick, buff coloured, 4-6 ridges, apex round.	Sweet pulp	Popular in Andhra Pradesh. Suitable for distant marketing.
Long	Narrow, small leaves	Long, small, medium	Sweet	Grown in Maharashtra, poor bearer.
Pala	Light green foliage	Small- medium, oval/egg shaped. Fruit apex broadly pointed.	Very sweet, good flavour	Grown in Andhra Pradesh, Tamil Nadu heavy bearer, borne in clusters.
Gavarayya	Dark green foliage	Small-medium and oval. 8-10 ridges on the fruit surface. Fruit shoulder dropping on a side, other side raised.	Soft, melting and very sweet	Grown in Andhra Pradesh and Tamil Nadu
Pot sapota	Light green foliage	Small and oval. Fruits with pointed apex and no ridges.	Very sweet, flavor good	Suitable for homesteads, dwarf stature
Tagarampudi	Light green foliage	Medium, round or oval. Fruits with flat base, skin thin and rough.	Flesh soft, juicy melting, very sweet and buff coloured	Tolerate distant transportation.
Oval	Light green leaves	Small and oval	Soft, melting, sweet	Shy bearer

Table 2: Characteristics of improved varieties of sapota

Places	Varieties	Parents	Characters
Selections			
TNAU, Coimbatore	CO 2	Baramasi	Tree medium to large, fruits medium and round, fruit weight 100-120 g, TSS 23^0 Brix.
TNAU, Periakulam	PKM-1	Guthi	Dwarf trees, fruits oblong, fruit weight 80-85 g, TSS 23^0 Brix
Periakulum	PKM-4	OP of PKM-1	The fruit and tree is in spindle shape which is suited for processing.
Periakulum	PKM-5	OP from Virudhunagar	Oval shaped attractive fruits with smooth, light brown peel. Pulp retains coppery brown colour, tolerance to leaf spot and leaf webber incidences under field conditions.
Hybrids			
TNAU, Coimbatore	CO 1	Cricket Ball x Oval	Trees large, fruits medium and oval, shaped, fruit weight 100-125 g, TSS 18^0 Brix
TNAU, Coimbatore	CO 3	Cricket Ball x Vavilavalsa	Trees large, fruits medium to large, round, fruit weight 120- 140 g, TSS 24^0 Brix
Periyakulam	PKM-2	Guthi x kirtibarthi	Trees medium, spreading, fruit oblong, fruit weight 80-85 g, TSS 26.2^0 Brix
Periyakulum	PKM-3	Kalipathi X Cricket Ball	Trees erect, oblong fruits and weight 80-85 g, TSS 26.1^0 Brix
UAS, Dharwad	DHS-1	Kalipathi x Cricket Ball	Round, large fruits, weight 200-210 g, TSS 23^0 Brix
UAS, Dharwad	DHS-2	Kalipathi x Cricket Ball	Oval medium fruits, 150-160 g, weight, TSS 26^0 Brix

Floral biology, flowering and fruit set

Seedlings take seven or more years and vegetatively propagated plants require 2 to 3 years for flowering. Economic yield is obtained after 7 or 8 years in grafted plants. Sapota gives flowers almost all months under tropical climate, but there are two main flowering seasons i.e. February-March and August - November. According to Nalawadi *et al.,* (1977) there are three flowering seasons; main season is in June while in October and March, flowering is less. Sambamurthy and Ramalingam (1954) reported that in Sapota, bud to flower opening takes 36 days. Flowers open during early hours of the day and the anthesis occurs between 4.00 AM to 10.00 AM. Peak anthesis is at 4.00 AM and maximum stigma receptivity on the day of opening of flower. In nature sapota appears to be wind pollinated. Flowers are protogynous; style and stigma grow out of the bud about two days before actual opening of the flower. Self incompatibility and self sterility are seen in sapota. It is a cross pollinated crop and pollination is mainly effected by wind. Because of these problems, fruit set may be very low when a single sapota tree is planted in homesteads.

Farooqui and Rao (1976), observed Metaxenia in sapota (effect of pollen used on fruit development). The fruit set, size, weight of the fruit, number of seeds per fruit, colour of the peel, scurf and maturity were influenced by pollenizer. The self pollinated fruits were oblong to oval as compared to crossed ones in Cricket Ball and Kalipatti cultivars.

In sapota, fruit set is relatively high but fruit retention is poor. It was observed that maximum fruit drop occurred immediately after fruit set. The fruit drop was very less during fruit development period (Patil and Narwadkar, 1974). Usually, the flowers emerged at base of inflorescence opened and set earlier and developed rapidly while remaining flowers in the same inflorescence may develop slowly or drop off. Fruit set can be improved by using bioregulators. Spraying NAA (25 to 100 ppm) at flowering and pea stage increased fruit set in sapota.

Sapota fruits exhibited a double sigmoid pattern of growth with two periods of active growth and one period of lag phase. Time taken from fruit set to maturity varies with variety, climate and other factors. It requires four months and can be extended to 10 months. Fruits matured during March-May exhibited better fruit quality. Fruit weight and fruit shape are positively correlated with number of seeds and its distribution around the placenta. Round shaped fruits have well distributed seeds and number of seeds increases fruit weight in sapota.

Genetical studies

Genetic variability among 20 Indian sapota cultivars was studied using RAPD for the first time by Meghala *et al.,* (2005), and observed wide genetic variability. Earlier studies by Dinesh and Reddy, (2000) on morphological parameters revealed variability in many characteristics. In a study using 20 cultivars, fruit length ranged from 5.3 to 7.9 cm, fruit size from 43-160 g, TSS from 15.6 to 23.7 and seed number from 1-6. In another study by Ponnuswamy and Irulappan (1987) involving 12 sapota cultivars, percentage coefficient of variation (CV) varied from 11-51 for various characteristics. Genetic correlation and cluster analysis made by Rekha *et al.,* (2011) classified 20 sapota cultivars based on fruit shape Round and Oval. The available variability in sapota based on fruit characters were used to conduct cluster analysis which revealed four definite clusters. High variability was observed for fresh- fruit weight, fruit length, fruit girth, fruit weight at ripening, pulp weight, peel weight, number of seeds and TSS. Correlation studies among fruit parameters indicated positive relationship between all the parameters involved, except TSS which was negative relationship with rest of fruit parameters.

Future lines of work

- Sapota being a slow growing tree crop, there are very less studies on genetics and breeding. With help of molecular markers, there is hope of identifying better cultivars with dwarf trait and good fruit quality. The dwarf stature will help in accommodating more number of trees per unit area and harvest would be easy.
- There is a need to regularize flowering and fruit set for a uniform development of fruits and harvest at one time.
- There is also a need to create variability by raising large number of half sib progenies as the number of progenies, produced out of controlled pollination is less.
- Molecular markers should be developed for some of the quantitative traits like dwarf stature so that progenies can evaluated at the seedling stage itself.

References

Bose. T.K. and Mitra, S.K. 1990 (Ed.). Fruits: Tropical and Sub-Tropical. Naya Prokash, Calcutta.

Cheema, G. S., Bhatt, S.S. and Naik, K.C. 1954. Commercial fruits of India. MacMillan and Co., 422 p.

Dinesh, M. R. and Reddy, B.M.C. 2000. Fruit evaluation studies in sapota (*Achras zapota* L.). *J. Appld. Hort.* 2:19-20.

Farooqi A.A. and Rao, M.M. 1976. Studies on Metaxenia in Sapota (*Achras sapota* L.). Mysore *J. Agric. Sci.* 10: 413-423.

Meghala.R, Ravishankar, K.V. Lalitha Anand and Rekha, A.2005. Genetic diversity of Indian Sapota (*Manilkara zapota*) cultivars characterized by RAPD markers. PGR-Newsletter FAO-IPGRI issue No.142 pp 43-46.

Nalawadi, U.G. Dasappa and Sulikeri, G.S. 1977. Floral biology of some varieties of sapota (*Achras sapota* L.). *Prog. Hort.* 9(1):27-32.

Patil, V.K and Narwadkar 1974. Studies on flowering, pollination and fruit drop in chiku. *Punjab Hort. J.*, 14:39-42.

Peter. K.V. (Ed.) 2007. Horticultural Science Series. Fruit Crops, Vol 3.

Ponnuswamy, V. and Irulappan, I. 1987. Evaluation of the chemical composition of fruits of different varieties and hybrids of sapota. *South Indian Horticulture*. 35:446-47.

Purseglove, J.W. 1974. Tropical crops.719 p.

Rekha.A, M.R.Dinesh, Venugopalan, R. and Murthy, B.N.S. 2011. Genetic correlation and cluster analysis in sapota (*Manilkara zapota*). *J. Hort. Sci.* 6(2):101-104.

Sambamurthy. K and Ramalingam, V. 1954. A note on hybridization in sapota (*Achras sapota* L.). *Indian J. Hort.* 57-60.

17

Jackfruit Breeding

Prakash Patil

Jackfruit (*Artocarpus heterophyllus* Lam) is a multi-purpose tree species that provides food, timber, fuel, fodder and medicinal and industrial products. It is one of the species that has been brought into wider cultivation through a farmer-driven domestication process because its products meet the household needs of families with limited land, which require wood and animal feed to ensure family food security. Despite such vast potential and usefulness, jackfruit remains an underutilized fruit species and deserves to be given the necessary thrust for research and development. There is little progress on the genetic improvement of the species, including study on its reproductive biology, even though it demonstrated substantial use and importance. Therefore, a brief background on the reproductive biology of jackfruit is covered here before dealing with its genetic improvement.

Reproductive biology

The jackfruit tree is monoecious, having male and female inflorescences or spikes on the same tree. Male and female spikes are borne separately on short, stout stems which sprout from older branches and the trunk.An overview of the reproductive biology of jackfruit is given here as per Bawa and Krugman (1990) to include flowering and fruiting morphology and flowering and fruiting phenology, which are essential for effective management of the species for both conservation and production purposes. Pushpakumara *et al.* (1997, 2010) and Pushpakumara (2006; 2006/07) conducted a detailed study on the reproductive biology of *A. heterophyllus,* collecting qualitative and quantitative data from both field and laboratory experiments. This information, along with other results reported, are summarised here to facilitate appropriate tree improvement, utilization and conservation of the species.

Male inflorescences and flowering

Male inflorescences occur mainly on the terminal shoots and branches of the tree's crown and also in the main stem (Fig. 1). They are noticed above the female spikes. Based on a detailed phenological experiment on the development of male inflorescences, Pushpakumara (2006) suggested six stages of male inflorescences. At the beginning, male inflorescences are light green with a smooth surface, becoming yellowish and rough when anthers appear. They are cylindrical or barrel-shaped, 2–11 cm long and 1–5 cm wide at the time of maximum pollen presentation. The flowers are tiny and pale green when young, and turn darker with age.

The stamen has four anthers and emits a sweet scent (Purseglove, 1968). It splits to expose a mass of sticky, spherical, yellow pollen grains. A single anther produces a large number of pollen grains (a mean of 931 ± 44 and range from 500–1,100). This equals approximately five to six million pollen grains per male inflorescence (Pushpakumara, 2006). The pollen grains are smooth, spherical in shape, 15–20 μm in diameter with three pores.A few inter-floral bracts are shed immediately after anthesis. The peak of anthesis is between 2 and 3 pm (Samaddar and Yadav, 1982). However, Pushpakumara, (2006), reported that anther dehiscence of male flowers occurs late in the morning (10.30 am) or early in the afternoon (2.30 pm). After they open, the anthers shed pollen over a mean period of approximately six days (6.3 ± 0.04), before the empty pollen sacs wilt and abscise.

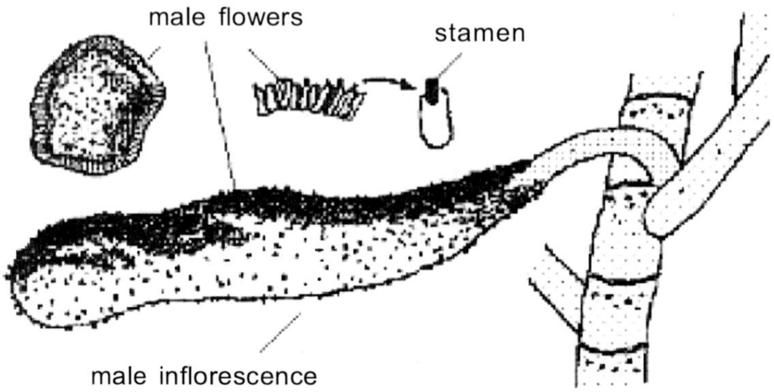

Fig. 1: Branch of *Artocarpus heterophyllus* male Inflorescence

Among the male flowers, sterile ones are also found. The sterile male flower has a solid unlobed perianth, whereas the fertile male flower is tubular and bi-lobed. The abscission of male spikes starts with a change in colour from green to yellow, turning to brownish yellow, and finally brown to black. Spikes generally fall off two to four weeks after emergence (Querijero, 1988). However, they may remain on the tree longer, before falling to the ground as a black mass.

Female inflorescence and flowering

Female inflorescences are solitary and are produced on short axillary leafy twigs called footstalks (Fig. 2). Footstalks are borne either on the tree trunk or on older branches (Acedo, 1992; Corner, 1981; Jarrett, 1959; Moncur, 1985; Pushpakumara, 2006; Soepadmo, 1991). Based on detailed phenological investigations of the development of female inflorescences, Pushpakumara (2006) suggests nine phenological stages.

Female flowers are larger, elliptic or rounded, with a tubular calyx. The flowers are reportedly pollinated by insects and wind, with a high percentage of cross-pollination. Individual female flowers are borne on fleshy receptacles. They are characterized by a small, greenish perianth and a unilocular ovary with a relatively long style measuring 9.46 ± 0.03 cm Fig. 2 (range 6–13 mm).

Fig. 2: Branch of *Artocarpus heterophyllus* with Inflorescence (A), Longitudinal Section of Female Inflorescence (B), Longitudinal Section of Female Flowers (C), Separated Female Flower Parts (D).

Creamy white stigmas protrude from the surface of the spike when receptive, usually 4–6 days after opening of the spathes. Stigmas tend to be receptive for 28–36 hours (Sambamurthy and Ramalingam, 1954; Azad, 1989). Observations from previous studies, for example by Moncur (1985) and Primack (1985), also suggest a long duration of stigmatic receptivity (about two weeks) in A. *heterophyllus.* In contrast, Sambamurthy and Ramalingam (1954) suggested that stigmas are receptive for only 36 hours. Lengthy stigmatic receptivity may be a common feature of the genus *Artocarpus,* along with other *Artocarpits* species.

Plate 1: Branch of *Artocarpus heterophyllus* with male and female Inflorescence

Flowering habit

In Asia in general, flowering in jackfruit starts in December and continues until March (Samaddar, 1985; Azad, 1989). However, flowering in off season, which is between September and October, is not uncommon (Singh, 1992) and geographical differences are certainly seen to affect flower initiation. Further, attempts have been made to induce flowering in jackfruit. For example, chemicals such as Ethephon (550 ppm), when sprayed onto the foliage in early October improved the number of female spikes (Das *et al.*, 1981) and the application of potassium nitrate (1%) by trunk injection also induced flowering (Angeles, 1983).

Male and female inflorescences are borne on footstalks, which appear on branches or stems initially as yellow-green, bud-like structures. The ratio of male-to-female inflorescences varies and most of the studies indicate a higher proportion of male than female inflorescences. Samaddar and Yadav (1982) reported 96% male inflorescences, whereas Srivastava (1961) reported 80% male inflorescences per tree. Azad (1989) reported very low number of female inflorescences (3–5%), whereas Sahadevan *et al* (1950) reported 3–40%.

The spike can easily be identified when small, because the length and diameter of female inflorescences are larger than the male inflorescences. The surface of a young male spike is smooth, whereas that of the female is granular. Male and female inflorescences develop more or less simultaneously but male spikes tend to reach maturity 3–5 days before the female ones. On any one tree, male and female flowers open over a long period (Soepadmo, 1992).

Pollination

The production of more male flowers than female flowers in jackfruit is a characteristic of the typical anemophilous species (Purseglove, 1968). Pushpakumara *et al.* (2010) suggested that wind is a primary vector in the pollination of *A. heterophyllus*, given the conditions in Sri Lanka. This is supported by observations of the deposition of *A. heterophyllus* pollen grains on pollen traps and pollen grains and/or grains with tubes on open-pollinated stigmas. Pollen grains are not shed readily if there is slight physical disturbance, as would be expected from a wind-pollinated plant (Moncur, 1985).

The flowering male inflorescences have the sweet scent of honey and burnt sugar, which attracts small flies and beetles and other insects. Their activity promotes the release of pollen grains from the protruding anthers. Nonetheless, very a few insects visit the female flowers and this reduces the chances of direct pollination by insects. The wind and insect activity together may act as a pollination mechanism. If the flowers of the female spikes are not adequately pollinated, fruits do not develop normally; the fruit size is small and the shape may be irregular. Further, lack of pollination may lead to the dropping of female spikes (Azad, 1989), but this should not be confused with early fruit drop, which appears to be a natural physiological thinning mechanism.

Previous reports ascribe syncarp (the aggregated fruits of several flowers) abscission principally to the lack of pollination (Acedo, 1992; Moncur, 1985). Observations of the abscission of female inflorescence buds and non-receptive female inflorescences suggest that factors other than inadequate pollination are involved. The possibility of syncarp production without pollination also suggests that pollen limitation is not the only factor controlling inflorescence abscission in *A. heterophyllus* (Pushpakumara, 2006). Sinha (1973) reported that pollen viability in jackfruit is affected by different ranges of relative humidity and temperature. Gopinath *et al.* (1983) reported on pollen fertility of 94.5% when pollens were kept on 1% agar and 10% sucrose.

Developmental irregularities, such as abnormal female inflorescences, also result in abscission. Pests and diseases attacking the inflorescences and young fruits can also have a direct or an indirect effect on inflorescence abscission. However, it is seen that pollination can be improved by hand pollination.

Fruiting

A. heterophyllus produces a large, round, cylindrical to pear-shaped, greenish fruit that is technically termed syncarp, which is a type of fruit consisting of several achenes (fruitlets or true fruit). Syncarps mature approximately three months after the emergence of inflorescences from the stipules. Footstalks usually produce more than one inflorescence and in some cases, young syncarps and inflorescence buds are produced on the same stalk. In these situations, maturation of syncarps generally occurs chronologically, giving early-appearing inflorescences a higher expectation of producing syncarps, although occasionally the later-appearing inflorescence also develops as a mature syncarp before the first. The syncarps are composed of individual fertilised ovaries, each of which develops into a fruitlet with soft, yellow, fleshy perianths that develop into pulp and surround the seeds. The fruitlets, or true fruits, are sometimes called bulbs, pods or pips.

Unfertilised female flowers develop into tough, strap or string-like structures, called perigones, white pulp or rags. They occupy the spaces between the fruitlets. Whereas the perianths of fertilised flowers form edible pulp, the ovary wall does not become fleshy and it develops into a pericarp. Initially, fruitlets appear as tape-like structures. The ovary enlarges and the embryo becomes distinct after one week of fertilisation. In some genotypes, the fruitlet becomes highly visible and differentiated only at the eighth week of flower emergence (Laserna, 1988). Each fruitlet develops 3–11 cm long, 2–4 cm wide, and weighs 5–62 g. It is composed of the fleshy aril and the seed. The aril is the part that is usually consumed fresh and is sometimes referred to as flake. It can be waxy, firm or soft. Its colour ranges from yellow to yellow orange or coppery red.

Depending on the genotype, the whole fruit varies from 20–100 cm in length and 15–50 cm in diameter, and weighs 0.5–50 kg (a fruit weighing 81 kg was also reported from Panrutti in Tamil Nadu). Each fruitlet is large and oblong or reniform. The overall fruit's shape may be oval, oblong or oblong-ellipsoid. The white pulp is firmly attached to a core. The core or receptacle is pithy and 4.2–10.5 cm thick. The colour of the fruit is pale or dark green at the immature stage but it becomes greenish yellow, yellow or brownish when ripe. The fruit has a green stalk or peduncle, which is 2–10 cm long and 1.2–3.5 cm thick. It is covered with a rubbery rind and spines that are hard, pyramidal, and either pointed or blunt. The spine is formed from the tip part of the perianth. Spine lengths vary from 1.5–10 cm, with around 3–12 spines per cm^2 surface area. The rind is formed from the hardening of the distal tip of the tubular perianth (Haq, 2006).

Reproductive morphology in jackfruit

Fruit Set: The female inflorescences in *A. heterophyllus* are generally highly successful in syncarp production. Normally, around 70-75% of the fruits are produced through open pollination. The fruits develop during spring (February) or summer (June). A fruiting season of about four months means ripe fruits are generally harvested from June onwards in most Asian countries. Geographical differences in harvest periods tend to range from April to August or September to December in Malaysia, January to May in Thailand, July to October in Nepal and June to October in India (Soepadmo 1992). The fruit matures in 79–163 days or may even take as long as 180–240 days after the emergence of flowering spikes (Yap, 1972; Pushpakumara, 2006).

Fruit maturation can be divided into 10 stages based on specified growth parameters (Narasimham, 1990). Increase in fruit size and its different parts during maturation indicate that the growth rate from stage to stage is more or less uniform. However, there is also some indication of slow growth in fruit length between stages five and six. At the stage of maturity, which is between stages 9 and 10, the whole fruit and its edible region increase in size at a faster rate. The perianths of non-fertilised flowers have similar growth rates as fertilised ones but they do not increase much in breadth and thickness, whereas the perianths of fertilised flowers attain a size of 3–5 cm in breadth and 0.5–1 cm in thickness.

Seed Set: *Artocarpus heterophyllus* has a very low seed-set, with a mean of 3% of the flowers in an inflorescence producing seed as a result of open pollination. The low seed-set in *A. heterophyllus* may be the result of competition rather than pollen availability, and the difference in seed-set levels between different pollination treatments and the open pollinated inflorescences may be due to different levels of competition (Pushpakumara *et al.*, 1997). The seeds originate from a meristematic tissue of the ovary wall, which is called ovule primordium (Copeland, 1976). The seed is firm and waxy, oval, oblong or oblong ellipsoid in shape (Corner, 1938).

There are numerous seeds in each fruit. The development of seeds occurs in parallel with the development of the ovary, which begins with the successful pollination of the flower. A seed-like structure coated with a white, loose covering starts to fill the ovary a month after pollination. Its subsequent growth in weight, length and thickness follows a linear pattern up to maturity. The seed and testa may become very distinct in two months after pollination. The seed may become fully developed within four months of pollination. The developed single large seed remains free from the fruit wall and is attached to the pericarp at only one point, (the funicle).

Varietal improvement

Use of available genetic resources can be a major part of any crop improvement programme. Bearing this in mind, we can therefore consider genetic criteria for an overall improvement strategy of *A. heterophyllus*.

As a cross-pollinated and seed-propagated species, jackfruit population diversity results from the breeding system and natural selection associated with local environmental differences (evolution) or from human selection and the preferences of the local community that is cultivating them (domestication). As an underutilized crop, jackfruit has escaped intensive selection and cultivation. For these reasons, a wide range of genetic and morphological variation have been reported for jackfruit (Azad *et al.*, 2007; Hossain 1996; IPGRI 2000; Jagadeesh *et al.*, 2006; Saha *et al.*, 1996; Schnell *et al.*, 2001; Shyamalamma *et al.*, 2008; Ullah and Haque 2008). All these studies indicate that both within and between the jackfruit populations, variations exist. These studies have, however, been limited in geographical scope and in the number of individuals examined, because very a few germplasm collections of jackfruit are known (IPGRI 2000).

For any crop improvement programme, it is important that ideotypes are selected for different uses and products that could be marketed, either locally or internationally, and for environmental adaptations and cultural practices. Haq (1995) believed that the most important characteristics of jackfruit ideotypes are:

a. Easy manageability and requiring less or no pruning for fruit production. In the case of timber, long straight trunks would be appropriate. This would be acceptable for crop diversification and agro-forestry systems.

b. Vigorous and prolific plants compatible with one or more rootstocks with early flowering and regular bearing.

c. Good quality fruits with acceptable pulp colour and texture, good flavour and sweetness.

d. Fruits with symmetrical form and acceptable size.

e. Fruits that are borer, die-back and flood resistant.

f. Wider geographical and environmental adaptation.

g. A range of times of fruit maturity.

h. Off-season fruiting.

i. Long post-harvest life.

j. Above all, high yield.

Breeding objectives

In jackfruit, the yield is not similar every year. Trees are erratic in bearing. The first objective is to develop a variety that is regular and prolific in bearing high quality fruits. According to Bhatia *et al.* (1955), the edible pulp (bulbs) forms

29% of the ripe fruit, the seeds 12% and the outer rind 59%. The outer rind consists of a soft structural matrix (unedible internal tissues) and fleshy yellow-green spines on the skin. However, the share of this mass in a fruit needs to be reduced through breeding. In fact, sincere breeding effort is required to enhance the weight of the edible pulp from about 30 to 45% of the fruit. The seeds should constitute about 5% of the fruit and the outer rind along the internal structural matrix should not exceed 50% of the total fruit weight. Usually big-size fruits are inconvenient to handle. Thus, to improve yield, the number of fruits per tree should be increased instead of getting a lower number of bigger-sized fruits.

The jackfruit tree has a long juvenile phase (7–8 years). Reducing this period is another important breeding objective. According to Richards (1950), the Singapore variety comes into bearing in about 18 months under favourable conditions in Sri Lanka. Nangka Salak is another early bearing variety in Java (Yaacob & Subhadrabandhu, 1995). Such varieties can be used as one of the parents for inducing precocity. Reduction in tree height is desired for high density plantings and also for convenience in cultural operations. In this regard, efforts on the use of dwarfing root stocks may also need to be explored.

Little is known about the breeding of jackfruit. This may be due to its status as a minor fruit although it has high nutritive value and a wide range of uses. Any attempts to produce improved jackfruit cultivars would be targeted at commercial production but would also be of value for home gardens. To initiate improvement, it requires a basic understanding of the existing clones. Farmers have selected the clones from natural populations for their desirable characters but the selection has not been rigorous. Hence, some trees produce sweet aromatic fruits; others are nearly dry and sour. Better selection and vegetative propagation of clones is practicable and efforts should be made to also extend the fruiting season. Although some work has been done on rootstock and scion compatibility, the evidence so far is that there is wide variability in scion performance with different rootstocks (Azad *et al.*, 2000).

Inheritance pattern

There seems to be very little work done on improving jackfruit through crossing (hybridization). Hence, there is hardly any information on the heritability of economic characters such as the texture of edible flesh, sweetness, seed size, number of fruits per tree, size of fruit and thickness of rind. In breeding for improvement of qualitative traits, knowledge of genetic correlations between traits is helpful in deciding parents. Unfortunately, very little information is available on this aspect too. Genetic investigations need to be initiated on priority, particularly related to mode of inheritance of important characters and identification of genes for seedlessness, dwarfness, resistance against fruit fly, fruit rot and soil salinity.

Selection

Varietal improvement in jackfruit has so far been limited to selection of high yielding, better quality genotypes. Whatever varieties are grown commercially, they are all selections made by enlightened growers or enthusiasts based on their own ideas of suitability.

Jackfruit has a high nutritive value and a wide range of uses but it is still recognized as a minor and under-utilized fruit. As a result, research investment to improve this crop is very limited and thus breeding strategies have not been developed for it. Genetic improvement of a tree species is time-consuming because it requires 15–20 years to produce a cultivar for farmer use. There are two reasons for this: first, jackfruit trees require 4–8 years to reach the fruiting stage and second, the germplasm must undergo several stages of screening, including on-station and on-farm evaluation. Farmers' participation must also be considered during the screening process because farmers are producers as well as consumers. As mentioned earlier, jackfruit trees are cross-pollinated and have been propagated through seeds. The progeny then segregates and produces different types. Because of this, heterozygous trees are found in homestead farms and small orchards. Farmers want uniform and good quality jackfruit planting materials.

Several authors (Cambell *et al.*, 2004; Reddy *et al.*, 2004; Haq, 2006; Jagadish *et al.*, 2007) have reported the selection of desirable types through germplasm characterization and evaluation. Table 1 shows a list of cultivars that have been selected through evaluation (Haq, 2006). Azad *et al.,* (2007) evaluated the diversity of jackfruit in five regions in Bangladesh and selected 10 trees as potentially superior tree types based on farmers' criteria, which included high yield, fruit quality, sweetness, early fruiting types and off-season types. These types were recommended for cloning.

Table 1: List of Selected Cultivars in Different Producing Countries

Countries	Cultivars
Bangladesh	Topa, Hazari, Chala, Goal, Koa, Khaja
India	Khujja or Karcha, Ghila, or Ghula, Hazari, Rudrakshi, Gulabi, Hazar, jackfruit NJ-1, jackfruit NJ-2, jackfruit NJ-3, jackfruit NJ-4, Koozha Navarikka or Pazam Varikka, Safeda, Khaja, Mammoth, Ever-bearer, Rose-Scented, Kooli, Varika, Gerissal, Barica, Ghila, Karcha, Rudrakshi, Champa, Swarna Manohar, Swarna Poorti, Safeda, Khawja, Bhusila, Bhadaiyan. Handia, T-Nagar jack, Velipala, Gulabi, Champa, Singapore or the Ceylon Jack
Malaysia	J-30, J-31, NS-1, Na2, Na29, Na31
Myanmar	Talaing, Kala
Philippines	J-01, J-02, TVC, Torres
Sri Lanka	Vela, Varaka (Waraka), Peniwaraka, Kuruwaraka, Singapore Jack/ Ceylon Jack
Thailand	Dang Rasimi, Kun Wi Chan, Kha-num Nang, Kha-num Lamoud
USA:Florida	Black Gold, Cheena, Dang Rasimi, Galaxy, Golden Nugget, Honey Gold, Lemon Gold, J-30, J-31, NS-1, Tabouey, Delightful

Sources: www.fairchildgarden.org/research/jackfruit-cultivars.html;Morton (1987), Samaddar (1990), Soepadmo (1992), APAARI. (2012)

Farmers have also propagated clones from selections taken from natural populations for their desirable characters, but the selection has not been rigorous. However, because the tree is heterozygous, any improvement requires a basic understanding of the existing clones. An effective selection of clones with desirable characters and vegetative propagation of these clones is practicable (Haq, 2006). Although little work has been done on rootstock and scion compatibility, the evidence so far is that there is wide variability in scion performance when grafted to different rootstocks (Haq, 2006). The selection of superior mother trees and their clonal development may be faster and may have greater impact than conventional breeding.

Haq (2006) has reported on the major effort in selection that has been made in nine countries in Asia. This was carried out by characterization, evaluation and identification of trees with desirable characteristics through farmer participation. Table 2 shows the selection of "superior" mother plants in the second column. These were propagated by grafting to produce relatively large numbers of plants for home gardens and commercial use. These trees can be grown with ease and with little or no care and it should prove to be profitable for farmers.

Table 2: Characterization, Evaluation and Selection of Promising Lines of Jackfruit

Countries	No. of Accessions	Selection
Bangladesh	70	10
India	281	54
Nepal	350	47
Pakistan	10	5
Sri Lanka	77	3

Source: Haq (2006)

A few reliable reports are available on the true varieties of jackfruit that were developed through breeding programmes in various countries. It seems that cultivars are broadly classified on the basis of a few characteristics of the fruit. Classifications were done based on the scent of the pulp, the yield of the trees, and the seasonal availability of fruits. However, there are other characters that were overlooked in this classification, because there are many other local names of cultivars of jackfruit (Azad, 2000; Haq, 2006).

Some local selections was named Gulabi (rose-scented), Champa (flavour like that of *Michelia* sp.) and Hazari (bearing more number of fruits in a tree). In fact, some better types are said to have been collected as a result of a local survey. Because there is a good demand for raw jackfruit in India as a vegetable for culinary purposes, emphasis is also given on fruit characters such as thickness of rind and softness of flesh at a premature stage of fruit development.

Samaddar (1985) reported that there is no distinct cultivar of jackfruit. Hossain (1996) also reported that there is no recommended variety of jackfruit for cultivation. Different types of jackfruit have come into commercial cultivation with different local names for the cultivars. In reality, there is little comprehensive or reliable information on any true variety of jackfruit that has been developed through any breeding programme anywhere. However, the cultivars, Rudrakshi and Singapore Jack (Ceylon Jack), are known to breed more or less true to type from seed (Soepadmo, 1992) in India. These produce jackfruit weighing 5–20 kg within 2–5 years of planting the seedlings.

In most jackfruit-producing countries, there are many local names given to the clonal selections that are propagated (for example, in India, the cultivars are known as Gulabi, Champa and Hazari; in Bangladesh, as Gala, Khaja, Hazari and so on). But how these are to be distinguished from each other is not yet established. Selections from one country that have moved to another have been given separate names, leading to some confusion about true types. In Sri Lanka, a variety called Vela with characteristic soft pulp is the same as the one described by Da Costa as Gerisssa in India. He classified Barica as a firm and harder pulp type. Under these two names, there are many subtypes which differ in taste, colour, fibrousness and sweetness and they have their own local names. Many of the local names relate to attributes rather than places, for example, Gulabi refers to a rose-scented type; Champa refers to a type with taste like champedak; Hazari refers to a type that produces a greater number of fruits (Samaddar, 1990). Manjunath (1948) also classified jackfruit into two types on the basis of size: (i) Barica, a smaller fruit and (ii) Kapa, a large fruit. Barica is locally called Katahali in Uttar Pradesh and yields small fruits that are somewhat sour in taste and weigh 2–8 kg each. Kapa is called Kathal in UP and its neighbouring regions, yields large fruits that weigh up to 40 kg and at ripening has fleshy sweet bulbs.

Sreenivasan (1970) reported a local cultivar in Tamil Nadu, Muttam Varikka, which has an average weight of 7 kg and produces fruits at ground level. A large fruit of this variety weighs up to 55 kg. Singh (1986) also reported local selected strains, including Hairialyalva, Bhadunha, Zarda and Bhusala from Northern India. Morton (1987) reported selected cultivars of jackfruit and these are Safeda, Khaja, Bhusila, Bhadaiyan, Handia and T-Nagar jack. She also reported Velipala as a local selection from the forest having large fruits with superior quality large bulbs.

Recently a crop improvement programme at the University of Agricultural Sciences, Dharwad, India was initiated (Jagadish *et al.*, 2007), which evaluated and selected 65 types from the hilly zone and 30 from the coastal zone. Further, breeding involving these selected types is in progress. The Indian Institute of Horticultural Research, Bangalore, characterized 70 accessions, which are being conserved.

Farmers' participatory methods are useful for varietal selection of any crop because farmers are the best placed to know the preferences of the market. Farmers also play an important role as retailers in the local market. It has been seen that farmers prefer high-yield and longer-fruiting tree types, with a desirable pulp colour, texture and sweetness of fruit. Therefore, awareness of farmers' preferences has made the selection process more acceptable, because they are involved as producers, consumers and retailers of jackfruit. However, farmers use seeds of the mother trees for propagation, which fail to maintain trueness to type. Therefore, little impact of this type of selection was found in the development of cultivars. Selection is seen to be effective only by propagating from superior mother trees by using asexual means.

Gunasena *et al.,* (1996) stated that breeding by using hybridization of perennial trees is cumbersome, time-consuming and sometimes impossible. Therefore, selection of superior mother trees is more convenient than other breeding methods. Jackfruit can live up to 100 years and have a very long juvenile period before coming into fruit bearing. Thus improvement of the trees by hybridization is not suitable for jackfruit. The procedure of selection of superior mother trees from the existing population saves time, money and effort.

The current large number of varieties have apparently evolved through natural crossing. These varieties are selected by professional tree breeders, principally based on fruit characteristics. However, varieties or hybrids of jackfruit that were selected for their timber characteristics have not been reported. In this regard, various selections are being made from the characterised germplasm and have been released by the respective regions.

Varieties

India

Palur-1 Jack or PLR.1: TheTamil Nadu Agricultural University (TNAU) developed this improved variety in 1992. It is a clonal selection from Panikkankuppam of Panrutti taluk, in South Arcot district of Tamil Nadu. It is a high-yielding tree with medium height, which is less spreading and suitable for high-density planting. In addition to the regular bearing season (March–June), it produces fruits during off season (October–December) also. The annual yield per tree is about 80 fruits, weighing around 900 kg. The average fruit weight is 12 kg and contains 115–120 flakes. The off-season bearing is a unique trait of this variety and the fruits fetch premium market price. The fully ripe fruits have flat stigmatic surface instead of a spiny surface. The fruit quality is good with attractive golden yellow firm flakes. The flakes are very sweet with high consumer appeal and good keeping quality. The seeds are also edible with high palatability. Total Soluble Solids (TSS) is 19°brix.

Plate 2. PLR-1 Jackfruit

PPI Jack: The Tamil Nadu Agricultural University (TNAU) developed this improved variety in 1996. It is a clonal selection from Ulagumoodu near Pechiparai. It is a medium tall tree and yields 105 fruits per tree per annum (each weighing 17 kg), which accounts for 40.8 per cent more yield than the local type. The fruit bearing occurs twice annually (April–June and November–December) and produces high quality crisp flakes with more TSS and ascorbic acid content. The flakes are sweet and tasty with a pleasant aroma. Maximum bearing takes place on the tree trunks. This variety is suitable for commercial planting and home gardens.

Plate 3: PPl-1Jackfruit

PLR (J)-2: TheTamil Nadu Agricultural University (TNAU) developed this improved variety in 2007. It is a clonal selection from Pathirakkotai. This variety has good quality and bigger-sized fruits. It has highly palatable and edible flakes, which fetch more prices due to their attractive characters and keeping quality. This results in increased income for farmers. This variety also has reported lesser incidence of major pests and diseases. It can be planted during June–December and is suitable for growing in tropical climates with deep well-drained loam soil. This variety yields 95–110 fruits per tree per year, with an average weight of 1,600–1,950 kg.

Plate 4. PLR (J)-2 Jackfruit

Swarna: The University of Agricultural Sciences (UAS) Bangalore, Karnataka released a jackfruit selection by the name of Swarna in 2010. Swarna is a very good yielder that bears fruits in bunches on the trunk, on primary branches and also on secondary branches. The medium-sized fruits weigh around 6–8 kg. The fruitlets are golden brown in colour, juicy with a TSS of 25 to 26^{0}B, with thick flakes and a very thin rind (< 1 cm). The parent tree of Swarna is located at the back of the University quarters near its Cauvery Guest House.

(a)

(b)

(c)

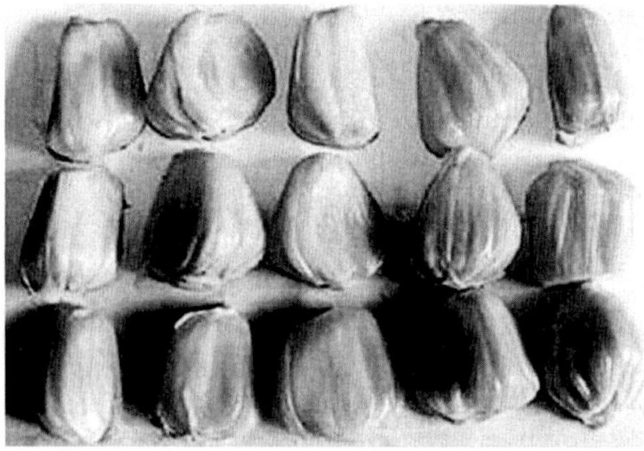

Plate 5 a, b & c: Swarna Jackfruit

Konkan Prolific

This variety was released in 2004 by the Regional Fruit Research Station (RFRS), Vengurle of the Dr Balasaheb Sawant Konkan Krishi Vidyapeeth, in Dapoli, Maharashtra. It is a seedling selection with a potential yield of 450–550 kg per tree and an average yield of 420.6 kg per tree. It produces 73 fruits per tree per year with individual fruits weighing 5.70 kg (fruit length of 34.4 cm and breadth of 15.2 cm). It is a medium-sized fruit, which is oblong in shape, with firm flesh even in the monsoon season. The pulp is golden yellow in colour and the fruit bears in bunches. Its surface is spiny and the tree starts bearing within 6 to 7 years of planting. The fruits have a TSS of 25°B, seed weight of 132 g and flower in January–February.

Plate 6: Konkan Prolific Jackfruit

Registration and GI facilitation for jackfruit

A three hundred year old elite jackfruit tree has been located in Kachahalli village (Thubagere taluk of Doddaballapur district, Karnataka, India), which is documented for its excellent fruit qualities. The fruits are ellipsoid or elongate and weigh 20–25 kg, with 300–350 flakes per fruit. The flakes are deep coppery red and have recorded a TSS of 32°B.

a

b

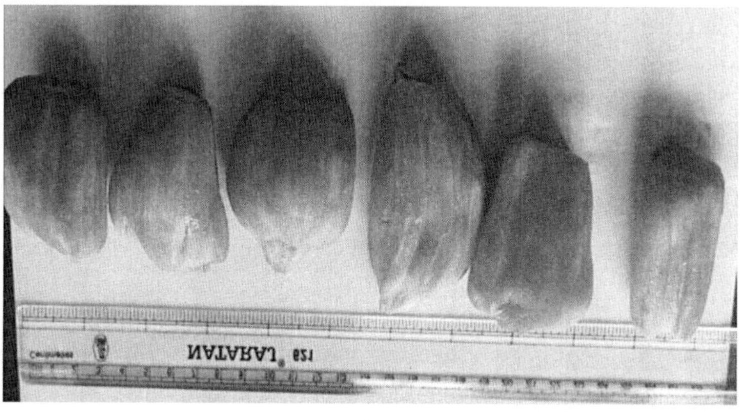

c

Plate 7 a, b & c: Kachahalli Jackfruit

Bangladesh

BARI Kanthal-1

Developed by the Horticulture Research Centre, Bangladesh, in 2008, this is a high-yielding, regular-bearing, early variety that has medium-sized and uniformly shaped fruits, which weigh 9.5 kg each. The pulp is medium soft, has a very sweet (TSS 22°B) taste with less fibre, and is very juicy. The edible portion accounts for 55%. The variety yields11,800 medium-sized fruits per ha. The fruit colour is light green. It is suitable all over Bangladesh and can be sown during June to September. The harvesting time is May to June.

Hybridization

There is no published report on hybridization in jackfruit. Because wide genetic variability exists in this crop, there is an apparent need to exploit the heterosis for yield gain and improvement in fruit quality. Inter-specific hybridization also deserves attention to know the species that are compatible with jackfruit and their contribution in a particular direction, for example, in tree size (dwarfness), precocity, hardiness for salt, low water stress and so on. Jackfruit and breadfruit have considerable potential. Planned breeding with the aid of biotechnological skills is required to exploit their potential.

The Fruit Experimental Station at Kallar began breeding work in 1952 to develop short, compact, many-branched trees, which are precocious and productive and bears large, yellow, high quality fruits, half in the main season and half in the late season. Singapore Jack was chosen as the female parent because of its early and late crops and as the male parent, Velipala, a local selection from the forest that has large fruits with large flakes of superior quality and is borne regularly in the main summer season was chosen. After 25 years of testing, one hybrid was rated as outstanding for precocity, fruit size, off-season as well as main season production, and yield that excelled its parents. It had not been named when reported on by Chellappan and Roche in 1982.

Biotechnology

Biotechnological techniques could be useful in breeding for seedlessness, disease, insect resistance, salt tolerance and germplasm storage. Jaiswal and Amin (1992) reviewed the work done on this aspect and described the developments made in the field. So far selection from seedling population is the only breeding method used in jackfruit because the tree is monoecious in nature and commonly raised from seed. Biotechnological approaches such as cell hybridization, organelle transfer, genetic transformation (that is, generation of new genetic combinations), protoplast fusion, anther culture (haploid production) and embryo culture (rescue) are yet to be employed to produce both intra- and inter-specific hybrids. Tissue culture techniques have been developed for faster multiplication of improved genotypes by using nodal segments (Roy *et al.*, 1990) and terminal and axillary buds (Roy *et al.*, 1991) as explants. Dumet and Berjak (1995) suggested that cryopreservation of embryogenic axes can be helpful in creating a gene bank.

Current improvement efforts

A major effort has been made to initiate selection by carrying out standard characterisations and identification of trees with desirable characteristics. This was done under the auspices of Underutilised Tropical Fruits in Asia Network (UTFANET). The effort was intended to be participatory in association with farmers. Different participating countries of UTFANET have characterised and evaluated the germplasm. Further selections of superior mother plants were then used for vegetative propagation by grafting to produce relatively large numbers of plants for home-gardens and commercial use.

References

Acedo, A.L. 1992. *Jackfruit biology, production, use and Philippine research, Multipurpose Trees Species Network Series*. Winrock International Institute for Agricultural Development, Thailand, p. 51.

Angeles, D.O. 1983. Cashew and jackfruit research. *Research Technical Report*. Laguna, Philippines. Department of Horticulture, University of Philippines at Los Banos (UPLB).

APAARI. 2012. *Jackfruit Improvement in the Asia-Pacific Region – A Status Report*. Asia-Pacific Association of Agricultural Research Institutions, Bangkok, Thailand. 182 p

Azad, A.K. 1989. *Studies on floral biology, pattern of fruit set and quality of fruit borne in different sections of jackfruit plants*. M.Sc. (Ag.) Thesis. Faculty of Agriculture, BAU, Mymensingh, Bangladesh.1-77.

Azad, A.K. 2000. *Genetic diversity of jackfruit in Bangladesh and development of propagation methods*. Ph. D Thesis, University of Southampton, UK. pp. 200.

Azad, A.K., Jones, J.G. and Haq, N. 2007. Assessing morphological and isozyme variation of jackfruit (*Artocarpus heterophyllus* Lam.) in Bangladesh. *Agroforestry Syst.* 71: 109-125.

Bawa, K.S., and Krugman, S.L. 1990. Reproductive biology and genetics of tropical trees in relation to conservation and management. *In:* Gomez-Fompa, A., Whitmore TC and Hadley M (eds.) *Rain forest regeneration and management*. MAB series 6, UNESCO, Paris and Parthenon Publishing Group. pp. 119-136.

Bhattia, B.S., Siddappa, G.S. and Lal, G. 1955. Composition and nutritive value of jackfruit. *Ind. J. Agric. Sci.* 25: 303-306.

Cambell, R.J., El-Sawa, S., Wasilekwski, J. and Ledesma, N. 2004. Breeding and selection of jackfruit for south Florida. *Proc. Fla. Hort. Soc.* 117: 193-194.

Copeland, L.O. 1976. *Principles of seed science and technology*, 15. Burgués Publishing Co., Minneapolis, Minnesota, USA.

Corner, E.J.H. 1938. Notes on the systematics and distribution of Malayan phanerogams II. The jack and the chempedak. *Gardener. Bull.*10:56-81.

Corner, E.J.H. 1981. *Moraceae*. Dassanayaka, MD and Fosberg, FR (eds.). *A revised handbook to the flora of Ceylon*. Amerind Publishing, New Delhi, India. pp. 213-220.

Das, G.C., Rana, S. and Konhar, T. 1981. Effect of phosphoric acid on induction of flowering and gynoecium in jackfruit (*Artocarpus heterophyllus* Lam.). *Orissa Journal of Horticulture*. 9:49-52.

Dumet and Berjak 1995. *Desiccation tolerance and cryopreservation of embryonic axes of recalcitrant species. (In) Basic and applied aspects of seed biology* (Ellis RH, Black M, Murdoch AJ and Hong TD -edts.), Kluwer Academic Publishers, Dordrecht, Netherlands, pp. 771-776.

Gopinath, R., Kumar, S.P. and Kumaran, K. 1983. Spike characteristics anthesis and pollen viability in jack. *South Indian Horticulture*, 31: 178-180.

Gunasena, H.P.M., Ariyadasa, K.P., Wikramasinghe, A., Herath, H.M.W., Wikramasinghe P and Rajakaruna SB 1996 Manual of Jack Cultivation in Sri Lanka. Forest Information Service, Forest Department: 48.

Haq, 1995. *Analysis of questionnaire to select priority species for research and development in Asia.*(In) Arora RK and Rao VR (eds.), *Proceedings of Experts Consultation on Tropical Fruit Species 01 Asia*. IPGRI. Singapore

Haq, 2006. *Jackfruit*, *Artocarpus heterophyllus*, Southampton Centre for Underutilised Crops, University of Southampton, Southampton, UK. p. 192.

Hossain, 1996. *Status report on genetic resources of jackfruit in Bangladesh*. IPGRI Regional Office, Singapore, 30.

IPGRI, 2000. *Descriptors for Jackfruit* (*Artocarpus heterophyllus*). International Plant Genetic Resources Institute (IPGRI), Rome.

Jagadeesh, S.L., Reddy, B.S., Swamy, G.S.K., Gorbal, K., Hegde, L. and Raghaven, G.S.V. 2006.Chemical Composition of Jackfruit (*Artocarpus heterophyllus* Lam.) Selections of Western Ghats of India. *Food Chemistry* 102:361–365.

Jagadeesh, S.L., Reddy, B.S., Basavaraj, N., Swamy, G.S.K., Gorbal, K. and Hegde, L. 2007. Inter tree variability for fruit quality in jackfruit selections of Western Ghats of India. *Sci. Hort.* 112:382-387.

Jaiswal, V.S. and Amin, M.N. 1992. *Guava and jackfruit*. (In) Hammerschlag F.A., Litz R.E. (eds), Biotechnology of perennial fruit crops, Biotechnology in agriculture. Vol. 8. CAB International, Wallingford, UK, pp. 421-431.

Jarrett, F.M. 1959. Studies in *Artocarpus* and allied genera, III. A revision of *Artocarpus* subgenus *Artocarpus. Journal of Arnold Arboretum*,40: 113-155, 298-368.

Laserna, J.C. 1988. Fruit abscission and its effect on yield and quality of jackfruit(*Artocarpus heterophyllus* Lam.) B.Sc. Thesis. UPLB Laguna, Philippines.

Manjunath, 1948. *Wealth of India*. Vol.1 CSIR Publications. New Delhi, 1-240.

Moncur, M.W. 1985. Floral ontogeny of the jackfruit, *Artocarpus heterophyllus* Lam. (Moraceae). *Australian Journal of Botany*,33: 585-593.

Morton, 1987. *Fruits of warm climates*. Creative resources system, Inc., Winterville, North Carolina, USA. pp. 58-64.

Narasimham, P. 1990. Breadfruit and jackfruit. (In) Nagy S, Shaw PE and Wordowski WF (eds.). Fruits of Tropical and Sub-tropical origin - Composition, Properties and Uses. *Florida Science Source, Inc*: 193-259.

Primack, R.B. 1985. Comparative studies of fruits in wild and cultivated trees of chempedak *(Artocarpus integer)* and terap *(Artocarptis odoratissimus)* in Sarawak, east Malaysia with additional information on the reproguctive biology of the Moraceae in southeast Asia. *Malayan Nature Journal* 39:1-39.

Purseglove, 1968. *Tropical crops Dicotyledons*, Longman, London pp. 377-390.

Purseglove, J.W. 1968. *Tropical crops: dicotyledons* 2. John Wiley and Sons Inc., New York.

Pushpakumara, D.K.N.G. 2006. Floral and fruit morphology and phenology in *Artocarpus heterophyllus* Lam. (Moraceae). *Sri Lankan Journal of Agricultural Science*,43: 82-107.

Pushpakumara, D.K.N.G. 2006/07. Agamospermy in *Artocarpus heterophyllus* Lam. (Jackfruit). *Tropical Agriculturist*,156: 87-98.

Pushpakumara, D.K.N.G. 2011. *Reproductive biology*, (In) *The Jackfruit* - (eds) Thottappilly G, Peter KV and Valavi, SG (Publisher- Studium Press India Pvt. Ltd).

Pushpakumara, D.K.N.G., Boshier, D.H. and Simons, A.J. 2010. Pollination vectors of *Artocarpus heterophyllus* Lam. (Moraceae) in Sri Lanka. (Web:http://2010.botanyconference.org.engine/search/index/pnp?fune=detail&aid=487).

Pushpakumara, D.K.N.G., Boshier, D.H. and Harris, S.A. 1997. Mating system in *Artocarpus heterophyllus* Lam. *Tropical Agricultural Research*, 9: 1-14.

Querijero, N.L.B. 1988. The flowering of jackfruit (*Artocarpus heterophyllus* Lam.). BSc Thesis. UPLB, Laguna, Philippines.

Reddy, B.M.C., Patil, P., Shashikumar, S. and Govindaraju, L.K. 2004. Studies on Physico-chemical characteristics of jackfruit clones of South Karnataka: *Karnataka Journal of Agricultural Sciences* 17(2): 278-282.

Richards, 1950. A note on the cultivation of Singapore jack. *Trop. Agiculturist*. 105: 12-13.

Roy, S.E., Rahman, S.L. and Majumdar, Rita 1990. *In vitro* propagation of jackfruit (*Artocarpus heterophyllus* Lam.). *J. hort. Sci.* 65 (3): 355-358.

Roy, S.K. and Hadiuzzaman, S. 1991. Micropropagation of two species of *Artocarpus* through *in-vitro* culture. *Bangladesh Journal of Botany* 20(1):27-32.

Saha, M.C., Saha, M.G., Rahman, M.A., Nazrul, M.I., Quasem, A., Halder, N.K. and Haque, A.F.M.E. 1996. *Viability in jackfruit*. A paper presented at the Internal Research Review Workshop. Horticulture Research Centre, BARI, Gazipur, Bangladesh.

Sahadevan, P.C., Abdul, S.A. and Menon, K.C.P. 1950. Floral abnormalities in Jack. *Madras Agricultural Journal*, 37:497-498.

Samaddar, H.M. 1985. Jackfruit, (In) Bose, T.K. and Mishra, S.K. (eds.), *Fruits of India: Tropical and subtropical* (pp. 638-649). Calcutta, India: Naya Prokash.

Samaddar, H.N. 1990. Jackfruit. (In) Bose, T.K. and Mitra, B. (eds.) *Fruits of India*, Tropical and Sub-tropical. Naya Prokash, Calcutta, India: 638-649.

Samaddar, H.N. and Yadav, P.S. 1982. Flowering, floral biology, pollination, fruit set and fruit growth of jackfruit. *Hort. J.* 1:54-60.

Sambamurthy, K. and Ramalingam, V. 1954. Preliminary studies in blossom biology of the jack (*Artocarpus heterophyllus* Lam.) and pollination effects. *Indian Journal of Horticulture* 11:24-29.

Schnell, R.J., Olano, C.T., Campbell, R.J. and Brown, J.S. 2001. AFLP analysis of genetic diversity within a jackfruit germplasm collection. *Scientia Horticulturae*; 01(3-4): 261-274.

Shyamalamma, S., Chandra, S.B.C., Hegde, M. and Naryanswamy, P. 2008. Evaluation of genetic diversity in jackfruit (*Artocarpus heterophyllus* Lam.) based on amplified fragment length polymorphism markers. *Genetics and Molecular Research*, 7(3): 645-656.

Singh, A. 1986. *Fruit physiology and production.* Kalyani publishers, New Delhi, India. p.210.

Singh, S.P. 1992. Budding in some fruit crops: a review. *Advances in Horticulture and Forestry* 2:84-97.

Sinha, M.M. 1973. Pollen viability and pollen storage studies in jackfruit (*Artocarpus heterophyllus* Lam.) *Progressive Horticulture*, 3:69-73.

Soepadmo, E. 1991. *Artocarpus heterophyllus* Lamk. Verheij, E.W.M. and Coronel, R.E. (eds.) *Plant Resources of Southeast Asia No. 2: Edible fruits and nuts*. Pudoc, Wageningen. pp. 86-91.

Soepadmo, E. 1992. *Artocarpus heterophyllus* Lam. (In) Verheij E.W.M. and Coronel R.E. (eds.) *Plant Resources of Southeast Asia No. 2: Edible Fruits and Nuts*. PROSEA, Wageningen, Netherlands. pp. 86-91.

Sreenivasan, K. 1970. Muttam Varikka -A promising jackfruit variety. *Agric. Res. J. Kerala*, 8: 51-52.

Srivastava, R.P. 1961. Preliminary studies in the sex determination, growth behaviour and the influence of α-napthalene acetic acid sprays on the size, seed number and weight of jackfruit (*Artocarpus heterophyllus* Lam.) *Journal of Science Research Banaras Hindu* University, 11:213-221.

Ullah, M.A. and Haque, M.A. 2008. Studies on Fruiting, Bearing Habit and Fruit Growth of Jackfruit Germplasm. *Bangladesh Journal of Agricultural Research*. 33:391–397.

Yaacob, O. and Subliadrabandhu, S. 1995. *The Production of Economic Fruits in South-East Asia*, Oxford Univ. Press, Kuala Lumpur, Malaysia. pp. 122-128.

Yap, A.R. 1972. *Cultural Directions for Philippine agricultural crops*. Fruits. BPI, Department of Agriculture, 1:137-141.

18

Breeding Sugar Apple/Sweetsop

T. Sakthivel

The Annonaceae or custard apple family comprises about 120 genera and more than 2000 species (Leboeuf *et al.,* 1982). The genus *Annona* is the most economically important one, containing 120 species. The major commercial species throughout the world are: the cherimoya (*Annona cherimola*), sweetsop (*Annona squamosa*) and *atemoya* (a hybrid of *Annona cherimola* and *Annona squamosa*). The custard apple species name '*squamosa*' refers to the knobby appearance of the fruit. Custard apple is a small tropical tree originating in the New World tropics, probably Central America. This species is the most widely grown in the tropical regions of the America, Africa, Asia and the Pacific. Sweetsop is synonym also named sugar apple and has many other regional names, such as Custard apple (India), anon (Spanish), ata (Portuguese), noi-na (Thailand), stis (The Philippines) and fan-li-chi (Taiwan). Sweetsop is the mostly grown Annona species. The fruit is frequently found in village markets but has not shown much potential for large commercial cultivation due to the small fruit size, frequent cracking at maturity and poor shelf life. The perishable nature and supply shortages make marketing localized or air shipment essential.

Fruit is usually harvested from July to October and can be extended to March if summer pruning is conducted. In India, the peak of production occurs later between August to November and in Thailand, Florida and the Caribbean between July and September. Taiwan's peak of production occurs between July and March.

Sweetsop is a good source of carbohydrate, potassium, calcium, phosphorus and ascorbic acid. The fruit is usually consumed fresh and also can be used to make juices, shakes and ice creams. Proximate composition of sweetsop fruit is given below:

Proximate sweetsop fruit composition per 100 g (USDA, 2002)

Proximate	**(%)**
Water	73.32
Energy (kcal)	94
Protein	2.06
Fat	0.29
Carbohydrate	23.64
Fibre	4.40
Ash	0.78
Minerals	**mg**
Calcium	24
Iron	0.6
Magnesium	21
Phosphorus	32
Potassium	247
Sodium	9
Vitamins	**mg**
Ascorbic acid	363
Thiamine	0.111
Riboflavin	0.113
Niacin	0.883
Vitamin E	0.590
Panthothenic acid	0.226
Vitamin B_6	0.200
Vitamin A	6 IU

Cytogenetics and breeding

Understanding cytology of a crop is useful in planning breeding programme. The chromosome number of *Annona squamosa* is 2n=14. Cross-pollination between species is conducted primarily to determine compatibility for increasing fruit set. Inter – and intraspecific cross combinations among *Annona squamosa*, *A. atemoya*, *A. reticulata*, *A. glabra*, *A. cherimola* and *A. muricata* were investigated by Kumar and Jalikop (2000).They reported that *A. squamosa* was the most effective pollen parent of *A. atemoya* and *A. cherimola*. However, *A. squamosa* was not an effective female parent with A. *atemoya*, whereas *A. cherimola* was compatible with both *A. squamosa* and *A. cherimola*. A. *atemoya* can be crossed freely with *A. reticulata*, *A. glabra* and *A. cherimola*, *A.* a*temoya* is the only hybrid that has gained importance and it has inherited the glabrate leaf character of *A. squamosa* and a leaf size almost as large as that of *A. cherimola*. Flowering and fruiting seasons are similar to those of sweetsop. Skin, pulp and seed characters of both parents are inherited in varying degrees by each plant. A desirable hybrid would be between the cherimoya and soursop that combines the larger fruit size and acidity of the soursop and the cherimoya's sweetness, flavour and texture. Attempts to cross the soursop with cherimoya, ilama, bullock's heart or sweetsop have not been successful and may reflect a considerable genetic distance of soursop from the other species (Paull and Duarte, 2011). Jalikop (2010a) investigated the inheritance of foliage characteristics in trispecies hybrids of atemoya (*A. cherimola* x *A. squamosa*) and *A. reticulata*. Segregation for leaf colour (green or dark green), leaf angle (erect or drooping), leaf apex shape (rounded or pointed) and time of leaf fall (early or late or intermediate) into discreet phenotypic classes revealed that their inheritance followed simple Mendelian genetics. Duplicate dominant gene interaction governed the leaf colour and leaf position. Individuals with single or both the dominant genes (A-B-, A-bb,aaB-) produce green colour leaves and those with recessive genes (aabb) dark colour leaves. Likewise erect leaf individuals are C-D-, C-dd, ccD- and drooping leaves are ccdd.A single gene determined shape of leaf apex that was either rounded (Rr) or pointed (rr).Segregation of progenies for leaf fall into early , intermediate and late suggested that codominant alleles were responsible for time of leaf fall. The pattern of segregation of tree and fruit traits in trispecies hybrids (*A. cherimola* Mill. x *A. squamosa* L.) and *A. reticulata* L. have been investigated by Jalikop (2010b) with the objective of salvaging useful genes from the three edible annonas and determining the extent of variation in the progeny. The heterogenous nature of *Annona* species and simultaneous segregation of three distant genomes resulted in a tremendous heterogeneity in the progenies and thus demonstrated that *A. reticulata* is potentially a valuable source of a variety of novel traits. However, the F_1 trihybrids exhibited reduced fertility with pronounced undesirable traits of A. *reticulata*, and the survival of F_2 trees was poor, with those that did survive being extremely weak and failing to flower or set fruit. Consequently, backcrossing may be more suitable approach than advancing trihybrid generation for transferring desirable genes from A. *reticulata*. Backcrossing F_1 progeny of [(*A. cherimola* x *A. squamosa*) x *A. reticulata*] to *A. squamosa*, a highly fertile species, is expected to yield productive introgressants in which the contribution of the *A. squamosa* genome will be extended from 25 to 62.5% in F_1BC_1 generation.

Problems in breeding

Wester (1913) was the first scientist to realize the possibilities for genetic improvement of annonas and initiated breeding programmes in Florida and in the Philippines. The long reproductive cycles, higher levels of heterozygosity and the costs associated with evaluating large populations of crosses limits breeding programmes. Existing commercial cultivars show considerable variation in growth, fruit set, fruit size and quality. No single variety has all the desirable characteristics. The length of the juvenile period varies, with the earliest production occurring in 2 years and full production in 5–6 years. This juvenile period is extremely variable with scions on seedling rootstocks. The seedling rootstocks are derived from extremely heterogeneous, open-pollinated seeds; hence it is difficult to fix specific characters in a short period. Breeding programmes have focused on selections from seedling populations. Seedling rootstocks of annonas are generally derived from heterogeneous open-pollinated plants; hence, it is often difficult to fix specific characters in a short period. Early maturity, better fruit appearance and long post-harvest life for tropical annonas, and in the subtropics, greater cold tolerance, are objectives for cherimoya breeding (Nakasone and Paull, 1998).

Varietal wealth

India

Arka Sahan: It is a progeny of Island gem (*Annona atemoya* Hort.) X Mammoth (*A. squamosa* L.). Arka Sahan fruits come to harvest in September-October and mature fruits (210 g) take about 6-7 days to ripe. The skin has a waxy bloom, light green in colour, moderately thick (0.5 cm) with large, flat eyes. The creamy white colour flesh is juicy with mild pleasant aroma and tender with scarce seeds (9/100 g) and large segments. The edible pulp is remarkable for its sweetness with 22.8% total sugars and measures more than 30°B as against 24°B in Mammoth. A 100 g pulp of Arka Sahan contains 2.49 g of crude protein, 42.29 mg P and 225 mg Ca against 1.33 g, 17.05 mg and 159 mg respectively in Mammoth. On an average, 12 tonnes of fruits can be harvested/ha and are resistant to drought (http://www.iihr.res.in/varieties).

Balanagar: Tree grows upto 3 m height. Fruits are spherical, pyramidal or cordate in shape; medium to big in size. Average fruit weight is 200 g but recorded upto 640g; rarely 8.3 cm long and 3.5 cm in diameter. Areoles are tuberculate. Very rough, pitted, forming deep furrows; rind greenish, mesocarp white with coarse and medium granules, pulp white, buttery sweet, with moderate to plenty of juice, flavor excellent; seeds 40–80. Highly productive, fruits remain green when ripe; rich in reducing sugars and proteins.

Barbados: Tree grows upto 2.4 m height. Fruits are spherical to cordate in shape; medium in size. Average fruit weight is 230 g. Fruits are 7.0 cm long and 7.7 cm in diameter; Areoles tuberculate. Very rough; rind fine; mesocarp with medium granules; pulp creamy white, moderately juicy, with agreeable flavor, sweet; seeds 40-60. Flowering late in June to early September, prolific yielder.

British guinea:Tree grows upto 1.8 to 2.4 m. Fruits are spheroid to cordate in shape; medium in size. Fruit weight ranges from 170 to 335 g. Fruits are generally 6.8cm long and 6.8 to 7.5 cm in diameter; Areoles tuberculate, acute, rough, rind grayish or grayish green; mesocarp white, soft; pulp white, soft buttery, crisp, moderately juicy, sweet, seeds 30-50. Flowering is late in June to mid September. High yielder and fruits remain green at ripening.

Crimson custard appple: Tree grows upto 1.8 to 2.0 m. Fruits are spherical to oval in shape; small to medium in size. Areoles medium-tuberculate, truncate, slightly rough, rind crimson, with grayish blue bloom, mesocarp reddish with medium to big granules, pulp with light red tinge, soft, moderately juicy, sweet, seeds 40-60. Flowering early June to end of August; trunk and branches dark green, new growth pinkish green.

Local custard apple: Tree grows upto 1.8 to 2.4 m. Fruits are spherical or cordate or conical in shape; Fruit weight ranges from 140 to 330g and recorded upto 615 g. Areoles small to big. Smooth to very smooth, round or mamillate; rind green, becoming light yellow on ripening. Mesocarp white, with fine, small to medium granules; pulp dull white, juiciness little to plenty, insipid to very sweet, seeds 30-80. It is an average yielder.

Mammoth: Tree grows upto 1.5m. Fruits are irregular, triantrous to spherical in shape; average in size. Fruit weight ranges from 170 – 250 g, recorded up to 450 g. Areoles medium to big, very smooth, round, rind light russet; mesocarp white with small to medium granule; pulp white and translucent, soft bright, juicy, sweet, seeds 20-40. It has tendency to branch profusely; prolific bearer; rind becomes wrinkled on ripening; possesses low acidity; fruits have a good keeping quality and stand transport better.

Red sitaphal: Tree grows up to 1.8 m. Fruits are spherical and cordate in shape. Fruit weight ranges from 170-250 g. recorded upto 450 g; areoles medium to big. tuberculate, truncate, rough and pitted; rind pink; grayish blue; mesocarp pinkisk with coarse and big granules; pulp with pink tinge, moderately crisp, juicy, slightly acidic, sweet, seeds 40-60. Trunk and stem dark green; section of the stem shows the presence of pink coloured parenchyma in a circular band; mesocarp becomes deep pink on ripening; seeds small as in *Annona reticulata*. It contains maximum reducing sugars.

Red specked custard apple: Tree grows upto 1.50 to 2.0 m. Fruits are spherical to cordate in shape; small to medium size. Fruit weight ranges from 170 – 335 g.Areoles medium, tuberculate, truncate, very rough, ridged, pitted; rind light russet-green; splashed with red colour; mesocarp white with red colour; with medium to big granules; pulp dull white , moderately juicy, slightly stringy and just sweet, seeds 40–70.Flowering till the end of August; poor yielder; appears to be a cross between 'Red Custard apple' and Local Custard apple.

Washington 98797: Tree grows upto 1.8 to 3.0 m. Fruits are spheroid to cordate in shape, medium to big size. Average fruit weight is 690 g, areoles medium, truncate, tuberculate, ridged, rough, rind light russet-green; mesocarp with medium to big granules; pulp creamy-white, soft, moderately juicy, pleasant, moderately sweet. Seeds 50–60. Flowering till early September. Heavy yielder.

Washington 107005: Tree grows upto 1.8 to 2.4 m. Fruits are spherical to short conical in shape. Average in size; fruit weight ranges from 170 to 335 g, recorded up to 560 g. Areoles tuberculate, with small pits; rind grayish green, mesocarp white with small to medium granules ; pulp soft, bright white, buttery; moderately to very juicy, sweet, flavor, delightful, seeds 30-50. Rich in reducing sugars and also in total acidity and protein.

APK-1: It is a clonal selection from a high yielding type in the State Horticultural farm, Courtallam. It bears higher yield in rainfed vertisol (black soil). It is a drought tolerant variety which bears sweet fruits (TSS of 24.5 Brix with an acidity of 0.2 per cent). The Fruit weight ranges about 207.5 g with a mean of 72 fruits per tree per year. The first bearing commences in a graft/budded plant at 3-3 ½ years. The Optimum productive life is 25 yrs. The best season

for planting is May – June / August – September. It yields about 7300 kg fruits / ha (14.90 kg / tree which is 30.7 percent higher than Balanagar). It is suitable for cultivation in plains of Tamil Nadu especially in semi arid regions and marginal soils of both vertisol and alfisol in dry tracts. It is suitable for both rainfed and irrigated conditions (http://www.agritech.tnau.ac.in/horticulture/horti_TNAU_varieties_fruitcrops.html#custard).

Other custard apple varieties

Some cultivars are identified based on colour of the fruit, origin and plant part: Yellow custard apple are named after the colour of the fruit; Cultivars like Kakarlapahad, Mahaboobnagar, Local Hyderabad and Saharanpur Local are named after the place of origin; White stemmed custard apple is named after the part of plant.

Cultivars from Alexandria and Egypt

Beni mazar

It is round, large, fruit weight ranges from 150 to 180g; 56-60% flesh; 15-30 seeds per fruit (Morton,1987).

Abd El razik

This has light-green or reddish rind; nearly round, large, average fruit weight is 236.3 g; 69.5% flesh; 14 seeds per fruit (Morton,1987).

Cuban seedless and Brazilian seedless

It is a seedless form ('Seedless') with seemingly identical selections from Brazil and Cuba ('Cuban Seedless' and 'Brazilian Seedless'). This form is not likely to be widely grown because the fruit splits very badly as it matures on the tree and fruit quality appears to be slightly inferior to most-seedling fruit. It is not as productive as the seeded types, is nearly identical in quality and flavor.

Another "seedless" sugar apple was introduced from Brazil, which appeared to be identical to the Cuban variety (http://university.uog.edu/cals/people/PUBS/Sweetsop/MG33000.pdf)

Some selected cultivars of custard apple (Chen and Paull, 2008)

Country/region	Cultivar name
Taiwan	Ruan-zhi, Cu-lin , Da-mu , Xi-lin and Tai-nong no.1
India	Balanagar, Mammoth, Arka sahan, barbadose, Washington and Red Sitaphal
Thailand	Fai Kaew, Fai Krung, Nang kaew, nang Sir Krung and Nang Thong
Florida	Lessard, kampong Mauve, Red Sugar, Cuban seedless
Egypt	Abd El-Razik

Flower biology

The flowers of custard apple are hermaphrodite and are produced singly or in small clusters on the current season's growth. The flower buds start appearing on new shoots in the spring, immediately after the commencement of the vegetative growth. The flowering period in custard apple was from the middle of April to end of June. The average time from visible flower initiation to full flowering was 30.8 days (Nalawadi *et al.,* 1975). Anthesis took place continuously throughout the day and night. Anthesis commenced at 6.00am and continued upto 6.00 pm, with its peak at 6.00 am under Dharwad condition, between 5.30 and 6.30 am under Bhubaneshwar (Sahoo *et al.,* 2000) whereas anthesis occurs at 5.00 am in Brazil (Ribeiro *et al.,* 2007). Kiill and Costa (2003) reported from Brazil that the anthesis takes place around 5.00 pm, the flowers last for approximately two days, and show female phase in the first twenty hours and male phase in the following twenty hours, characterizing dichogamy. Anther dehiscence commenced at 12.00 pm and was completed by 4.00 am with a peak at 2.00 am (Nalawadi *et al.,* 1975).

The flower exhibits a protogynous dichogamy nature that limits self-pollination and cross-pollination can increase fruit set. Pollen grains appearing early in a flowering season have lower germination rates than the pollen from late flowers (Chen and Paull, 2008).The highest pollen germination was recorded between 18.9 and 20.4% in 20% sucrose for green and red type respectively .The stigmas were receptive from one day prior to anthesis till 2-3 days after anthesis indicating a protogynous condition (Thakur and Singh, 1965; Sahoo *et al.,* 2000). Nitidulid beetles, *Carpophilos domidiahus* and *Carpophilos hemipterous* were identified as pollinators in terms of visitation frequency, pollination potential score and pollination efficiency. However, *Carpophilos hemipterous* was relatively more effective pollinator than *Carpophilos domidiahus*. Custard apple clearly exhibit obligate specialization by filtering only *Carpophilos hemipterous* and *Carpophilos domidiahus* as pollinators and thereby explicitly indicating canthrophilous pollination syndrome (Kundan Kishore *et al.,* 2012).

Artificial pollination is frequently practiced to ensure pollination and good fruit shape. Hand pollination is normally carried out before 8.00 am using a small brush. Pollen can be collected in the morning between 5.00 and 8.00 am from fully open flowers, when the sacs have turned from white to cream. The collected pollen is used to pollinate half-open flowers whose pistils are already receptive (Jalikop and Ravindra Kumar, 2007).

Varietal evaluation

India

Several varietal evaluation studies were carried out in various regions of India using different cultivars. Evaluation of sweetsop genotypes (APK-1, Mammoth, Balanagar) under sodic soil at Trichy, Tamil Nadu by Krishnamurthy (2009) showed superiority of APK-1 with regard to yield in nine years old trees which produced 47.78 kg per plant, followed by Balanagar (27.05 kg per plant).Thirty five germplasm of custard apple have been evaluated for scarce rainfall zone under rainfed at Anantapur, Andhra Pradesh by Dhanumjaya Rao and Subramanyam (2011) in 10 years old trees and the maximum yield have been recorded from MMR-10 (15.0 kg per plant). Girwani *et al.* (2011) evaluated sweetsop hybrids under Sangareddy condition and revealed that all the hybrids varied in fruit shape (round, conical and cordate), fruit colour (yellowish green, grayish green, light green and red), pulp colour (creamy white, light pink and white), areole shape (tuberculate, hexagonal) and texture (course, soft and meaty). Time of fruit maturity varied from 2nd fortnight and lasted till the end of December. Hybrid-1 (17/4 Atemoya × Balanagar) was September earliest to mature. Maximum fruit weight (250 g) was recorded in Hybrid-1 (17/4 Atemoya × Balanagar) followed by Hybrid-6 (15/3 Red Sitaphal × Atemoya) (225 g) and Hybrid-4 (1/6 British Guinea × Atemoya) (220 g). Among all the hybrids, maximum number of fruits per tree (94) was recorded in Hybrid-1 (17/4 Atemoya × Balanagar) followed by Hybrid-6 (15/3 Red Sitaphal × Atemoya) (67). The TSS ranged from 22 to 28°B and was maximum in Hybrid-2 (15/2 Red Sitaphal × Pondapple), while the seed content per fruit was also minimum (20) in this hybrid. Based on overall performance, the Hybrid-1 (17/4 Atemoya × Balanagar), Hybrid-3 (15/3 Red Sitaphal × Atemoya) and Hybrid-2 (15/2 Red Sitaphal × Pondapple) were found excellent in quality with pleasant aroma, smooth pulp texture with less seed, good sugar acid blend and shelf life.

Breeding objectives

The first objective is the determination of the specific characteristics that are important for the new cultivar-the ideo type. These characteristics may help a breeder to select parental groups to be used in an annona breeding programme, in order to obtain desirable progenies.These progenies may not have all desirable characteristics, but at least will have those most important to growers, retailers and consumers. For instance, a sweet sugar apple cultivar with long shelf life fruits (>5days), which is important from a consumer's viewpoint, may not be selected, if it presents low yield (<20kg/tree/year) and fruit weight lower than 400 g. Genetic improvement for yield is the most difficult and expensive of all breeding objectives, due to the complex nature of its inheritance and the numerous environmental factors influencing the trait. Quality is another important characteristic for improvement of annonas, as important or some times more so than yield, since market value is based on the fruit's appearance and its organoleptic characteristics (Pinto *et al.,* 2005).

Maincharacteristics of sugarapple ideo-types(adult plants)

Characteristics	Sugar apple
Plant	
Vegetative vigour	Medium to low
Reproductive vigour (Yield)	High (>20 kg per plant)
Bearing	Regular
Rootstock / scion compatibility	High
Flower number	Abundant (>180 flowers per tree)
Pollen fertility (20°C)	High (>76%)
Fruit	
Size / Weight	
For industry	Large/>400 g
For fresh consumption	Medium/300-400g
Shape	Rounded to Heart
Fruit set (naturally)	High (>25%)
Skin	Tuberculate
Pulp	Sweet, low fibre
Flavour	Pleasant acidulous
Seed number per 100g of pulp	Low or absent (<10 seeds)
Shelf life (15 to 30°C)	Long (>5 days)

Breeding strategies

Most *Annona* species and cultivars differ in environmental adaptation, productivity and fruit quality. Therefore, different conventional methods can be used in their breeding. According to Fehr (1987), there are three requirements for the development of an asexually propagated cultivar: a) a suitable source of genetic variability; b) evaluation of individuals from the population; c) a sexual multiplication of a new cultivar for commercial use.

Introduction of superior genotypes and/or cultivars to establish a germplasm collection is, basically, the first requirement of any breeding programme. This can be complemented by the introduction into the collection of some wild *Annona*species with useful genes, mainly for resistance to diseases. All accessions require comprehensive characterization and documentation, followed by evaluation and selection. Several types of populations can be developed by hybridization, from which superior clones are selected.

Selection and cultivar development

India and Taiwan have selected a few named cultivars of sugarapple from seedling population that are propagated vegetatively. In Cuba, researchers developed 'Cuban Seedless', which is a seedless cultivar with medium-sized fruits, and another cultivar with low fibre content that is very important for the commercial market (Nakasone and Paull, 1998). In the Philippines, there are 3 forms of sugar apple fruits: (a) a green-fruited seedy form, which is grown in most parts of the country; (b) a purple-fruited seedy form, reportedly introduced from India; (c) and the green-fruited seedless form, which is a recent introduction and whose origin is unknown. The selection of superior strains is aimed in the direction of a green-fruited seedy form (Coronel, 1994).

Some selections and cultivars of sugar apple that are currently grown in various countries (Pinto *et al.*, 2005)

Cultivar/Selection	Country of origin	Cultivar/Selection	Country of origin
IPA selections	Brazil	Leone	Italy
Libby	USA	Madeira	Portugal
Lisa	USA	Mateus 1	Portugal
Mossman	Australia	Negrito	Spain
Ott	USA	Reretai	New Zealand
Whaley	Australia	White	USA
Molate	Philippines	Cuban Seedless	Cuba
Lobo	Philippines	Noi	Thailand
Red Sugar Apple	USA/Florida	Mammoth	India
Balanagar	India	Red Sitaphal	India

Hybridization

Tri-species hybrids between *A. atemoya* x *A. squamosa* have been attempted by Jalikop and released one hybrid for commercial cultivation during 1997 in India which is a cross *A. atemoya* cv. Island gem x *A. squamosa* cv. Mammoth and named as Arka Sahan (Jalikop and Ravindra Kumar, 2007). Likewise, Girwani *et al.,* (2010) also developed sugar apple hybrids at Sangareddy.

Modern biotechnology

The first haploid plants induced by anther culture in fruit trees were reported by Nair *et al.* (1983) with sugarapple. The availability of haploids is very important for fruit breeding, because of the long generation intervals, the highly heterozygous nature of most fruit species and the presence of parthenocarpy and self-incompatibility. These researchers obtained callus differentiation, and formation of triploid roots and shoots from sugar apple endosperm (Nair *et al.,* 1986). Their aim was development of seedless fruits, but a complete plantlet was not obtained.

Future thrust

1. Development of seedless sugar apple varieties
2. Development of sugar apple varieties resistant to Anthracnose and fruit fly
3. Production of double haploid in sugar apple

References

Chen,C.C. and Paull, R.E. 2008. *Annona squamosa*-sweetsop. (In) The Encyclopedia of fruit and nut, Janick, J. and Paull, R.E.(Eds.).CABI, Oxfordshire, UK: p:52.

Coronel, R.E. 1994. Atis.In: Promising fruits of the Philippines. Coronel,R.E., Zuno,J.C. and Sotto, R.C.(Eds.). College of Agriculture, University of the Philippines, Los Banos, Laguna, The Philippines: pp.1-18.

Dhanumjaya, Rao, K. and Subramanyam, K. 2011. Growth and yield performance of custard apple germplasm under scare rainfall zone. *Indian J. Agric. Res*. 45(2):156-160.

Fehr, W.R. 1987. Principles of cultivar development - Theory and Technique.Vol.1, MacMillan Publishing Company, New York.

Girwani, A., Madahvi, A., Kumar, T.S. and Reddy, G.S. 2011. Evaluation of custard apple hybrids for fruit yield and quality attributing characters. *Acta Hort*. 890: 251-254.

Jalikop, S.H. and Ravindra Kumar. 2007. Psedo-xenic effect of allied Annona spp. Pollen in hand pollination of Cv.'Arka Sahan' [(*A.cherimola* x *A.squamosa*) x *A.squamosa*]. *Hort Sci*. 42(7):1534-1538.

Jalikop, S.H. 2010a. Inheritance of foliage characteristics in (*A. cherimola* Mill. x *A. squamosa* L.) and *A. reticulata* L. hybrids. *Euphytica*. 171:175-180.

Jalikop, S.H. 2010b. Tree and fruit traits of progenies from the cross between (*A. cherimola* Mill. x *A. squamosa* L.) and *A. reticulata* L. and approaches for the introgression of valuable genes from *A.reticulata* L. *Euphytica*. 173: 161-171.

Kiill, L.H.P. and Costa, J.G. da. 2003. Floral biology and reproductive system of Annona squamosa L. (Annonaceae) in Petrolina, P.E., Brazil. *Ciencia Rural*. 33(5):851-856.

Krishnamoorthy,V. 2009. Studies on the performance of different varieties of aonla,pomegranate and annona under sodic soil. *Asian J. Hort*. 4(1):149-151.

Kumar, P.S. and Jalikop, S.H. 2000.Cross-compatibility among *Annona* species. *Indian J. Hort.* 57(4): 309-313.

Kundan Kishore., Shukla, A.K., Naresh Babu., Sarangi, D.N. and Samiksha Patanayak. 2012. Pollination biology of *Annona squamosa* L.(Annonaceae): evidence for pollination syndrome. *Sci Hort.*144: 212-217.

Leboeuf, M., Cave, A., Bhaumik, P.K., Mukherjee, B. and Mukherjee, R. 1982. The phytochemistry of the Annonaceae. *Photochemistry*. 21: 2783-2813.

Morton, J. 1987. Sugar apple. (In) Fruits of warm climates, Miami, Florida.p.69-72.

Nair, S., Gupta, P.K. and Mascarenhas, A.F. 1983. Haploid plants from *in vitro* anther culture of *Annona squamosa* L. *Plant Cell Reports*, 2: 198-200.

Nair, S., Shirgurkar, M.V. and Mascarenhas, A.F. 1986. Studies on endosperm culture of Annona *squamosa* L. *Plant Cell Reports*, 5:132-135.

Nakosone, H.Y. and Paull, R.E. 1998. Annonas. In: tropical fruits. Nakasone, H.Y. and Paull, R.E. (Eds.), CABI, Oxfordshire, UK. p 45-75.

Nalawadi, U.G, Sulikeri, G.S. and Singh, C.D. 1975. Floral biological studies of *Annona squamosa* L. under Dharwar condition. *Prog Hort.* 7(1):15-24.

Paull, R.E. and Duarte, O. 2011. Annonas: Cherimoya, Atemoya and sweetsop. In: Tropical fruits, Volume 1, CABI, Oxfordshire, UK, pp: 134.

Pinto, A.C.de Q., Cordeiro, M.C.R., Andrade, S.R.M.de., Ferreira, F.R., Filgueiras, H.A.de C., Alves, R.E. and Kinpara, D.I. 2005. Annona species. Williams, J.T., Smith, R.W., Hughes, A., Haq, N. and Clement, C.R. (eds). International Centre for Underutilized Crops, University of Southampton, UK.

Ribeiro, G.S., Jose, A.R.S., Reboucas, T.N.H. and Amaral, C.L.F. 2007. Floral biology aspects related to the production of sugar apple (*Annona squamosa* L.) fruits. *Acta Sci. Bio. Sci.* 29(4): 369-373.

Sahoo, S.C., Panda, J.M. and Mohanty, D. 2000.Studies on floral biology of custard apple (*Annona squamosa* L.) under Bhubaneshwar condition. *Orissa J.Hort.* 28(1):43-45.

Thakur, D.R. and Singh, R.N. 1965.Studies on floral biology of Annona. *Indian J. Hort.* 22: 238-253.

US Department of Agriculture (USDA). 2002. Sugar-apples (Sweetsop), Raw, USDA National Nutrient database for Standard References, Release 15, Available at: http://www.nal.usda.gov/fnic/foodcomp/egi-bin/list_nut_edit.pl.

Wester P. J. 1913. Annonaceous possibilities for the plantbreeder. *Philippine Agri. Review,* 6:312-321.

19

Breeding in Indian Jujube (Ber)

S.S. Hiwale

Indian jujube (Ber) belongs to the genus *Zizyphus* and family Rhamnaceae having more than 600 species. The family includes trees with spine, shrubs and climbers. The original home of Ber is Myanmar and India. Many workers have tried to classify it on the basis of leaf, flowers and fruits characteristics. Some tried to divide into two groups i.e. having smooth leaves on both surfaces and having leaves downy beneath. Ber as a tree has been described it as small, much branched or a large shrub with dense spreading habit, bark dark gray, branches elongated, leaves with glabrous shining upper surface densely covered beneath with whitish hairy growth.

Ber has attracted sustained breeding attention owing to its wide adaptability, good quality and nutritive value. However, problems like poor shelf life, incidence of powdery mildew disease, fruit fly and fruit borer have hindered the extension of the area under this crop. In Chinese jujube, the problem of fruit cracking and 'witches broom' disease are the major hindrances in its cultivation. There are two major cultivated species viz., *Ziziphus mauritiana* and *Ziziphus jujuba*, which are widely distributed covering vast regions in Asia. The wild population found is heterozygous and extremely variable. As a result, many local selections were made, some of which have become widely recognized as cultivars. The success of any breeding programme depends on the extent of genetic variability in the source population. The assessment of variability is therefore a basic requirement of breeding programme. Since most of the plant characters of economic importance are governed by the group of genes and are highly influenced by environmental variation, it is difficult to judge whether the observed variability is heritable or due to environment. This necessitates optimizing of phenotypic variation into its heritable and non heritable components, so that the selection of parents

Indian jujube is a polyploid with chromosome number being 12, 20.24,30,36 or 48. Khushoo and Singh (1963) studied the range of cultivars and found n=24 in most, but in two cultivars it was observed to be n=48 and in one it was n=30. Mehetre and Dahat, (2000) reported a range of diploids, triploids, tetraploids, pentaploids and octoploids in Indian jujube. Nehra *et al.* (1983) in some wild varieties and in naturalized wild varieties observed n=48. *Z. lotus*, *Z. numularia* and *Z. oenoplia* appear to be diploid, *Z.numularia* shows a polyploidy series. Chinese jujube also represents a polyploid series with chromosome counts tending to represent 2n=45, 60, 90.The level of ploidy appears to be important for some cultivars of Indian jujube. Knowledge of basic cytology is needed to plan introgression from wild species.

The majority of cultivars of both Indian and Chinese jujube are selections from heterogeneous population; superior types protected and used over years by local people so that location of specific cultivars is wide spread. Those that are used now a days tend to be propagated clonally and some have become very wide spread eg., the variety Umran in Indian jujube or Li in Chinese jujube. As per literature over 180 named cultivars are mentioned (Pareek, 2001).

Morphological variability

A wide range of variation due to cross pollination is exhibited by jujube with regard to vegetative, floral, fruit and quality traits. Variation in morphological and physico- chemical characters of ber cultivated types are reported by Hiwale, *et.al.* (2008) and Shobha *et. al.,* (2001).Out of fifty *Z. mauritiana* genotypes, 23 were erect types and 27 spreading types under semi-arid rainfed conditions of Gujarat in India. Variation in plant height, spread and nature of branching was observed in different cultivars at the 'Central Horticultural Experiment Station, Godhra Gujarat, India under rainfed conditions. The most appropriate vegetative character for classification is leaf area, while most dependable fruit character is the fruit length. A non destructive method of determining leaf area using linear parameters has been described (Hiwale and Raturi, 1991). Metroglymph and index score analysis showed wide morphological variation in 80 germplasm accessions of *Z. mauritiana* and eight distinct groups were recognized on the basis of variation in 23 morphological and fruit characters (Pradeep and Jambhle, 2002). Paired spines were observed in many cultivars of *Z. mauritiana* like Umran, Kathaphal, Gola Gurgaon No.3, Bahadurghdhia, Gola, Dankan Gola and Kakrola Gola (Gupta *et al.,* 2003).

Flowering, fruit set and fruit retention

The peak period of flowering and fruit set in most of the cultivars of *Z. mauritiana* was September – October under semi-arid rainfed conditions of Gujarat. Saran (2005) found the longest period of bloom in Katha Rajasthan (78 days) and the maximum period of bloom in Safeda Rohtak (23 days). Flowering occurred from 57 to 75 days, depending on cultivar (Dhaliwal and Bal, 1998). Only one flowering season was observed in Hyderabad from May to July and its total duration varied from 68-94 days, (Babu and Kumar, 1988). The variety Umran had the highest number of hermaphrodite flowers (22.2%) followed by Gola Gurgaon (20.1%) (Darbara and Jindal, 1982).

The pollen of ber varieties has three zonicoloporate apperture, pistillate exine pattern and circular endosperm.The pollen grains were sub prolate to prolate and spheriodial. Pollen germination in 25 percent sucrose solution was observed to be more than 50 percent in most of the ber cultivars. The variety Illachi was sterile. Size and shape of pollen with in the cultivars were quite uniform , differences in exine pattern, size and P/E ratios could be used for identification of *Ziziphus* genotypes (Diwakar *et al.,* 1996). Size of pollen of Z *mauritiana* cv. Illachi, Umran and four wild forms was different (Nehra *et al.,* 1984). Pollen diameter in seven cultivars of ber (*Z. mauritiana*) ranged from 20.05 μ in Darakhi to 32.04 u in seedless. Pollen stainability ranged from 63.69 percent in seedless to 87.12 percent in Darakhi -1. *In vitro* germination was a better indicator of fertility than stain test (Hulwale *et al.,* 1995). Similarly considerable differences were also observed in pollen diameter among different ploidy levels (Pradeep and Jambhale, 2000). Phenotypic relationship among 32 genotypes of *Ziziphus mauritiana* and one of *Z. numularia* studied using data of 54 characters did not show a close similarity with *Z. mauritiana* genotypes (Diwakar *et al.*, 1992). Studies in India indicated that in *Z.mauritiana*, some cultivars have cross incompatibility (Teotia and Chauhan, 1964). Ackermann (1961) showed that many jujube cultivars fruit poorly without cross pollination with fruit cracking during rainy season being another problem.

Fruit set

The presence of cross incompatibility and pollen sterility showed that fruit set depends on physiological and environmental conditions. Umran was compatible as female parent with Sanuar -2 but Sanuar-2 did not set fruit after pollination by Umran (Mehrotra and Gupta, 1985). The time of anthesis was also observed to be cultivar specific. Peak receptivity of the stigma appeared to be just at the time of flower opening (Dhaliwal and Bal, 1998). Some Chinese jujube clones developed fruits through self fertilization and very a few set appreciable crop by self fertilization and fruits developed from self pollination usually were observed to be smaller than normal and tend to drop pre-maturely. Cross fertilization was found necessary for development of viable seeds and many of the aborted seeds were result of self fertilization (Akerman, 1961).

Fruit set in *Z. mauritiana* cultivars took place at different times. In the early varieties like Gola and Goma Kirti flowering was early leading to early fruit set and early maturity by third week of December whereas late varieties matured in the 2nd week of January under Gujarat conditions (Hiwale *et al.*, 2008). Neeraja *et al.,* (1995) reported higher fruit set in hand pollinated than open pollinated and 60 % fruit set was observed in Umran x Seb. In Chinese jujube, fruit set was uniformly high in Ya and Nikitskii, Melmoplonyi (Sin Ko,1974 a). Yakobashvili (1973) in a six year study of fruit set, yield and other characters, showed the varieties Ta Yan Tszao, Seedling -2 , Seedling 1 and 3 to be promising.

Fruit drop studied in seven *Z mauritiana* cultivars indicated the lowest drop (24.1%) in Ponda and the highest (68.6%) in Illachi (Vashistha and Pareek, 1979). Kakrola Gola had considerably higher drop (up to 80.6 %) than Kaithali and Umran which did not differ significantly, showing 7.2 and 12.1 % drop respectively (Panwar, 1980). Fruit maturity in different cultivars of *Z. mauritiana* was reached in 174 to 200 days (Kumar *et al.,* 1986). Gola was the earliest (109 days) and Umran took 147 days to ripen under Gurgaon conditions (Singh *et al.,* 1983a). Fruit growth in terms of length and diameter showed three distinct phases (double sigmoid growth curve) in all the fifty cultivars. Hiwale *et al.*, (2008) based on maturity period and harvesting time classified them into three categories as early, medium and late.

Fruit and seed variability

Fruit maturity varied right from September in South to March in North India. High variability was observed for all fruit and seed characters. There is considerable variation in length, breadth and fruit weight in different cultivars of *Z. mauritiana,* with maximum length in Banarasi (4.82cm), and breadth in Goma kirti (3.02 cm). The fruit weight reaches even 150g in Wugianzhong, a *Z. mauritiana* cultivar (Li *et al.,* 1999). Significant genotypic differences were found in the seeds of *Z. mauritiana* varieties (Babu and Kumar, 1986). Saran (2005) observed variation in stone size of 35 cultivars of ber and reported maximum stone length in Ponda (3.72cm) and minimum (0.53 cm) in Illachi.

Yield variability

Variation in yield of *Z. mauritiana* cultivars (55.5 to 116.11kg/tree/year) have been observed by Kumar *et al.* (1986). In the semi arid subtropical climate of Northern India under irrigated conditions, the fruit yield per tree ranged from

80 to 200 kg depending on the varieties and management practices during the prime bearing age of 10-20 years (Bakshi and Singh, 1974). The cultivar Pewandi is considered the best for commercial cultivation in Uttar Pradesh, India (Teotia *et al.,* 1974).

Variability for quality traits

A wide range of variability for quality traits was detected in fifty cultivars of ber (*Z. mauritiana*) at CHES, Godhra under semi-arid rainfed conditions. The maximum TSS was recorded in the cultivar Katha (26.09%). Maximum sugar and minimum acidity content were recorded in Chuhara (Hiwale *et.al.*2008). Under rainfed conditions of Bawal, Haryana cv. Nazuk had the highest total soluble solids (28.9%) followed by Illachi (Yamadagni *et al*., 1985 b). Acidity varied from 0.228-0.78 % and ascorbic acid from 80.85 -178.04 mg/100g pulp (Godara, 1980). Ascorbic acid content of several cultivars ranged from 70-165 mg /100g of pulp (Jawanda and Bal, 1978). Bisla *et al.* (1980) observed maximum Vitamin C content (120.15 mg /100g) in the cv. Illachi. The Vitamin C content was maximum in Narikeli, followed by Kaithali, 165 and 125 mg/100g fruit pulp, respectively (Gupta, 1977). Godhan was the most nutritive with regard to total soluble solids and ascorbic acid content, where as Kharki was the sweetest at Hoshngabad, Madhya Pradesh, India. Singh and Jindal (1980) found the highest content of pulp (97.2%) TSS (21.4%) and total sugar (10.7%) in Gola Gurgaon and Kaithali had the maximum (113.5 mg/100g) ascorbic acid content. Cultivars grown in China, United States and Pakistan contained 500-600, 300-500 and 45.2-160.8 mg/100g Vitamin C respectively. Variation in ascorbic acid content of 10 varieties was reported by Ahmed and Mallik (1971). The soluble protein in fruits and leaves on fresh weight basis ranged from 9.37 to 26.90 mg and 25.40 to 56.98 mg /gram respectively in 42 cultivars of *Z. mauritiana* (Sudhir *et al*., 1999). Umran was noted for high sugar and protein content, followed by Kathaphal (Khera and Singh,1976).

Genetic variability

At CHES Godhra, the highest Genotypic Co-efficient of Variation (GCV) and Phenotypic Coefficient of Variation (PCV) values were observed in leaf number /shoot and leaf area in respect of morphological parameters, (Hiwale *et al.,* 2008). Praveen and Patil, (1998) reported maximum GCV and PCV values for stomatal frequency. GCV and PCV values were the lowest for relative water content (RWC). High values of heritability and genetic gain were observed for stomatal index. Significant and positive correlations were observed between yield per plant and stomatal frequency, leaf area and stomatal index. Selection of these traits may be effective for the improvement of ber. The total variation accounted for by all the characters was 86 percent. In a study of 30 cultivars of ber at Hissar, India, fruit weight, size, seed weight and pulp-stone ratio were significantly correlated with yield (Bisla and Daulata, 1987). Yield was positively correlated with total sugar and disease intensity (Bisla and Daulata, 1988 b). TSS had positive and high correlation with total sugar while yield had negative correlation with quality traits except ascorbic acid and non reducing sugars (Saran, 2005). Traits such as TSS, fruit breadth and fruit length showed comparatively low genetic coefficient of variation, where as weight/fruit, pulp stone ratio and yield (kg/plant) exhibited high genetic coefficient of variation (Bisla and Daulata, 1988a). Liu (1996) studied nine major quantitative characters of Chinese jujube. They observed significant differences among the characters, the coefficients of variation was the highest in fruit weight (51.95) and the lowest in edible part of the fruit (2.67%). Saran (2005) studied 10 quantitative characters in 35 genotypes. Saran (2005) also observed high estimates of genetic coefficient of variation, phenotypic coefficient of variation, heritability and genetic advance for stone size, pulp- stone ratio, fruit weight and yield. This indicates the effectiveness of improvement through selection. He observed that fruit yield had positive and significant correlation with tree spread and fruit size (0.319), fruit weight (0.515) and stone size (0.353). He also observed that fruit size (0.580) and flesh thickness (0.811) were indirectly contributing through fruit weight. Fruit weight (0.998) contributed to yield mainly through its direct effect followed by indirect effect of stone size and number of flowers per cyme.

Heritability

The estimation of heritability enables selection programme to be based on phenotypic performance. Heritability in 30 ber cultivars ranged from 54.2% for total soluble solids to 99.63% for disease intensity and the value for acidity was 91.61 % (Bisla and Daulata, 1988a). To improve the yield through hybridization in *Z mauritiana,* individual trees with more no of leaves/shoot, higher leaf area, low fruit acidity, low disease intensity and high reducing sugar should be selected. Heritability was noticed to be high for fruit weight (97.2%) and size (87.9), pulp stone ratio (87.5%) and seed weight (84.6) in *Z. mauritiana* at Hissar (Bisla and Daulta, 1988b).

The highest heritability values were seen for days from fruit set to ripening (99.2%), fruit set (94.7%), days from pruning to sprouting (93.6%), fruit drop (85.0%), shoot length (82%). High heritability coupled with genetic advance for yield (kg/plant), pulp-stone ratio indicated that high heritability obtained for these characters was probably due to additive gene effect. Expected genetic gain was the highest for fruit set (142.8%) and lowest for days from pruning to flower initiation (3.5%). Selection for higher fruit set and lower fruit drop was recommended for improving the yield by Bisla and Daulata, (1988a).

High genetic advance was reported in fruit retained /shoot, percent powdery mildew incidence , yield, acidity and pulp to stone ratio in different cultivars, the genetic advance expressed as percentage of mean ranged from 3.28 percent (plant spread) to 128.58 percent (percent powdery mildew incidence -Hiwale *et.al.*, 2008). Fruit weight, fruit size and seed weight showed high genetic advance (Bisla and Daulata, 1988b). The fruit length, fruit breadth and weight per fruit had low heritability along with low genetic advance, indicating that dominance of epistatic effect is of considerable value for these characters. Hence, improvement in these characters is possible through selection. Individual plant selection for incidence of powdery mildew, yield (kg/plant) and pulp- stone ratio would be effective in ber as these characters exhibited high heritability as well as high genetic advance, where as for other characters selection may not be so effective.

Ideotype for ber breeding

- Canopy spread, low headed
- Leaf area -High (> 50cm^2)
- Internodal length- Short (< 5cm)
- Leaf colour - Dark green Chlorophyll less than 0.04mg/kg fresh wt.
- Leaf surface- Glossy dorsal and tomentose ventral
- Leaf size- Large
- Fruit size- 30g
- Fruit colour- golden yellow
- TSS-15 ° Brix
- Acidity-0.3%
- Pulp stone ratio- > 20

The inflorescence consists of male and hermaphrodite flowers. It is highly heterozygous because of cross pollination. The existence of polyploidy and presence of self and cross incompatibility in *Ziziphus* species has resulted in wide hybridization. Current ber cultivars were evolved through selection of superior types from the wide natural variability. There is a need to improve the crop for commercial cultivation in different agro- ecological regions. Introduction, selection hybridization, polyploidy, mutation, biotechnological methods and genetic engineering are some of the techniques.

Important characteristics for initiating breeding

Despite wide cultivar diversity of the two major jujubes, only a few cultivars are of commercial importance. The main objectives are:

- Higher productivity.
- Ability to withstand transport and storage.
- High TSS, good sugar acid blend.
- Crisp flesh and good flavor.
- Resistance to common insect pests and diseases.
- Earliness is a desirable trait in cultivars meant for dry, rainfed region.
- Time of maturity is also an important characteristic to be considered.
- Salt and low temperature tolerance.
- Better eating and processing qualities.

Breeding

Ber breeding is hampered by high heterozygosity, inadequate knowledge of inheritance pattern and lack of well established pre-selection criteria such as incompatibility patterns, sterility and fruit drop and the small size of flowers. Ber exhibits wide range of polyploidy and incompatibility in its cultivars (Pareek, 1996). Most of the ber cultivars are tetraploids showing segregation; many cultivars have been reported to be reciprocally cross incompatible (Taeotia and Chauhan, 1964). In ber the main breeding methodologies are same as those adopted in other crops. These are enumerated below;

Selection

Most of the cultivars of ber grown in tropical, subtropical and temperate regions of the world are selections with major emphasis on clonal selection especially of early maturing clones. A wide range of variability exists for ber in India for all important characters suggesting substantial scope for improvement. Goma Kirti is a clonal selection from Umran which flowered 25 days earlier. A clonal selection for earliness with better quality of fruits and better storage life with consistent productivity from Umran was identified at Central Horticultural Experiment Station, Godhra Gujarat, India. After evaluation trials, the selection was early compared to Umran, it matured with the early cultivar Gola. It was released as "Goma Kirti "(Hiwale, 2005).

Hybridization

In order to find superior hybrids with a wide range of fruit traits, as well as adaptation to diverse climate, several cultivars have been used in India as parents. They are Umran, Banarasi, Karaka, Kakrola, Gola, Mundia Murhara, Sanuar, Illachi, Safeda selected, Kaithali, Rashmi, Chhuhara and Seb. At Hissar, Umran and Banarasi karaka were used for high yield, Gola for quality and earliness, Illachi for profuse bearing, BS 75-3 for drought tolerance and fruit fly tolerance, Rashmi for fruit quality (softness). Fruit setting in four *Z. mauritiana* cultivars (Umran, Karaka, Popular Gola and Chhuhara) after pollination with powdery mildew resistant cultivars Villiati, Darahi-1, Darakhi-2 and Guli was observed.The best female parent was Umran and Gola both setting fruits with each of the pollinators, Darakhi-2 was the best male parent for fruit setting with each of the female cultivars (Karale *et al.,* 1992).

Mutation

Mutation can be induced by treating with chemical mutagens such as ethyl methane sulphonate, diethyl sulphonate, nitrous compound (Nitrous acid, nucleoside analogs and Sodium azide) and by ionizing radiation. Somatic mutations of spontaneous origin within vegetatively propagated material are commonly referred as sports. Desirable mutations occurring in adapted asexually propagated plants may result in an immediate improvement. Tainung No.1, a bud sport of Kaolang No1 was found in Pintung, Taiwan (Chang *et al.,* 2001) and it was confirmed that this new cultivar was a hereditary cultivar Luyuan Xiaozao bud sport of Xiaozao (Li *et al.,* 1996).

The induction of mutation with colchicine treatment (0.15%, /18h) was 50% in cv. Linyilizao and 43.3% in Dongzao and Lajiaozao as reported by Jing an Liu (2004) .The chromosome number of the mutated plants (2n= 4x= 44) and DNA content were both twice that of the control. Useful mutants have been produced by gamma irradiation and colchicine treatment in *Z. jujuba* (Akundova and Agaev, 1989). Variation in seedling of *Z. jujuba* after treatment with ethyleneimine, ethyl methane sulphonate, and diethyl sulphate and gamma radiation was observed and some were propagated vegetatively for further trials (Sin· ko and Chemarin, 1979b, 1982). Sin· ko and Chemarin (1979a) reported that when seeds of *Z. jujuba* were irradiated at 0.1 to 100 Krad, 60-98% germination was obtained from seeds treated with low dose (0.5-2.0 Krad). The 30 Krad dose was critical, the treatment had no effect on the rate of emergence. Colchicine and ethyl methane sulphonate treatments of *Z. mauritiana* increased crossability and reduced pre mature fruit drop (Gupta and Minhas, 1991). Mutants were selected following N-methyl-N-nitrosourea treatment (0.2 -0.4 % for 12 h) of pre germinated seeds of *Z. mauritiana* cultivars Ma Hong, and Gio Loc> They maintained early maturity of cultivars, allowing two harvests per year. Cultivar Dao Tien, which produces round fruits which are large (25 vs. 20 g) with better flavor than those of its parental cultivar Thien Phien. It is also one month earlier in maturity, allowing two harvests/ year rather than the one produced by Thien Phien (Hoang and Tuynh, 1989).

Polyploidy

Transfer of powdery mildew resistance from diploid to tetraploid cultivars requires doubling of chromosome in resistant genotypes in order to overcome the possible post fertilization barrier. Attempts were made to induce polyploidy in some of the resistant diploid genotypes of *Ziziphus* that were confirmed cytologically (Pradeep and Jambhale, 2003). The behavior at meiosis e.g. bivalents and quadrivalents can be shown in some cultivars of *Z. mauritiana* and are cytologically alloploid (Pradeep and Jambhale, 2002). The number of stomata and chloroplasts in diploid, tetraploid, pentaploid and octoploid *Z. mauritiana* genotypes indicated significant differences in stomatal dimensions, frequency of their occurrence and chloroplast number per two stomatal guard cells among the different ploidy levels. However, no significant differences were observed between pentaploid and octoploid genotypes for these attributes, indicating the deleterious effect of high gene dose beyond the pentaploid level (Pradeep and Jambhale, 2000). In a triploid *Z jujube* (*Z. sativa*) cultivar the water saturation deficit, transpiration rate and daily maximum transpiration rate were lower than two diploid cultivars (Wan, 1994).

Biotechnological Methods

Tissue culture

Tissue culture is one of the most widely used techniques for rapid asexual in *vitro* propagation. This technique is economical in time and space, it gives greater output and provides disease free and elite propagules .It also facilitates safer movement of germplasm between nations. Protocol is developed for successful clonal propagation of *Z. mauritiana* and *Z. numularia* (Rathor *et al.,* 1992) and *Z.jujuba* (Zhao *et al*, 2001). Methods for rapid *in vitro* multiplication of ber using stem explants of mature trees have been developed (Yan *et al.,* 1990 and Mathur *et al.,* 1995). Benzyl adenine promoted bud differtiation and IBA promoted root growth when added to MS medium on which stem segments of *Z. jujuba* lines A-17, A-27 and A-80 were cultured (Yan *et al.,* 1990).Gola and Seb cultivars of *Z. mauritiana* took 150 days from initial culturing to transplanting (Goyal and Arya, 1985).

Genetic engineering

Use of molecular markers can facilitate proper identification of genotypes and help to overcome the problem of duplicity in gene bank. Raja (2004) worked out the techniques to identify 12 genotypes at Hissar, India for categorization and identification by molecular markers. Transgenic Chinese jujube plants have been obtained using young stems precultured for one day and infected with *Agrobacterium tumefaciens,* which contained anti ACC synthase gene (He, 2004b).

Through molecular markers, it is possible to score the plants for different traits e.g. disease resistance at the seedling stage itself. The use of RFLP (Restricted Fragment Length Polymorphism), RAPD (Random Amplified Polymorphic DNA), AFLP (Amplified Fragment Length Polymorphism) and isozyme markers in plant breeding are numerous. The genetic relationship between 27 cultivars, strains and related species (*Z. acidojujuba* and *Z. mauritiana*) of *Z. jujuba* were studied using RAPD techniques by Li *et al.,* (2003). They reported a total of 92 DNA bands amplified with 15 primers screened from 80 arbitrary 10-mer primer, 77 of which (83.7%) were polymorphic beside RAPD fingerprint of 11 excellent strains of Zanhuangzao were established using 5 primers. The S154-780 bp band was regarded as a molecular marker linked to the stoneless character according to the study of 17 samples of two closely related groups viz. Jinsixiaozao (with stones) and Wuhexiaozao (without stones).The DNA based marker systems namely RAPD and RFLP have been used in ber. RAPD analysis of germplasm resources on Chinese jujube was also reported by Peng *et al.* (2000). Genomic DNA was isolated from leaves of *Z. mauritiana* suitable for RAPD analysis without liquid nitrogen, making it advantageous over other common protocols (Sharma *et al.* 2003). Polymerase chain reaction restriction fragment length polymorphism (PCR-RFLP) can be used to determine the relationship among the phytoplasm infecting various woody host species. Amplification of the DNA with primers specific to the 16s rRNA gene generated a 14 kb band in *Z. jujuba* (Han *et al.*, 1996). *Ziziphus jujuba* was reported infected with phytoplasm a in Korea. Isolation and PCR amplification of genomic DNA from *Z. mauritiana* phytoplasma were reported by Shukla *et al.* (2000).

Embryo rescue

Culture of excised embryos at suitable stages of development can overcome problems encountered in post zygotic incompatibility .This technique is highly significant in intractable and long duration horticultural species. Most cultivars of Chinese jujube have serious abortion (Qi and Liu, 2004). Liu and Qi (2004) established an optimized system for embryo culture to get hybrids of four excellent cultivars, Dongzao, Jisifeng, Wudeng and Fuzao. Embryo rescue technique may help in developing hybrids of suitable combination in Indian ber especially in sterile and incompatible cultivars.

in vitro screening

The flower number per fruit branch, pollen number, time of anther dehiscence, pollen germination percentage, pollen vigour and storage of pollen collected at different stages of flowering in *Z. jujuba* var. *spinosa*, were studied *in vitro* by Liu *et al.* (2004b). The two taxa varied in flower number per fruit branch, pollen number and pollen germination percentage and these parameters. The pollen germination percentage after storage showed a drop. The enormous variability in *Z. mauritiana* and *Z. jujuba* and related species are not adequately exploited.

Future line of work

In ber, there is a need to make use of variability. Apart from hybridization, methods like mutation need to be tried for creating variability and developing better recombinants. Work can be attempted on some of the following subjects.

1. Overcoming polyploidy through embryo rescue to produce hybrids.
2. Breeding thornless varieties with higher pulp stone ratio and bigger fruit size.
3. Varieties having compact canopy with short internodal length.
4. Development of drought tolerant, pest and disease resistant varieties with high yield and longer shelf life.

References

Akerman, W.L. 1961. Flowering, Pollination, Self sterility and seed development of Chinese jujube. *Proceedings of American Society of Horticultural Science*, 77:265-269.

Ahmed, M. and Mallik, S.M. 1971.Canning of ber. *Journal of Agriculture Research*, Pakistan, 9 (3):210-217.

Boertjes , C. 1968. Mutation breeding in Vegetatively Propagated crops. In: Proceeding of research co-ordination Meeting on them use of induced mutations in Plant breeding ,held in Vienna, 11-15 September, 1967, pp 59-62.

Bakshi, J.C. and Singh, P. 1974. The ber –a good choice for semi-arid and marginal soils, *Indian Horticulture*, 19:27-30.

Babu, R.H. and Kumar, P.S. 1988.Studies on growth and flowering of certain cultivars of ber of Hyderabad. *Haryana Journal of Horticultural Sciences*, 17 (1/2):24-27.

Babu, R.H. and Kumar, P.S. 1986. A note on the seed characters certain varieties of ber (*Ziziphus mauritiana* Lam.), *Progressive Horticulture*, 18(3-4):297-298.

Bharad, S.G., Mali, D.V., Ingle, V.G. and Sagane, M.A. 2002. Effect of time and severity of pruning on physico-chemical characters and yield in ber varieties. *PKV Research Journal,* 26(1/2):31-35.

Bisla, S.S., Chauhan, K.S. and Godara, N.P. 1980. Evaluation of late ripening germplasm of ber under semiarid regions. *Haryana Journal of Horticultural Science*, 9(1/2):12-16.

Bisla, S.S. and Daulata, B.S. 1988a. Variability studies for some growth characters in ber. *Indian Journal of Horticulture*, 45 (1-2):29-33.

Bisla, S.S and Daulata, B.S.1988b. Variability studies for some fruit and seed characters in ber. *Agriculture Science Digest, Karnal*, 8 (2):112-114

Bisla, S.S and Daulata, B.S 1987. Correlation and Path analysis studies for some fruit and seed characters in ber. *Agriculture Science Digest, India*, 7 (3):170-172.

Dahliwal, J.S. and Bal, J.S. 1998. Floral and pollen biology studies in (*Zizyphus mauritiana* Lamk). *Journal of Research, Punjab Agricultural University*, 35(1-2):36-40.

Darbara Singh and Jindal, P.C. 1982. Studies on flowering and sex ratio in some ber (*Zizyphus mauritiana* Lamk) cultivars. *Haryana Agriculture University Journal of Research*, 12 (2): 292-294.

Diwakar-Hegde, Sharma, V.P. and Hegde, D. 1992. Numerical taxonomic studies of the ber (*Zizyphus.* spp). *New Botanist*, 19(1-4):21-26.

Diwakar-Hegde, Sharma, V.P. and Hegde, D. 1996. Palynological studies in Ber (Z. Spp). *New Botanist*, 23(1-4):141-151.

Dhingra, R.P., Singh, J.P and Chitkara, S.D. 1973. Varietial variation in Physico-chemical characters of ber(*Zizyphus mauritiana* Lamk). *Haryana Journal of Horticultural Science*, 2(3/4):61-65.

Daulata, B.S and Chauhan, K.S. 1982. Ber- a fruit with rich food value. *Indian Horticulture,* 27(3):7-9

Godara, N.R., Chauhan, K.S. and Bisla, B.S. 1980. Evaluation of mid season ripening ber (*Zizyphus mauritiana Lamk.*) germplasm. *Haryana Journal of Horticultural Science*, 9(3/4):101-105.

Godara, N.R. 1980. Studies on floral biology and compatibility behavior in ber (*Zizyphus mauritiana Lamk.*). Ph.D.Thesis. Haryana Agriculture University, Hissar, India.

Goyal, Y. and Arya, H.C. 1985. Tissue culture of desert trees. II. Clonal multiplication Zizyphus *in vitro*. *Journal of Plant Physiology*, 119:399-404.

Gupta, R.B., Sharma, S, Sharma, J.R. and Sareen, P.K. 2003. Cytological studies in some varieties of genus Zizyphus. *National Journal of Plant Improvement,* 5(1):19-21.

Gupta, M.R. 1977. Physico-chemical characters of some promising ber cultivars grown at (Patiala) Bhadurgarh region. *Punjab Horticulture Journal,* 17(3-4):131-134.

Hiwale, S.S. 2005. Goma kirti –a clonal selection in ber. *Indian Horticulture*, 50(1):24.

Hiwale, S.S. and Raturi G.B. 1991. Determination of leaf area jujube (*Zizyphus mauritiana*) using linear parameters. *Indian Journal of Agricultural Sciences*, 61(5):335-336.

Hiwale, S.S, More, T.A. and Bagle, B.G. 2008. Evaluation and Characterization of ber germplasm under semiarid rainfed conditions. First International Seminar on Jujube held at Agricultural University of Hebei, Baoding, China Conducted by International Society for Horticulture Science (ISHS) dated on 21-25 Sept. 2008.

Hulwale, T.D Desai, U.T., Kaulgaud, S.N. and Keskar, B.G. 1995. Pollen studies in ber (*Zizyphus mauritiana* Lamk) cultivars. *Annals of Arid Zone*, 34(1):47-50.

Han Museok, Noh Yuan Woon , Lee Jae Soon, Chung, Hoosup, Youn Yang, Youn Jeong Koo and Han M.S.1996. Differtiation in Phytoplasm by PCR-RLFP technique in woody plants. *Research Report of the Forest Genetics Research Institute, Suwon,* 32:72-79.

He YeHua, Xiong-Xinhua, Lin Shun Quan,Wu-HuiTao, Lin-Liang Bin, He-Gang, and Chen – JianHua 2004. Transformation of *Ziziphus jujuba* Hetau with antisense ACC synthase gene. *Journal of Hunan Agricultural University*, 3 (1):31-36.

Jawanda, J.S and Bal, J.S. 1980. Comparative study on the growth and development of ZG-2 and Kaithali cultivars of ber. *Punjab Horticulture Journal*, 20(1/2):41-46.

Jawanda, J.S. and Bal, J.S. 1978. The ber- Highly paying and rich in food value. *Indian Horticulture,* 23(3):19-21.

Kumar, S.S, Babu, R.S. and Reddy, Y.N. 1986. Duration of fruit maturity seasons and yield of certain cultivars of ber (*Zizyphus mauritiana* Lamk) at Hyderabad. *Journal of Research, Andhra Agricultural University*, 14(1):88-89.

Khera, A.P. and Singh, J.P. 1976. Chemical composition of some ber cultivars *(Zizyphus mauritiana* Lamk). *Haryana Journal of Horticulture Sciences,* 5 (1/2):21-24.

Liu, M. 1996. Studies on the variation and probability grading of major quantitative characters of Chinese jujube, *Acta Horticulture Sinica,* 23(2):105-109.

Liu, Meng Jun., Liu, M.J., Frosline, P.L. Fidegheli, C., Richard, K., Meerow, A., Nienhus, J., Williams, D., Thorn, E., Tombolato, A.F.C. Knupffer, H. and Stoner, A. 2003. *Acta Horticulture,* 623:351-355.

Liu, M.J. and Qi, Y.F. 2004. Embryo rescue of Chinese jujube (*Ziziphus jujuba* Mill) *Acta Horticulture*, 663(1):479-482.

Mathur, N., Ramawat, K.G. and Nandwani, G. 1995. Rapid in vitro multiplication of jujube through mature stem explants. *Plant Cell and Tissue Organ culture,* 43(1):75-77.

Mehrotra, N.K. and Gupta, M.R. 1985. Pollination and fruit set fruit studies in ber (*Zizyphus mauritiana* Lamk) *Journal of Research, Punjab Agricultural university*, 22(4):671-674.

Madalgiri, B.B., Reddy, P.N. and Baswaraj, B. 1977. Ber cultivars HB-1, HB-2 and KB-1 have commercial possibilities. *Current research,* 6(7):113-114.

Nehra, N.S. Chitkara, S.D. and Singh, K. 1984. Studies on morphological characters of some wild types of cultivated varieties of ber. *Punjab Horticultural Journal,* (1/4):49-59.

Neeraja, G., Reddy, S.A. and Babu, R.S.H. 1995. Fruit set, fruit drop and fruiting behavior in certain ber (*Zizyphus mauritiana* Lamk) cultivars. *Journal of Research, Andhra Agricultural University*, 23(3/4):17-21.

Panwar, J.S. 1980. Pattern of fruit drop in ber (*Zizyphus mauritiana* Lamk) in the arid conditions of Haryana. *Journal of Research, Haryana Agricultural University*, 10(1):57-59.

Pareek, O.P. 2001. Fruits for the future 2: Ber International Center for Underutilized Crops, University of Southampton, Southampton, U.K.

Pareek, O.P. 1996. Indian jujube and pomegranate. (In) 50 years of Crop science research in India (eds). Paroda R.S. and Chadha, K.L. ICAR, New Delhi, India, pp 557-563.

Praveen, J. and Patil, P.B. 1998. Variability, heritability and correlation studies of some physiological characters in ber cultivars. *Karnataka Journal of Agricultural Sciences*, 11(2):450-452.

Pradeep, T. and Jambhale, N.D. 2002. Cytomorphological studied in relation to powdery mildew resistance in ber (*Zizyphus mauritiana Lamk.*) *Indian Journal of Genetics and Plant Breeding,* 62(1):46-51.

Pradeep, T. and Jambhale, N.D. 2000. Ploidy level variation for stomata, chloroplast number, pollen size and sterility in ber (*Zizyphus mauritiana Lamk.*). *Indian Journal of Genetics and Plant Breeding*, 60(4):519-525.

Peng –Jian Ying, Shu-HaaiRui, Sun Zhong Xu, Peng ShiQi, Peng, J.Y., Shu, H.R., Sun Z.X. and Peng, S.Q. 2000. RAPD analysis of germplasm resources on Chinese date. *Acta Horticulture Sinica,* 27(30):171-176.

Qi , Y.F. and Liu, M.J. 2004. Embryo abortion and Embryo culture in Chinese jujube. *Acta horticulture Sinica,* 31(1):78-80.

Rathore, T.S., Singh, R.P., Deora, N.S. and Shekhawat, N. S. 1992. Clonal propagation of Zizyphus species through tissue culture. *Scientia Horticulture*, 51(1-2):165-168.

Raja, S. 2004. Morphological and molecular characterization of ber germplasm. M.Sc. Thesis , CCS Haryana Agriculture University, Hissar, India.

Reddy, Y.N. 1998. Potential of dry land fruit crops to make social forestry programme more productive. National Seminar Dryland Horticulture, Hyderabad, India Abst 37.

Saran, P.L. 2005. Studies on genetic divergence in ber (*Zizyphus mauritiana* Lamk) germplasm Ph.D. thesis, CCS HAU, Hissar, India.

Sharma, R., Mahla, H.R, Mohapatra,T., Bhargava, S.C. and Sharma, M.M. 2003. Isolating plant genome DNA without liquid nitrogen. *Plant Molecular Biology Reporter*, 21(1):43-50.

Singh, R.R. Jain, R.K and Chauhan, K.S.1983a. Flowering and fruiting behavior in ber under Gurgaon condition. *Haryana University Journal of Research*, 13(1):112-114.

Singh, J.P and Singh, I.S. 1973. Some promising varieties of ber, *Indian horticulture*,18 (2):28.

Singh, P Bakshi, J.C. and Singh, R.1972a. Identification of ber (*Zizyphus mauritiana* Lamk.) cultivars through fruit characters, *Punjab Horticultural Journal*, 12(2/3):120-133.

Singh, D. and Jindal, P.C. 1980. Physico –chemical characters of some promising ber cultivars grown at Gurgaon (Haryana), *Haryana Journal of Horticultural Science*, 9(3/4):114-117.

Sin'ko, L.T. 1974. Economic and technological evaluation of Ziziphus fruits, *Bylleuten'-Gosudaratbvennogo- Botanicheskogo Sada,* 2:28-31.

Sudhir-Kumar, Sharma, V. P. and Kumar, S. 1999. Biochemical studies of ber cultivars in relation to protein. *Annals of Agricultural Research*, 20(3):266-269.

Shukla Laxmi, Singh, N.K. and Rao, P.B. 2000. Isolation of PCR and amplification of genomics from *Zizyphus numularia* Mill. *Physiology and Molecular Biology of plants,* 6(1):75-80.

Sin'ko, L.T. 1974. Economic and technological evaluation of Ziziphus fruits, *Bylleuten'-Gosudaratbvennogo- Botanicheskogo Sada,* 2:28-31.

Teotia, S.S. and Chauhan, R.S. 1964. Flowering, Pollination, fruit set and fruit drop studies in ber (*Zizyphus mauritiana* Lamk).II. Pollination, fruit set, fruit development and fruit drop. *Indian Journal of Horticulture,* 2:40-45.

Teotia, S.S., Dube, P.S. Awasthi, R.K. and Upadhay, N.P. 1974. Studies on physico-chemical characters of some important ber varieties (*Zizyphus mauritiana* Lamk), *Progressive Horticulture,* 5(4):81-88.

Tiwari, R.J. and Banafar, R.N.S. 1995. Studies on nutritive constituents, yield and yield attributing characters of ber (*Zizyphus jujuba.*) *Indian Journal of Plant Physiology,* 38(1):88-89.

Verma, M.K., Sharma V., Saxena, S.K. and Jindal, P.C. 2001. Stomatal density as influenced by stionic combinations of ber (*Zizyphus mauritiana Lamk.*). *Indian Journal Horticulture,* 58:350-353.

Vashishtha, B.B. and Pareek, O.P. 1979. Flower morphology, fruit set and fruit drop in some ber (*Zizyphus mauritiana* Lamk) cultivars, *Annals of Arid Zone*, 18(3):165-168.

Yan, R.L., Liu, G.R., Zhang, L., Wang, Z.X. and Yang, E.Q. 1990. Effect of exogenous plant hormones on rapid multiplication of Ziziphus jujuba test tube plants. *Journal of Plant Sciences*, 7(4):231-233.

Wang Yuan and Wang, Y. 1994. A preliminary study on the geographical provances of ber *Zizyphus mauritiana*). *Forest Research*, 7(3):334-335.

Yakobashvili, V.K. 1973. Some varieties and promising seedlings of jujube, *Sub-trop kul'tury*, 3:78-81.

Yamadagni, R. Gupta, A.K. and Ahalawat, V.P. 1985. Performance of different cultivars of ber (*Zizyphus mauritiana*) under rainfed conditions, *Annals of Arid Zone*, 24(2):175-177.

Zhao, J. and Liu, M.J. 2001. Optimization of RAPD analysis system of Chinese jujube (*Ziziphus jujube* Mill.). *Journal of Fruit Sciences,* 18(6):329-332.

20

Jamun Breeding

A. Rekha

Jamun is an evergreen tree commonly known as Java plum. It has its origin from Southeast Asia, India, Myanmar and East Indies. It belongs to the family Myrtaceae (which has flowers consisting of characteristic numerous stamens) and species *Syzygium cuminii* (L.) Skeels. Closely related to an important spice *Syzygium aromaticum* "Clove". It is cultivated in tropics as an evergreen tree with dense foliage and can provide lot of shade. It is a very popular avenue tree. A unique fast growing tree, known to be tolerant to high rainfall and also to water deficit conditions, it is present in both moist and dry situations, occurring in the tropical wet evergreen forests, tropical semi-evergreen forests, tropical moist deciduous forests, littoral and swamp, tropical dry deciduous, tropical dry evergreen, subtropical broadleaved hills, and subtropical pine forests. The tree favours moist, damp or marshy situations, where it tends to form gregarious crops. It tolerates prolonged flooding, and once established it can tolerate drought.

Taxonomy

There are about 1100 to 1120 species of *Syzygium* (http://www.theplantlist.org with variable chromosome number 2n=22 to 66. Some of the important species other than *Syzygium cuminii* are as follows:

1. *Syzygium aromaticum* L.: It is a conical evergreen tree of 8-12 m height, with grayish smooth bark, lanceolate leaves and flowers borne in terminal clusters. The flower bud is an important source of spice, the Clove.
2. *Syzygium jambos* L.: Commonly known as "Rose apple" was used as an ornamental tree. It is a small to medium tree in size, narrow lanceolate

leaves with glabrous bark. The flowers with numerous 5-8 cm long stamens are attractive. The fruits are greenish yellow in colour with a typical rose fragrance. Each fruit has two large round seeds.

3. *Syzygium malaccense* L.: Known as Malay rose apple. It is a small bushy tree with attractive red flower and fruit, borne in bunches.

4. *Syzygium samarangese* Blume: It is found in Andaman Islands. Tree is similar to *Syzygium malaccense*. The fruits are waxy; bell shaped resembling cashew apple also known as water apple.

Flowering and fruit development

Jamun trees flower from March to April. The flowers are in panicles, small, about 5 mm in diameter, hermaphrodite and light yellow in colour; borne in leaf axils. The fruits develop during May / June and are large berries. The fruit is oblong, ovoid, and green in the initial days, but turns crimson black at maturity. The fruit has a sweet, mildly sour and astringent flavor. Seed is recalcitrant; seeds germinate well when fresh, but viability is lost within 2 weeks of open storage at room temperature. Jamun fruits are a good source of iron and are useful in heart and liver problems. The seeds of jamun are effective medicine against diabetes.

Nutritional Value per 100 gm of Jamun Fruit

Protein:	0.7 gm
Fat:	0.3 gm
Minerals:	0.4 gm
Fiber:	0.9 gm
Carbohydrates:	14 gm
Energy:	62 kcal
Calcium:	15 mg
Phosphorus:	15 mg
Vitamin C:	18 mg
Sodium:	26.2 mg
Potassium:	55 mg
Oxalic Acid:	89 mg
Carotene:	48 mg

Cytology and genetics

The chromosome number is 2n=44, 66 which seems to be a hexaploid. The morphological variations exhibited by *Syzygium cumini* are correlated to the intraspecific chromosome variations found in it. The tree is represented in North West India, by intraspecific diploid, tetraploid and hexaploid cytotypes which could be differentiated based on the leaf size, such as large leaf type (6x), medium leaf type (4x) and small leaf type (2x) (Kaur 2013). Study on molecular characterization on Jamun by Shakya *et al.*, (2010) revealed good diversity among Indian accessions, which were grouped based on geographical region.

Breeding/selection

Considering Jamun as one of the important future fruits, work on Jamun collection and evaluation has been carried out in CISH, Lucknow, UHS, Arabavi and Konkan Krishi Vidypeeth, Vengula etc. Several region specific clonal selections have been made for bold fruit size, taste and seedlessness. Some of the important named varieties are Ra Jamun, Narendra Jamun, Seedless Jamun (CISH J-42), Kokan Bhadoli, Dhoopdhal and Goma Priyanka.

Some of the characteristic features of released/identified varieties of jamun are as follows:

1. *CISH J-42*: Fruits are unique, seedless with pulp content of 97-98% and TSS 14-15^0 Brix.
2. *CISH J-37*: It has bold fruits and high pulp quality (pulp content 90-92%) and TSS 16-17^0 Brix.
3. *Goma Priyanka:* It is semi-dwarf, spreading growth habit, dense foliage and drooping branches, early, precocious bearer (starts flowering in 4th year) and suitable for high density planting. It starts flowering in the month of March, ripens in the fourth week of May and recorded 19.86g average fruit weight, 85.06 per cent pulp, 16.80°Brix TSS.
4. *Konkan Bahadoli:* Fruits are bold sized, good keeping quality and more juice percentage.
5. *Ra Jamun:* This is a large-fruited type having oblong fruits of deep purple colour, having large- sized fruits is known as Paras in Gujarat.
6. *Narendra Jamun:* Another type in Varanasi has no seed. A selection with desirable trait which was isolated by the Agricultural University at Faizabad.
7. *Dhoopdal:* The jamun variety from Belgaum district with GI tag for its large sized sweet fruits.

Jamun Flower

Jamun fruits

Future thrust

- Urgent need for conservation of vast diversity, as trees are lost in developmental activities.
- Improved, dwarf and high yielding varieties.
- Long juvenile phase is major drawback for cultivating on large scale in spite of its multiple uses.
- Research work on proper propagation techniques.
- Due to high medicinal values, but poor shelf life, there is need to develop value added products.

References

The Plant List., 2010. Version 1. Published on the Internet; http://www.theplantlist.org/ (accessed September 2013).

Kaur. J., 2013. Exploration of cytomorphological diversity in woody species from shiwalik hills of North West of India. PhD thesis submitted to Department of Botany, Punjab University.

Shakya, R., Siddiqui, S. A., Srivastawa, N. and Bajpai, A., 2010. 'Molecular Characterization of Jamun (*Syzygiumcumini* L. Skeels) Genetic Resources. *International Journal of Fruit Science*, 10: 1, 29-39.

Sanjay Singh, Singh. H.P., Singh, A.K., and Sisodia P.S. 2011. Jamun – Potential fruit for future. Agrotech Publishing Academy, Udaipur. 96. pp.

Patil, S. J., S. M. Mutnal, V. Maheswarappa and Girish Shahapurmath. 2009. Evaluation of jamun (*Syzygium cumini*) collections for productivity on black clay soils. *Karnataka J. Agric. Sci.*, 22(4): (933-934).

Colour Plates

Chapter 6: Role of Insects in Fruit Crop Pollination

Apis florea

Apis cerana

Apis dorsata

Chrysomya megacephala

Eristalinus arvorum

An unidentified fly

Fig 1. Pollinator diversity on mango

Fig. 2. *Apisdorsata* foraging on guava flower

Fig 3. Honeybees (*Apisdorsata* and *A. cerana*) foraging on passion fruit

Chapter 10: Mango Breeding

Male flower

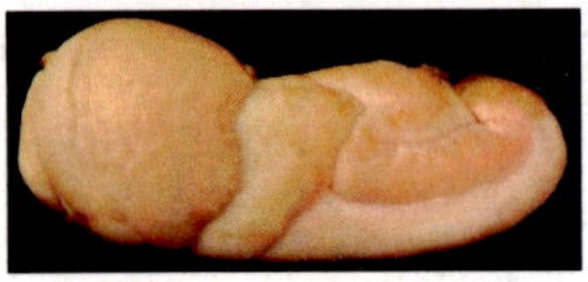

Polyembryony

Monoembryonic seedling

Pic 1

Pic 2

Pic 3

Pic 4

Pic 5

Pic 6

Flowering in mango tree

Variability for leaf colour and shape

M. andamanica

Parts of a Mango Fruit

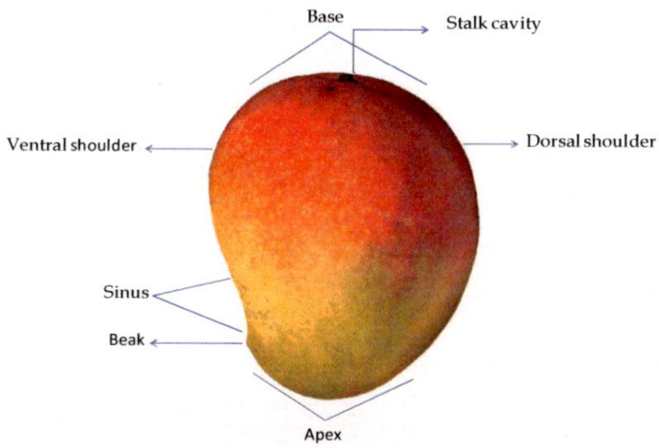

Variability for coloured varieties

Diversity for pickling mango

Diversity

Vanaraj a coloured variety

S Jehagir F

Chapter 11: Citrus Breeding

Variation in fruit characteristics in Rangpur lime

Low seeded clone of Nagpur mandarin

Early maturing clone of Nagpur mandarin

A view of variability in Nagpur

Early and late maturity in Nagpur mandarin

Fruits and leaves of a Rough lemon x Troyer citrange hybrid (centre) and the parents

The plant of Rough lemon x Troyer citrange

A block of citrus rootstock hybrids exhibiting segregation in the F_1 progeny

Chapter 12: Banana Breeding

Elakki bale AB type

FHIA-1 of AAAB

FHIA-2 of AAAA

FHIA-3 of AABB

Karpooravalli - ABB

M. acuminata

M. balbisisana

Mysore AAB

Nendran AAB

Robusta AAA

Seed Germination musa hybrid seedlings

Seeds

Chapter 14: Papaya Breeding

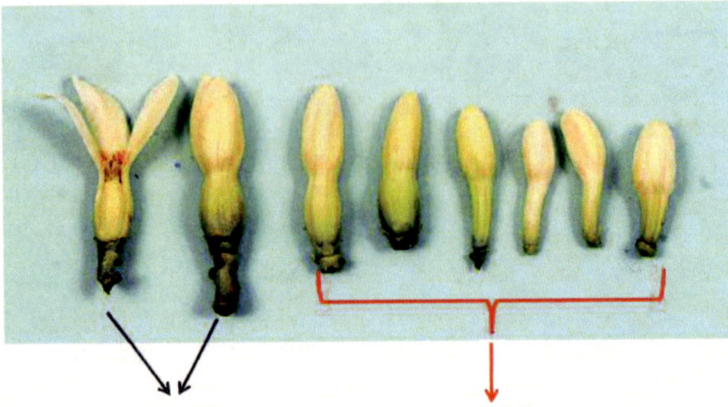

Fertile hemaphrodites   Sterile hermaphrodite flowers

Different flowertypes

Hermaphrodite flower

Male flower

Maleflower

Ringspotplant

Fruit from a female tree

Arka Prabhath

Fruit from bisexual tree

Ringspot

Arka Surya

Arka Surya (♀)

X

V. cauliflora (♂)

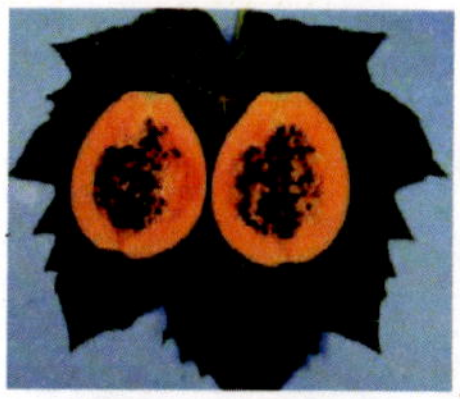

Yellow pulp

Single plant of intergeneric hybrid

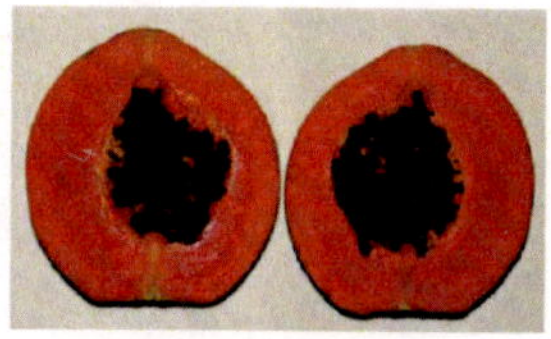

Pink pulp

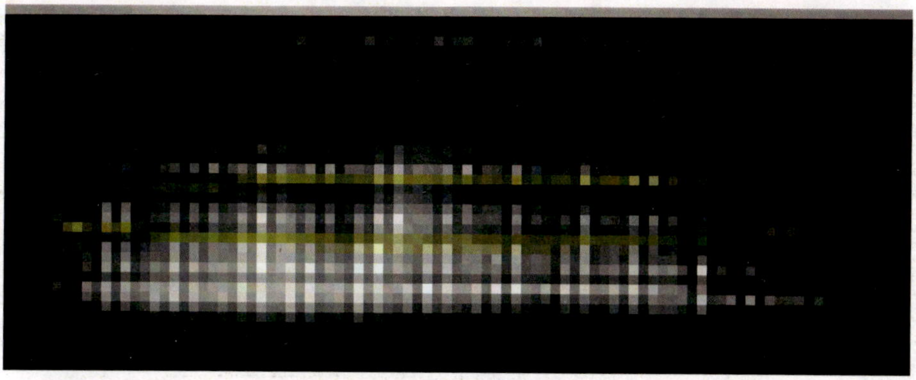

PCR amplified products showing polymorphism using ISSR primer UBC 834:1& M- 1kb ladder, 2- *A. Surya*, 3- *V. cauliflora*, 4 to 7- F1s & 8 to 46 Intergeneric hybrids

Chapter 15: Guava Breeding

Allahabad safeda

Apple Colour

Arka Kiran

Variability for fruit shape in guava

Guava flower

Mridula 4

Chapter 17: Jackfruit Breeding

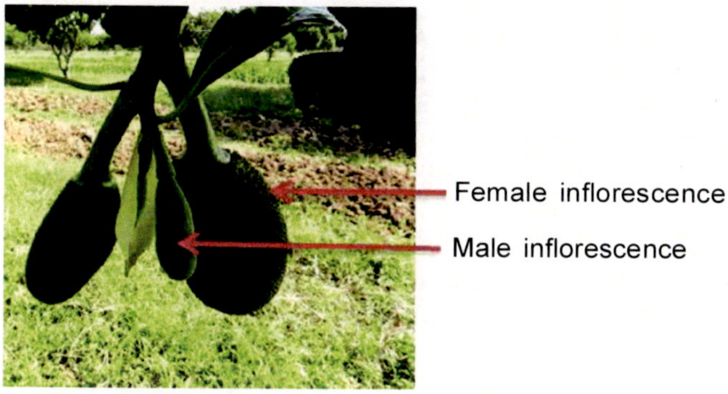

Plate 1: Branch of *Artocarpus heterophyllus* with male and female Inflorescence

Plate 2. PLR-1 Jackfruit

Plate 3: PPl-1Jackfruit

Plate 4. PLR (J)-2 Jackfruit

(a)

(b)

(c)

Plate 5 a, b & c: Swarna Jackfruit

Plate 6: Konkan Prolific Jackfruit

a b c

Plate 7 a, b & c: Kachahalli Jackfruit

Chapter 20: Jamun Breeding

Jamun Flower

Jamun fruits